# A QUANTITATIVE BIOLOGY OF THE PIG

# A Quantitative Biology of the Pig

*Edited by*

## I. KYRIAZAKIS

*Animal Biology Division*
*Scottish Agricultural College*
*Edinburgh, UK*

CABI *Publishing*

**CABI *Publishing* – A division of CAB INTERNATIONAL**

CABI *Publishing*
CAB INTERNATIONAL
Wallingford
Oxon OX10 8DE
UK

Tel: +44 (0)1491 832111
Fax: +44 (0)1491 833508
Email: cabi@cabi.org

CABI *Publishing*
10 E 40th Street
Suite 3203
New York, NY 10016
USA

Tel: +1 (212) 481 7018
Fax: +1 212 686 7993
Email: cabi-nao@cabi.org

A catalogue record for this book is available from the British Library, London,
UK.


**Library of Congress Cataloging-in-Publication Data**
A quantitative biology of the pig / edited by I. Kyriazakis.
    p.  cm.
  ISBN 0-85199-273-0 (alk. paper)
   1. Swine--Physiology. 2. Swine. I. Kyriazakis, I.
SF768.2.S95Q35   1998
636.4'0892--dc21                                   98–22769
                                                 CIP


ISBN 0 85199 273 0

Typeset by York House Typographic Ltd, London
Printed and bound in the UK at the University Press, Cambridge

# Contents

# Contributors

**G. Annison**  Australian Food Council, Barton, ACT 2600, Australia

**D. Bastianelli**  CIRAD-EMVT, Campus de Baillarguet, BP 5035, 34032 Montpellier Cedex 1, France

**J.L. Black**  John L. Black Consulting, Locked Bag 21, Warrimoo, NSW 2774, Australia

**R.D. Boyd**  Pig Improvement Company, PO Box 348, Franklin, KY 42136, USA

**H.J. Bray**  South Australian Research and Development Institute, Pig and Poultry Production Institute, University of Adelaide, Roseworthy, SA 5371, Australia

**A. Danfaer**  Danish Institute of Agricultural Sciences, Research Centre Foulum, PO Box 50, 8830 Tjele, Denmark

**J.-Y. Dourmad**  INRA Station de Recherches Porcines, 35590 Saint-Gilles, France

**F.R. Dunshea**  Victorian Institute of Animal Science, Sneydes Road, Werribee 3030, Australia

**G.C. Emmans**  Animal Biology Division, Scottish Agricultural College, West Mains Road, Edinburgh EH9 3JG, UK

**M. Étienne**  INRA Station de Recherches Porcine, 35590 Saint-Gilles, France

**H. Everts**  Department of Large Animal Medicine and Nutrition, Faculty of Veterinary Medicine, Utrecht University, PO Box 80.152, 3508 TD Utrecht, The Netherlands

**L.R. Giles**  NSW Agriculture, Elizabeth Macarthur Agricultural Institute, PMB 8, Camden, NSW 2570, Australia

**A.W. Jongbloed**   Department of Pigs and Poultry, ID-DLO, PO Box 65, 8200 AB Lelystad, The Netherlands

**P.A. Kemme**   Department of Pigs and Poultry, ID-DLO, PO Box 65, 8200 AB Lelystad, The Netherlands

**R.H. King**   Victorian Institute of Animal Science, Sneydes Road, Werribee 3030, Australia

**I. Kyriazakis**   Animal Biology Division, Scottish Agricultural College, West Mains Road, Edinburgh EH9 3JG, UK

**A.B. Lawrence**   Animal Biology Division, Scottish Agricultural College, West Mains Road, Edinburgh EH9 3JG, UK

**J. Le Dividich**   INRA Station de Recherches Porcine, 35590 Saint-Gilles, France

**J.P. McNamara**   Department of Animal Sciences, 233 Clark Hall, Washington State University, Pullman, WA 99164-6320, USA

**M.T. Mendl**   Division of Animal Health and Husbandry, Department of Clinical Veterinary Science, University of Bristol, Langford, Bristol BS18 7DU, UK

**C.A. Morgan**   Animal Biology Division, Scottish Agricultural College, West Mains Road, Edinburgh EH9 3JG, UK

**P.J. Moughan**   Institute of Food, Nutrition and Human Health, Massey University, Palmerston North, New Zealand

**Z. Mroz**   Department of Pigs and Poultry, ID-DLO, Postbus 65, 8200 AB Lelystad, The Netherlands

**B.L. Nielsen**   Animal Biology Division, Scottish Agricultural College, West Mains Road, Edinburgh EH9 3JG, UK

**J. Noblet**   INRA Station de Recherches Porcines, 35590 Saint-Gilles, France

**M.C. Père**   INRA Station de Recherches Porcines, 35590 Saint-Gilles, France

**S.M. Rutherford**   Institute of Food, Nutrition and Human Health, Massey University, Palmerston North, New Zealand

**D. Sauvant**   INRA Laboratoire de Nutrition et Alimentation, INAPG, 16 rue Claude Bernard, 75231 Paris Cedex 5, France

**A.P. Schinkel**   Department of Animal Sciences, 1151 Lilly Hall, Purdue University, West Lafayette, IN 47907-1151, USA

**J. Wiseman**   Faculty of Agricultural and Food Sciences, University of Nottingham, Sutton Bonington Campus, Loughborough, Leicestershire LE12 5RD, UK

# Foreword

This book is a good idea. It is about quantification, and taking a systems approach to the biological study of a particular organism – the pig. It is also full of good ideas. It is a commissioned challenge set by the Editor to the world's foremost scientific teams of this generation. They have risen to that challenge and produced an important book that will become an instant classic. It is both compulsive and compulsory reading.

Why is this book so badly needed?

First, because science must be understood in order to be correctly applied. This requires not empiricism and anecdote but rather rigorous underlying theory. Theory is difficult, much more difficult than doing an experiment and reporting the results.

Second, the application of science through the technology of pig production is coming to depend upon two principles: quality assurance and integrated management control systems. Both these require full and quantitative understanding of the whole of the process, and completion of all the links in the chain of knowledge into a closed loop.

It has taken many years for these needs to be recognized, and many more for them to be addressed. The work is by no means complete; but this book establishes the ground rules, points the way and gets well down the track.

Animal science has had insufficient theoretical analysis, and an excess of empirical experimentation. The latter is fine for short term fixes but the former is the only way forward for long-term advance. This is not to say that the next steps in science are just thinking about things – quite the reverse. Better and more penetrating experiments are required to yield quality data upon which unifying constructions of understanding can be built. It is a characteristic of this book that the authors have seen the need to provide the robust experimental knowledge-base so greatly needed if accurate scientific interpretation is to be achieved.

This book is of its time and ahead of its time. It will change the reader's perception of the issues, it will influence future thought and work. It is an important book, and I am delighted to be associated with it. I congratulate and thank the Editor and all the authors for this work.

**Colin Whittemore**, NDA, BSc, PhD, DSc, FIBiol, FRSE
Professor of Agriculture and Rural Economy, University of Edinburgh
1998

# Introduction

## I. Kyriazakis

*Animal Biology Division, Scottish Agricultural College, West Mains Road, Edinburgh EH9 3JG, UK*

## Why 'a Quantitative Biology of the Pig'?

The science and practice of pig production has changed dramatically over recent decades, so that the 'modern pig industry bears little or no resemblance to that of only a few decades ago' (Whittemore, 1993). These changes have arisen from the need to consider novel issues and alternatives regarding the quality of the end product, the welfare of the animals and the environmental impacts of pig production. They have been achieved through improvements and modifications in the breeding, feeding, housing and management of pigs. One of the greatest challenges that has arisen from these changes is the need for extremely accurate decision making, which can only be achieved by considering simultaneously the effects of many interacting factors which influence the outcome of a pig production system: 'Meat of a *given quantity* and *definable characteristics* requires to be produced from feedstuffs of *quantifiable* nutritional value when given to *specific* pig genotypes in *describable* environmental, managemental and economic circumstances' (Whittemore, 1993; my italics).

The tradition in the practice of pig science has been to focus mainly or only on the factors which appeared to have an overwhelming influence on the system in question. For example, it has been, and unfortunately still is customary to investigate the effect on a response by varying a small set of experimental variables; with all of the uncontrolled factors, and the way in which the components of the pig production system (such as pigstock and feedstuffs) change, such an approach has little or no chance of success in predicting the future, and in helping us to be accurate in our decisions

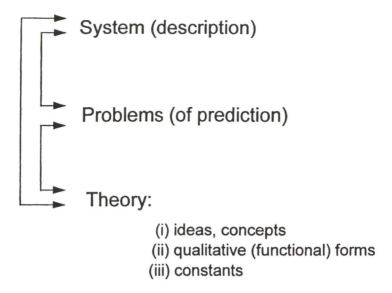

System (description)

Problems (of prediction)

Theory:

(i) ideas, concepts
(ii) qualitative (functional) forms
(iii) constants

**Fig. 1.1.** A process for theory construction and methodology for research. (Adapted from Emmans and Oldham, 1988.)

(Emmans, 1996). Fortunately, these complex interactions between the component factors can now be approached through simulation modelling, using biological principles and mathematics, which provides a means of focusing on the issues under consideration. Model building, however, requires a satisfactory theory which underlies the behaviour of the system to be simulated: 'In many senses a model *is* a theory' (Emmans and Fisher, 1986). The aim of this book is to develop theories or models dealing with the biological processes which underlie pig production. By doing so alternative theories in the field are inevitably considered and compared with the offered ones. These theories, like any good theory, are put into quantitative terms so that their predictions can be made explicit and compared with outcomes from the real world (see below).

## The Need for a Theory in Pig Biology

A theory arises from the need to be able to predict the behaviour of a system, and represents an idea or a concept of how a system operates. Although theories always form the basis of predictions in biological systems, this is not always obvious in agricultural systems, such as that of pig production. Emmans and Oldham (1988) suggested that the process of theory construction is done at three levels (Fig. 1.1). Once we have an idea of how a system operates, then the next step is to choose logical and functional forms which represent the theory qualitatively. The third level is to identify the variables, and find the values of the constants of the theory's functional forms, so that the theory is put in quantitative terms. The values of the constants are found by making measurements in the real world by experiments.

An application of the above process to an important task in pig production systems, that of describing quantitatively and being able to predict how the voluntary food intake of pigs changes from weaning to maturity, is provided here as an example. The concept is that food intake initially increases as the pig increases in size and the rates of gain of lipid-free body and fat are also increasing, and eventually reaches an equilibrium when the pig has achieved its mature size. The task then is to choose a function which describes adequately this pattern, and this may range from a simple regression equation to a more sophisticated set of functions (see Chapter 10). The choice should be based on the criterion of generality, i.e. how well the form describes various sets of data, and not how well some particular set of data is described by different functional forms without questioning the data. It may be found eventually that the chosen form is not adequate in describing the food intake of new breeds of pig, as these are introduced into the system, or of pigs manipulated by endocrine agents or immunological means. It would then need to be replaced with an alternative, better function(s) which would be able to describe quantitatively the change in the food intake of pigs.

The quantitative theories presented in this book have the above process inherent in their structure. The choice of the level at which these theories are developed (e.g. whole animal or metabolic, empirical or mechanistic) was offered to each individual author, but it was greatly dictated by their subject matter. Similarly, the form of presentation of the functional forms which represent each theory, varies among different chapters which deal with different biological processes. This is mainly a reflection of the diverse scientific backgrounds of the contributors of each chapter, but also of the fact that in some cases the specific values of the constants related to the functions representing a particular theory are yet to be determined experimentally. In these latter cases the value of the constructed theories will lie not in their immediate predictive ability, but in the identification of their existing shortcomings. Theories are both evolving and dynamic in their nature; their true value lies in the identification of the *problems* associated with their construction, rather than in the provision of the correct answers to these *problems*.

## The Book Structure

In order to be able to predict the behaviour of any system we need both a predictive theory and a description of the system. As Emmans (1996) pointed out the theory and the description of the elements of the system are not independent, since different theories call for different descriptions of the system. For this reason Part I of the book addresses explicitly descriptions of the elements of the system, which form the broad subject of this book.

In pig production systems their three obvious components are the pigs, the food offered to them and the environment in which they are kept. Chapter 2 examines the underlying biological traits which describe genetic populations of pigs, while Chapter 3 provides an overview of the chemical and physical characteristics of pig foods, and discusses how such a description may assist in developing a quantitative understanding of pig biology. A specific description of the foods offered to pigs, in relation to their 'energy' contents is in part revisited in Chapter 15. Chapters 4 and 5 deal with the descriptions of the environment in which pigs are kept while some conventional descriptions of it are provided, e.g. in terms of its thermal and physical characteristics; some other, less usual descriptions are also invoked. These include tentative descriptions of the infectious (Chapter 4) and the social environments (Chapter 5), and to a certain degree of their interactions, since it is becoming increasingly evident that these have a profound influence on the system under question. For example, it is now suggested that the reduced voluntary food intake seen in group housed pigs, relative to that of their individually kept counterparts (Chapple, 1993), cannot be accounted for through the effects of the climatic environment (i.e. ambient temperature, humidity, airspeed).

Part II of the book deals with the quantitative descriptions and predictions of the biological processes in the pig from its conception and birth, to slaughter. They are divided, in the traditional manner, into: mating, pregnancy and prenatal growth (Chapter 6), lactation and neonatal growth (Chapter 7), and (postweaning) growth and body composition (Chapter 8). The physiological controls which underlie these processes, and an attempt to define them in quantitative terms, are addressed in Chapter 9. This is done by attempting to answer questions such as how sensitive is a process responsive to a regulator (i.e. hormone)? and, what is the relative importance of each regulator within the whole animal? By doing so this chapter becomes the link between the whole animal and its underlying metabolic processes, which are addressed in Part III of the book.

In pig production systems the ultimate interest lies in the ability to predict the outcome of feeding a group of pigs in a particular way. This can be achieved through our knowledge of the amount of food either offered to or voluntarily consumed by the pig, and our ability to predict the fate of food within the animal (metabolism). Part III addresses these two issues by first describing quantitatively the voluntary food intake of pigs given access either to a single food or to more than one food as a choice (Chapter 10). The digestion and absorption of the food and its associated nutrients are described in Chapter 11 and subsequent chapters deal with the fate of the most important group of nutrients within a food, i.e. minerals (Chapter 12), amino acids (Chapter 13) and carbohydrates and lipids (Chapter 14). Chapter 15 addresses the fate of energy consumed within a pig, and by doing so it provides a concluding link with its preceding chapters.

## In Place of Conclusions

One can claim to understand how a system works only if one can successfully predict its behaviour. As discussed above this can only be achieved by developing theories which are then tested, criticized and eventually replaced. It has now been over two decades since the first attempt to represent quantitatively one of the biological processes in the pig, that of growth (Whittemore and Fawcett, 1974). The intervening twenty years have demonstrated, rather strongly, the importance of developing theories or models in making progress in pig science (as will be discussed in the final chapter). I do not, however, wish to give the false impression that the development of theories or model construction is relevant only to scientific endeavours. There is a tendency to consider theories as being appropriate only for the environment in which they have developed. However, it is the characteristic of good science that scientific and practical issues are or should be different aspects of the same thing.

## References

Chapple, R.P. (1993) Effect of stocking arrangement on pig performance. In: Batterham, E.S. (ed.), *Manipulating Pig Productions IV*. Australian Pig Science Association, Victoria, pp. 87–97.

Emmans, G.C. (1996) Problems of description and prediction. *Proceedings of the 15th Scientific Day of the Southern African Brand of the World's Poultry Science Association*, pp. 12–18.

Emmans, G.C. and Fisher, C. (1986) Problems in nutritional theory. In: Fisher C. and Boorman, N. (eds), *Nutrient Requirements of Poultry and Nutritional Research*. Butterworths, London, pp. 9–36.

Emmans, G.C. and Oldham, J.D. (1988) Modelling of growth and nutrition in different species. In: Karver, S. and van Arendonk, J.A.M. (eds), *Modelling of Livestock Production Systems*. Kluwer Academic Publishers, Brussels, pp. 13–21.

Whittemore, C.T. (1993) *The Science and Practice of Pig Production*. Longman, Harlow, Essex, p. 661.

Whittemore, C.T. and Fawcett, R.H. (1974) Model response of the growing pig to the dietary intake of energy and protein. *Animal Production* 19, 221–231.

# Describing the Elements of the System

# Describing the Pig

## 2

## A.P. Schinckel

*Department of Animal Sciences, 1151 Lilly Hall, Purdue University, West Lafayette, IN 47907-1151, USA*

## Introduction

Genetic selection for improved pork production efficiency has resulted in substantially different pig populations. In the past four decades, vastly different performance testing and selection programmes have been implemented by government agencies, regional cooperatives and seedstock companies (David and Johnson, 1982; Schinckel *et al.*, 1985; De Vries and Kanis, 1994). Initially, selection focused on the more highly heritable postweaning performance traits including growth rate, feed conversion and carcass lean percentage. Different testing procedures, selection criteria and selection intensities have resulted in substantially different absolute and relative rates of genetic changes for postweaning performance traits. Application of BLUP (Best Linear Unbiased Prediction), hyperprolific sow selection schemes and importation of Chinese breeds has resulted in renewed interest in the genetic improvement of more lowly heritable reproductive traits.

Several swine models have been developed which integrate our knowledge of the effects of genetic potential, nutrient intake, and environmental conditions on pig growth and reproduction. These models can be used to identify alternative means to improve the efficiency of pork production and to estimate daily nutrient requirements for pigs of different ages and genetic groups when managed under a range of environmental conditions (de Lange and Schreurs, 1995). For an effective application of swine growth models, the growth potential of pig genotypes must be accurately characterized (Schinckel and de Lange, 1996).

Researchers have primarily concentrated on the evaluation of commonly measured performance traits. To understand the biological changes which have occurred as a result of past selection practices and direct future genetic changes, genetic evaluation must be directed towards quantifying the underlying biological traits rather than conventionally measured performance traits. The differences observed between genotypes for measured performance traits, i.e. growth rate, feed conversion and per cent lean, are susceptible to nutrient intake and environmental interactions. Evaluation of the underlying biological traits, when used as inputs to swine growth models, allows the prediction of differences in performance traits given the nutritional and environmental limitations present. This chapter examines the underlying biological traits which describe genetic populations of pigs.

# Growth Performance

## Protein accretion potentials

Estimation of empty body protein accretion rates is the primary characteristic needed to describe the growth of pigs. The growth of all other non-fat body components including water, carcass lean and ash can be modelled as functions of protein accretion (Whittemore, 1994).

The growth of protein mass to time is best described as a sigmoid growth curve with respect to time. The sigmoid growth curve has two principal segments, the first of increasing daily growth rate or accelerating phase of growth and the second phase of decelerating growth (Brody, 1945). The point of inflection between the two growth phases is the point at which maximum protein accretion occurs. When protein mass ($Pt$) is fitted to a sigmoid function of time ($t$), daily protein growth is estimated as $\partial Pt/\partial t$. The inflection point is mathematically defined as the point in which the second derivative of protein mass with respect to time equals zero. Secondary parameters include the maximum protein accretion rate, the age at which maximum protein accretion was achieved, the protein mass at the inflection point as a ratio of mature protein mass and mature protein mass.

There are two primary methods in which protein accretion can be mathematically estimated. The first is to describe protein mass as a sigmoid function of age. The second method is to describe protein accretion as the product of the live weight growth rate and the rate of change in protein mass relative to live weight.

Protein mass data can be fitted to either fixed or variable inflection point functions. Fixed inflection functions assume that maximum protein accretion is achieved at a constant percentage of mature protein mass. The Gompertz equation, a fixed inflection point function, has been widely used to describe pig growth (Whittemore, 1994; Emmans, 1995a). The Gompertz equation has the form: $P_t = \hat{P}_t \cdot e^{-e^{-B(t-t^*)}}$ where $B$ is the rate parameter, $P_t$ is present protein mass, $\hat{P}_t$ is predicted mature protein mass, $t$ is time (age) and $t^*$ is the time at the point of inflection. The Gompertz equation assumes the maximum rate of protein accretion is achieved at a constant proportion (0.368) of mature protein mass. Although easily parameterized (Ferguson and Gous, 1993) the equation is relatively inflexible (Black *et al.*, 1995).

Several variable inflection point functions have been developed (Fitzburgh, 1976; Parks, 1982; Bridges *et al.*, 1986; Black *et al.*, 1995) which allow the maximum protein accretion rate to occur at different proportions of mature mass. Variable inflection point functions can also provide for either a sharper or a flatter rise in the protein accretion rates prior to and after the inflection point (Black *et al.*, 1995). For these reasons, flexible inflection point functions are more flexible to describe observed differences in the shape of protein accretion curves of different genotypes and sexes (Schinckel, 1994a; Black *et al.*, 1995; Schinckel and de Lange, 1996; Thompson *et al.*, 1996).

The problems with variable inflection point functions are: (i) difficulty to achieve convergence, especially when good compatible initial values of the parameters are not available; and (ii) unstable parameter estimates sensitive to the initial values used (Brown *et al.*, 1976; Fitzburgh, 1976). Numerical methods can be used to identify good initial values of the parameters in proximity to the global maximum by linearizing the function. Also, a weighted least-squares non-linear algorithm should be used to account for unequal variation in protein mass at different ages.

The second method to estimate protein accretion rates is to consider protein accretion to be a multiplicative function of live weight growth rate and partitioning of live weight gain into protein using the following function: $\partial(Pr)/\partial t = \partial w/\partial t * \partial(Pr)/\partial w$, where $\partial(Pr)/\partial t$ is the protein accretion rate $Pd$, $\partial w/\partial t$ is live weight growth rate and $\partial(Pr)/\partial w$ is the rate at which live weight is partitioned into protein (Whittemore *et al.*, 1988). This method divides the problem into two steps. The first step is to describe live weight as a sigmoidal function of time. Live weights need to be collected periodically before, during and after the weight interval of interest. For example, if the goal is to estimate protein accretion from 30 to 110 kg, pigs must be weighed starting at 15 kg and ending at no less than 125 kg.

The second step is to describe protein mass as a function of live weight. Allometric functions ($Y = aX^b$) have been used where $Y$ is the weight of a tissue or compositional component and $X$ represents the weight of the whole entity, i.e. empty body weight, live weight or carcass weight (Walstra, 1980; Moughan *et al.*, 1990). The equation is usually transformed to a linear log to log function ($\log Y = \log a + b \log x$). The log to log transformation equalizes the variation in component mass at each live weight, a required assumption for least-squares regression.

Allometric equations have several advantages including: (i) simple stable linear solutions after the log to log transformation; (ii) straightforward biological interpretation; and (iii) simple, stable derivatives. The allometric function ($Y=aX^b$) assumes that the ratios of the relative growth rates of $X$ and $Y$ are constant throughout growth (Evans and Kempster, 1979; Walstra, 1980; Moughan *et al.*, 1990). For every 1% change in empty body weight, the body component changes *b*%. Thus, the allometric function assumes that the body component mass expressed as a percentage of empty body mass uniformly decreases (*b* < 1), remains constant (*b* = 1) or increases (*b* > 1) as empty body increases. However, empty body protein and fat-free carcass lean percentages increase from birth to approximately 45 to 65 kg as per cent water decreases (Curtis *et al.*, 1967; Stant *et al.*, 1968; Shields *et al.*, 1983; Ferrell and Cornelius, 1984; White *et al.*, 1995) while after 65 kg, the percentage of protein and fat-free lean decreases as per cent lipid increases. Despite the fact that many researchers have used allometric equations, Evans and Kempster (1979) stated three objections to its use in relating body component mass to empty body mass (i.e. part to whole relationships), including: (i) it may be inapplicable at the earliest and later stages of growth; (ii) if the growth of a body component over time fits a flexible inflection point sigmoid growth curve, then the relationship of the body component to body weight is not in general allometric from birth to maturity; and (iii) there are difficulties in using the allometric function in a multicomponent growth model.

Research has been conducted with several functions (Walstra, 1980; Parks, 1982; Wagner *et al.*, 1995). Some are not stable due to a high correlation among the independent variables. A function known as an augmented allometric, $Y = aX^b(c - X)^d$, describes accretion of body components including protein, moisture and a fat-free carcass lean mass, which do not maintain the same relative growth rates in proportion to empty body weight (Schinckel *et al.*, 1995b; Wagner *et al.*, 1995). These and other body components expected to reach mature mass at a point less than ultimate potential mature empty body mass would be expected to be better described by an augmented or quadratic allometric growth function. Addition of the $(c - X)^d$ term is consistently significant ($P < 0.001$) when the data are from pigs with backfat depths greater than 18 mm at 115 kg live weight. The function provides a similar fit of protein or fat-free lean mass to live weight with only a slightly higher $r^2$ than the allometric function (Wagner *et al.*, 1995). However, the estimated derivatives, the amount of the component gained with each unit of live weight gain, are quite different.

The negative aspect of the augmented allometric function is that it is a more difficult and less stable solution in small datasets. The significance of the $(c - X)^d$ term of the augmented allometric function is dependent on the ratio of the relative growth rates of the body component to live weight or empty body weight changing and having adequate data over key live weight ranges. The increase in percentage protein from 30 to 70 kg live weight is dependent on essential amino acid : calorie ratios of the diets fed (Friesen *et al.*, 1996). The drop in percentage protein after 80 kg live weight is dependent on the feed intakes being above those needed for maximum protein accretion which results in a substantial increase in the ratio of lipid : protein accretion. When energy intakes, i.e. over 10 Mcal day$^{-1}$ from 90 to 110 kg live weight are

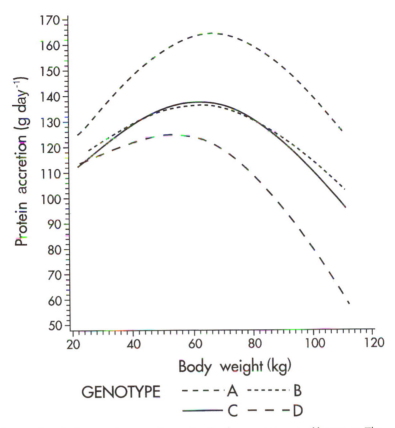

**Fig. 2.1.** Daily whole protein accretion rates for four genotypes of barrows. The genotypes are representative of high (A), average to high (C and D), and average (B) lean growth genotypes (Palmer *et al.*, 1993). The parameters of the Bridges function $Pt = PTM(1 - e^{-mta}) + 0.22$; where $Pt$ = protein mass, kg at time 't', $a$ = kinetic order constant, $m$ = exponential decay constant and $PTM$ = predicted mature protein mass. Values of $PTM$, $m$ and $a$ are 25.4, $3.44 \times 10^{-6}$ and 2.54 for genotype A; 15.9, $9.81 \times 10^{-6}$ and 2.40 for genotype B; 19.62, $7.74 \times 10^{-6}$, 2.39 for genotype c; 19.70, $1.66 \times 10^{-5}$ and 2.25 for genotype D.

achieved, it is likely that protein mass will be better explained by augmented allometric equations. Protein mass accretion of low feed intake, lean boars or gilts may not result in data which are better fitted by augmented allometric equations than allometric equations.

Protein accretion curves have been developed for several commercially available genotypes (Palmer *et al.*, 1993; Schinckel *et al.*, 1995b, 1996; Thompson *et al.*, 1996). Protein accretion curves for four genotypes of terminal cross barrows representing high (A), medium-high (C and D) and medium (B) lean growth genotypes are presented in Fig. 2.1. The major differences between genotypes are the overall mean protein accretion rate and the rate at which protein accretion declines after 90 kg live weight. In general, genotypes with overall high protein accretion rates have higher predicted mature protein mass and thus maintain higher protein accretion rates at heavier weights than average protein accretion genotypes.

Using conventional statistical procedures, it is not possible to assign confidence bands to either of the input functions ($\partial(Pr)/\partial w$ or $\partial w/\partial t$) or, more importantly, to the product of these inputs ($\partial(Pr)/\partial t$). Bootstrapping is a resampling procedure to estimate both bias and variation around a parameter estimate or predicted value (Efron, 1982). Bootstrapping procedures allow the development of confidence bands for each of the parameters of the non-linear equations, derivatives, and the resulting product of the derivatives (Thompson *et al.*, 1996). Bootstrapping procedures also allow the estimation of the confidence bands for component accretion curves and can be used to evaluate the number of animals, number and range of live weights and estimates of composition required for a desired degree of accuracy.

To develop the maximum protein accretion curves for a genotype, pigs must be fed diets not limiting in nutrients and must be reared in the most optimal environment possible. The dietary protein levels can alter the shape of the observed protein accretion curves. Feeding protein levels below requirements for the genotype will limit protein accretion. More importantly, if the restriction on protein accretion is removed, compensatory growth may occur (De Greef, 1992; Friesen *et al.*, 1996).

The pigs must be weighed periodically, and serial estimates of composition must be collected. The direct measurement of empty body composition through dissection, grinding and chemical analysis of serially slaughtered pigs is too expensive to be routinely conducted on an adequate number of pigs. Total body electromagnetic conductance (TOBEC), B-mode ultrasonics, and bioelectrical impedance are potential live animal measurements that can be taken repeatedly at different live weights. Pork processors are utilizing midline backfat depths, optical probes, and TOBEC to estimate carcass lean and value (Forrest *et al.*, 1989; Akridge *et al.*, 1992). Data from these instruments are less accurate than the data obtained from direct dissection but are obtained at a much lower cost. Part of the dissection measurements could be replaced with predicted values from live animal or carcass measurements to reduce the cost of the experiment without loss of precision via double sampling (Conniffe and Moran, 1972; Engel and Walstra, 1991).

One method to serially estimate body composition is via serial B-mode (real-time) ultrasonic backfat depth and loin muscle area measurements. Prediction equations from a large swine growth trial have been developed to predict fat-free lean, total carcass fat, empty body protein and empty body lipid mass for barrows and gilts from 20 to 155 kg live weight (Wagner *et al.*, 1995, Thompson *et al.*, 1996).

To evaluate accurately protein accretion for specified genotypes, prediction equations utilizing live or carcass measurements must accurately predict the actual genotypic differences in composition. Prediction equations should rank the genotypes correctly, that is there should be a high degree of relationship between the true and predicted genotype means for lean mass (Gu *et al.*, 1992). Also, the equations should be able to predict the true variation between genotypes. A number of prediction equations utilizing standard carcass measurements overestimate the lean content of genotypes with below average per cent lean and underestimate the lean content of the leanest genotypes (Rae *et al.*, 1985; Branscheid and Sack, 1988; Planella and Cook, 1991; Gu *et al.*, 1992; Wagner *et al.*, 1993). The systematic biases can be partially accounted for by statistical methods (Schinckel *et al.*, 1995a). However, the genotype-specific biases can only be accounted for by dissecting a number of pigs and recalibrating the equation for the specific genotype (Engel and Walstra, 1991).

### Lipid accretion

Empty body lipid growth can be described as a sigmoid growth curve with an accelerating phase, inflection point, decelerating phase and mature mass like that of protein accretion (Hammond, 1952; Whittemore, 1994; Emmans, 1995a). Daily lipid accretion rates increase from 25 to 90 kg live weight in most genotypes and reach a plateau from 90 to 120 kg live weight (Palmer *et al.*, 1993; Schinckel *et al.*, 1995b). In some genotypes, daily lipid accretion rates increase at a linear rate to 130 kg live weight (Fig. 2.2). Because maximum lipid accretion is achieved at 100–150 kg live weight, data from pigs 180–220 kg live weight or greater will be needed to fit the lipid data to a sigmoid function of time. Sufficient data beyond the inflection point, during the decelerating phase of lipid growth, are needed to allow the parameters of a sigmoid curve to be estimated.

### Energy partitioning

One of the most important factors determining protein accretion rates is energy intake. To evaluate alternative management or genetic changes that impact energy intake, the relationship between energy intake and protein accretion must be characterized for various populations of pigs. It is generally accepted that when other nutrient intakes are not limiting, a linear relationship exists between protein accretion and energy intake when energy intake is greater than the amount needed for maintenance and less than the amount required to maximize protein accretion (Whittemore and Fawcett, 1976; Campbell *et al.*, 1983; Dunkin and Black, 1987; Campbell and Taverner, 1988; De Greef, 1992; Bikker *et al.*, 1995; Quiniou *et al.*, 1995, 1996). The change in protein accretion per unit increase in energy intake is called the slope. The regression has the form: $PD = a + b\,ME$ where *PD* is daily protein accretion rate (g day$^{-1}$), *a* is the intercept, *b* is the slope and *ME* is either *ME* intake or *ME* intake above maintenance. The equation can estimate the protein accretion achieved at maintenance (zero net energy gain) or above maintenance. From a biological perspective, estimated positive protein accretion at low

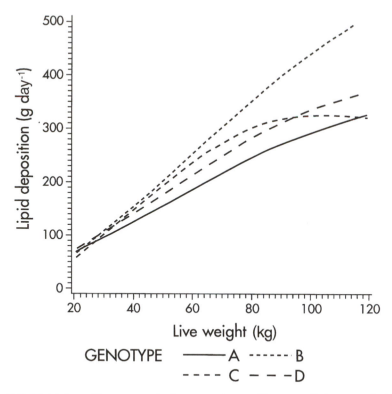

**Fig. 2.2.** Lipid deposition curves for four genotypes of barrows (Palmer *et al.*, 1993). For a description of the genotypes, see Fig. 2.1.

energy intakes indicates that protein accretion is a priority process associated with mobilization of lipid tissues when energy supply is limited (Quiniou *et al.*, 1996).

The slope of protein accretion on energy intake is the more common method of describing the partitioning of energy between lean and fat growth. The relationship between protein accretion and energy intake can also be described as the change in the lipid to protein deposition ratio per unit energy intake. These two methods are algebraically interchangeable (de Greef, 1992). The slope of protein deposition on energy intake quantifies the additional amount of protein which is deposited from each additional unit of energy intake (Fig. 2.3a). A constant amount of additional energy is available for lipid deposition from each additional unit of energy intake (Fig. 2.3b). In other words, a linear relationship also exists between lipid deposition and energy intake (de Greef, 1992; Bikker *et al.*, 1995). The steeper the slope of protein accretion on energy intake, the lower the overall lipid to protein deposition ratio. In addition, the steeper the slope of protein accretion on energy intake, the more slowly the lipid to protein ratio increases as energy intake increases up to the amount needed for maximum protein accretion (Fig. 2.3c).

Lean growth efficiency is maximized at the energy intake at which protein accretion is maximized (Fig. 2.3d). To estimate the energy intake required to achieve maximum protein accretion requires the estimation of both the slope and intercept of protein accretion on energy intake (Quiniou *et al.*, 1996). Lean efficiency declines more rapidly at energy intakes above that needed for maximum protein accretion in comparison to energy intakes below that needed for maximum protein accretion.

The magnitude and rate of the changes in the slope of protein accretion with respect to live weight or maturity have not been comprehensively evaluated. Bikker (1994) reported higher slopes in 20 to 45 kg pigs in comparison with 45 to 85 kg live weight pigs. Dunkin and Black (1987) reported that the slopes of nitrogen retention on ME intake rapidly declined from 30 to 46 kg live weight and remained constant at 46, 74 and 90 kg live weight. Small changes in the slopes at 65, 80 and 94 kg live weight (11.5, 10.9 and 9.0 g protein accretion per megajoule *ME* intake above maintenance) were reported by Quiniou *et al.*, 1995. In a second trial, slopes from 45 to 100 kg live weight were affected by pig genotype but not by stage of growth (48, 64, 79 and 94 kg live weight; Quiniou *et al.*, 1996). However, the intercepts became less positive as live weight increased, which caused the lipid to protein deposition ratios to increase with live weight.

Swine growth models require an estimate of slope of either protein accretion or lipid to protein deposition ratios on energy intake at different live weights. Economical, yet precise methods to establish these parameters have not been defined. One method is to feed various energy intakes at different live weight ranges and then establish the slopes for each live weight range. Protein accretion rates at the different energy intakes could be estimated by comparative slaughter or nitrogen balance techniques. The second method is to feed different energy levels above maintenance but below the amount needed for maximum protein accretion. By serially slaughtering animals and using allometric functions ($Y = aX^b$), the ratio of lipid to protein deposition curves as a function of live weight and energy intake can be established (de Greef, 1992).

Protocols which both economically and accurately estimate either parameter have not been developed and carefully evaluated. Two groups of pigs must be scale fed two energy levels and serial estimates of composition determined or nitrogen retention rates determined. The feed intake levels must be carefully considered as pig to pig variation likely exists for both the protein accretion rates and the energy partitioning parameters. The difference in energy intake levels must be far enough apart at each live weight so that the changes in protein and lipid mass at each serial live weight are large compared to the accuracy with which composition is estimated. Too high levels of feed intake may allow some of the pigs to consume more energy than that needed for maximum protein accretion. This will cause an

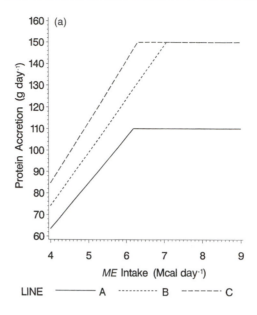

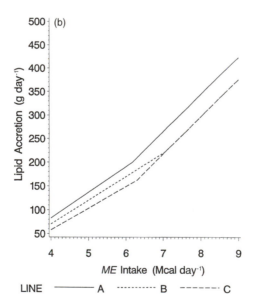

**Fig. 2.3.** (*and opposite*) Predicted growth for three genotypes at 53 kg live weight (50 kg empty body weight) at different energy intakes. Maximum daily protein accretion rates of 110, 150 and 150 g, slope of protein accretion (g) per Mcal *ME* intake of 21.3, 24.8 and 28.3 were used for genotypes A, B and C, respectively. (a) Relationship of daily protein accretion to energy intake. (b) Daily lipid accretion rates at different energy intakes.

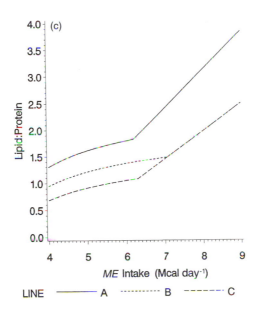

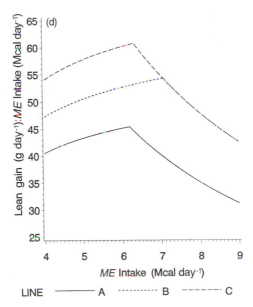

**Fig. 2.3.** *Contd* (c) Ratio of lipid accretion to protein accretion at different energy intakes. (d) Ratio of lean gain to *ME* intake at different energy intakes.

overestimation of the change in the lipid : protein ratio and an underestimation of the slope of protein accretion on energy intake at each live weight. Various protocols could be examined by simulating individual pigs which vary from a mean protein accretion curve and energy partitioning parameter. The number of pigs per feed intake level, alternative methods to estimate composition, and the number of live weights at which composition is estimated could be varied.

### Energy requirements for maintenance

The energy requirement for maintenance is defined as the total energy intake minus that needed for protein and lipid accretion. Maintenance energy requirements include a magnitude of physiological and physical activities. Maintenance requirements for energy can be expressed as a function of live weight or protein mass (ARC, 1981; Whittemore, 1983; NRC, 1988). Studies (Koong *et al.*, 1983; Noblet and Henry, 1991) indicate that such relationships only explain part of the variation of maintenance energy requirements that are observed between different groups of pigs.

Maintenance energy requirements are better expressed in relation to the distribution of the major tissue groups (viscera, muscle, fat) in the pig's body (Ferrell, 1988). The contribution of the activities of each tissue group in relation to the total maintenance energy requirements should be considered.

Recently, the concept of selection for reduced maintenance requirements through selection for residual feed intake has been examined (De Haer, 1992; Kennedy *et al.*, 1993; Mrode and Kennedy, 1993). Variation in residual feed intake can include variation in digestibility, energetic efficiencies, and variation in maintenance requirements such as physical activity, temperature regulation, protein turnover, and basal metabolic rate (De Haer, 1992). Genetic differences in maintenance requirements due to differences in physical activity have been found (Henken *et al.*, 1991). Future research on energy requirements may focus on genetic variation in protein turnover (Knap, 1996; Knap and Schrama, 1996).

### Energy intakes

Swine growth models require an estimate of daily energy intake to estimate other aspects of animal performance and nutrient requirements. Three non-linear equations have been primarily used to describe the curvilinear relationship between daily food or energy intake to either age or body weight. Both ARC (1981) and NRC (1988) suggested the relationship: $DE = C(1 - e^{-kw})$ where $w$ is live weight and $C$ is an estimate of mature feed intake. A slightly different non-linear equation is: $ME = C(1 - be^{-kw})$ where the additional constant 'b' is a scalar, usually close to 1, which can provide a slightly better fit in some datasets (Williams *et al.*, 1994). The third non-linear function is: $dF/dt = C(1 - e^{-t/t^*})$, where $dF/dt$ is daily feed intake (kg day$^{-1}$) at age $t$ (days), $C$ is mature feed intake and $t$ is the exponential decay constant or appetence (days) (Thompson *et al.*, 1996). Parks (1982) defined $t^*$ as the time required for the animal to increase its rate of feed intake to 0.63 of its mature feed intake $C$. In this function, both $C$ and $t^*$ can be considered as genetic parameters.

The negative aspect of these past feed or energy intake functions is that they provide no biological interpretation. An alternative approach with more direct biological interpretation assumes that a pig's desire to consume energy is driven by the pigs' protein accretion and lipid deposition potential (Emmans, 1995a). The pig's desire to consume energy becomes a function of its energy requirement for maintenance and growth of protein and lipid. Therefore, the important first steps toward characterizing energy intake capacities of different pig genotypes are to determine the rates of protein and body lipid deposition in a 'non-limiting' environment at the various body weight ranges (Ferguson and Gous, 1993; Emmans, 1995a, b).

If the actual body protein and body lipid accretion rates are determined for pigs given *ad libitum* access to feed, energy intake can be predicted largely based on the energetic cost of maintenance protein accretion and lipid deposition. Assuming that maintenance energy requirements are determined by empty body protein mass, daily energy intakes at each live weight were predicted within 4% of observed feed disappearance (Schinckel, unpublished data). A similar method of estimating feed intakes from predicted lipid and protein accretion rates accurately estimated mean feed intakes in groups of pigs (Whittemore *et al.*, 1995).

## Reproductive Performance

Until the mid-1980s, swine selection programmes concentrated on growth and carcass traits. Currently, there is renewed interest in the genetic improvement of sow productivity traits. Also, recent concerns have been raised that antagonisms exist between reproductive performance and continued intense selection for lean efficiency.

A number of reproductive traits exist, the majority of which are substantially impacted by environmental and management effects (Aherne and Kirkwood, 1985). Thus, the evaluation of different genotypes for reproductive performance must be carefully planned and should consider the impact of the environment in which the evaluation is done.

### Age at puberty in gilts

Age at puberty in the gilt is defined as the time of first oestrus and ovulation with continued regular oestrous cycles. Sexual maturity is the developmental stage at which the gilt will respond to exogenous gonadotrophin (LH and FSH) or GnRH with oestrus and ovulation and maintain either a regular cycle or pregnancy (Dyck, 1988). Hormone treatments can mimic natural pubertal

LH preovulatory surges and induce precocious ovulation (Lutz *et al.*, 1985; Burnett *et al.*, 1988). However, if the hormone treatments are done too early, before sexual maturity, a second ovulation will not occur. Thus, the pubertal LH surge is a necessary event but not the final limiting event to the natural onset of puberty.

In most mammals, the onset of puberty in females occurs shortly after the period of maximum growth (Brody, 1945; Joubert, 1963). A minimum body fat content permissive to pubertal onset, though not necessarily a trigger for ovulation, has been proposed (Beltranena *et al.*, 1991). Kirkwood and Aherne (1985) have suggested that minimum threshold values for live weight, fat mass and age must be achieved before sexual maturity can occur. However, although threshold levels for age, weight and fat composition are necessary, their achievement is not in itself sufficient to initiate puberty. Also, other factors including boar exposure, housing and social environment have stimulatory or inhibitory effects.

Age seems to be the more critical factor for the onset of puberty rather than live weight or body composition based on trials in which feed intake was limited after 60 kg live weight (Newton and Mahan, 1992, 1993). The amount of body tissue attained prior to dietary restriction may influence the onset of puberty. Etienne *et al.*, 1983, demonstrated that young gilts fed at 77% of *ad libitum* from 28 to 60 kg live weight had delayed onset of puberty, whereas when energy restriction was not imposed until 60 kg live weight, no effect was evident on age at puberty.

Age at puberty is moderately heritable with published estimates averaging 30% (Young *et al.*, 1978; Hutchens *et al.*, 1981; Rydmer, 1994). Differences between domestic breeds in age at puberty are relatively small, ranging from 180 to 220 days of age (Christenson, 1981; Hutchens *et al.*, 1982; Dyck, 1988). However, for Chinese breeds and crosses between Chinese and domestic maternal lines, age at puberty ranges from 80 to 120 days of age (Legault and Caritez, 1982; Legault, 1985; Christenson, 1993). These results and data from domestic crosses (Hutchens *et al.*, 1982) indicate that both additive and non-additive (dominance or epistasis) genetic variation exist for age at puberty.

## Pubertal development in boars

In boars, no distinct event has been defined as the achievement of puberty. To characterize pubertal development of boars, the age at which a number of developmental events occur must be evaluated.

Pubertal development in boars is characterized by sigmoidal growth of the testes. In domestic pigs, rapid prepubertal growth of the testes and epididymides is initiated at approximately 100 days of age (Allrich *et al.*, 1982). Testicular and epididymal growth is linear to 200 days after which time growth rapidly declines. The relative growth rate of the testes is very rapid from 100 to 130 days of age as testes weight quadruples and diameter of the seminiferous tubules doubles (Allrich *et al.*, 1982, 1983; Cameron, 1987).

Spermatozoa first appear in the lumen of the seminiferous tubules of domestic boars from 100 to 120 days of age. By 120 to 150 days of age, at a paired testes weight of approximately 200 g, 50% of the seminiferous tubules will have active spermatogenesis occurring (Schinckel *et al.*, 1983, 1984). Ejaculation of fertile spermatozoa is achieved in domestic boars between 170 and 200 days of age (Cameron, 1987).

Pubertal development in Chinese boars occurs substantially earlier than domestic boars (Lunstra *et al.*, 1993). Spermatids have been detected in the testes of Meishan boars at 64 days of age at a paired testes weight of 41 g. The pituitary function of Meishan boars is different from domestic boars. Meishan boars have much higher circulating FSH concentrations than domestic boars (554 vs 102 ng ml$^{-1}$; Wise *et al.*, 1996). However, the FSH concentrations are not different between domestic and Meishan females (Hunter *et al.*, 1993).

With the onset of rapid testicular growth, the production of testosterone (T) and oestradiol-17β ($E_2$) increases by five- to ninefold (Christenson *et al.*, 1984). Serum T and $E_2$ concentrations increase by fourfold from 100 to 200 days of age (Allrich *et al.*, 1982; Schinckel *et al.*, 1984). These gonadal steroids are needed to induce mating behaviour, the final step of pubertal development in the boar (Ford, 1990). From 200 to 250 days of age, most domestic boars will be able to mate naturally (Wilson *et al.*, 1977).

## Measures of prolificacy

A number of variables can be used to evaluate sow reproductive performance. The most encompassing single variable measurement of prolificacy is pigs weaned per sow per year. Pigs weaned per sow per year incorporates all other variables including pigs weaned per litter and litters per sow per year.

Pigs weaned per litter can be broken down into its component traits. Total born includes all fully formed pigs including live and stillborn pigs. Total born alive is the total pigs born minus the stillborn pigs. Number weaned is the product of number born alive times the preweaning survival percentage.

Litters per sow per year is a function of the days in the year divided by the interval between farrowings which is the sum of gestation length, weaning age and rebreeding interval. Litters per sow per year is also defined as the function:

Litters per sow per year = $(365 - NPD) / (GL + LL)$

where NPD is the number of non-productive sow days per year defined as all the days in which a sow is neither lactating nor gestating, GL is gestation length (115 days) and LL is lactation length.

Litter size, defined as either total born or total born alive, is a primary variable of importance. Litter size has been broken down into three biological component traits: ovulation rate, potential embryonic viability and uterine capacity (Bennett and Leymaster, 1989). Ovulation rate, the number of ova shed, is moderately heritable and responsive to selection (Lamberson *et al.*, 1991). Embryonic viability represents the reduction in number of ova shed to potentially viable embryos. Causes of this reduction include factors inherent to the ovum/embryo (i.e. fertilization failure, genetic lethals, abnormal blastocyst development) and not directly associated with limitations in maternal environment (Bennett and Leymaster, 1989). Uterine capacity, the second phase of prenatal mortality, consists of the reduction of potentially viable embryos prior to birth. This phase implies competition among embryo/fetuses for limited uterine resources or space (Christenson *et al.*, 1987).

An evaluation and understanding of the three biological component traits of litter size is needed to interpret the genetic responses of past selection experiments (Legault, 1985; Lamberson *et al.*, 1991; Gama and Johnson, 1993), to design more optimal future litter size selection programmes (Bennett and Leymaster, 1990) and to interpret the performance of various lines or crosses (Bolet *et al.*, 1986; Christenson, 1993; Gama and Johnson, 1993; Young, 1995). Nutritional treatments including level of feed restriction can have differential effects on the components of litter size (Dyck, 1974; Hughes and Varley, 1980; Dyck and Strain, 1983) and have different effects on different parities (Dyck and Cole, 1986).

Litter size generally increases with parity up to third to fifth parity and then gradually declines thereafter (Strang, 1970). However, in some cases, extended weaning to oestrus intervals in parity 1 and decreased litter size in parity 2 may occur (Brooks and Smith, 1980; Whittemore *et al.*, 1980; Clark and Leman, 1986; Mullan and Williams, 1989). Research results indicate that this effect may be due to inadequate protein intake during lactation and subsequent lean tissue catabolism (King and Dunkin, 1986; Brendemuhl *et al.*, 1987).

When evaluating litter size, it is important to realize the effects of sexual age (number of oestrous periods after puberty). When gilts are stimulated to reach puberty at young ages and light body weights, ovulation rates will increase with each oestrous period. This effect may not be primarily an effect of sexual age but rather an age, weight and body composition effect (Kirkwood and Aherne, 1985). The ovulation rate of different genotypes may increase at markedly differing rates from first to subsequent oestrous periods (Christenson, 1993).

## Longevity and lifetime performance

Sow longevity can be measured by several variables including number of litters per lifetime, average herd parity, sow age at death or culling and per cent of sows culled or removed by a given parity. Lifetime productivity of sows can be evaluated by measures such as pigs born alive per sow per lifetime or pigs weaned per sow per lifetime (Rozeboom, 1996).

Reproductive failure, degenerative problems, locomotor problems and death account for 70% of all removals in commercial sow herds (Stein *et al.*, 1990). Reproductive failure is the most common reason for culling across all parity groups except parity 7 and greater, in which degenerative problems were more common and reproductive failures ranked second in occurrence. Voluntary culling based on low performance (i.e. number born or number weaned) only accounted for 9.4% of the sow removals.

Research conducted evaluating the effects on longevity of nutrition during rearing (i.e. magnitude of scale feeding), age at first breeding or conception and backfat thickness at first breeding show a great deal of disagreement between studies (Rozeboom, 1996). The effects of age, live weight and backfat thickness on longevity may be related to and interact with other managerial and environmental factors. The only consistent result is that lean gilts, those with 12 to 13 mm or less backfat at breeding or farrowing, have increased parity one and two weaning to oestrus intervals and are more prone to culling (King *et al.*, 1984; Gueblez *et al.*, 1985; Yang *et al.*, 1989; Young *et al.*, 1991; Rozeboom, 1996).

## Genetic Selection and Populations

Several concepts are crucial to the discussion of genetic selection and the definition of genetic populations. First, the underlying biological growth traits, those needed as inputs for swine growth models, determine the differences observed in the measured performance traits (De Vries and Kanis, 1992; Schinckel, 1994b). Genetic evaluation must be directed towards quantifying the underlying biological traits rather than conventionally measured performance traits.

The second concept is that the choice of 'environment' in which the animals are tested will alter which animals are selected each generation. The type of feeding programme, i.e. 24 hour *ad libitum*, fixed time 2× per day *ad libitum*, or restriction with scale feeding, affects the genetic parameters and response to selection (Fowler *et al.*, 1976; Smith and Fowler, 1978; Webb and Curran, 1986; Cameron and Curran, 1994, 1995). The differences observed between genotypes for measured performance traits (i.e. growth rate, feed conversion and per cent lean) are susceptible to nutrient intake and environmental interactions (Fowler and Ensminger, 1960; Kanis, 1990; McPhee *et al.*, 1991; Cameron and Curran, 1995). The extent of feed restriction will substantially reduce the genetic expression of feed intake and partially reduce the variation in lean growth and fat growth. The amount of feed intake achieved relative to that needed for maximum protein accretion will affect the amount and direction of the genetic change in feed intake.

**Table 2.1.** Summary of performance differences observed between European and US terminal cross.[a]

|  | ADG (kg day$^{-1}$) | Feed intake (kg day$^{-1}$) | Live wt feed conver. | Fat depth (cm) | Fat-free lean gain (g day$^{-1}$) | Carcass fat gain (g day$^{-1}$) | Feed (kg) lean gain (kg) |
|---|---|---|---|---|---|---|---|
| Barrows |  |  |  |  |  |  |  |
| European | 0.957 | 2.19 | 2.30 | 2.05 | 345 | 236 | 6.36 |
| US | 0.975 | 2.46 | 2.54 | 3.15 | 304 | 313 | 7.95 |
| Gilts |  |  |  |  |  |  |  |
| European | 0.875 | 1.99 | 2.27 | 1.57 | 327 | 172 | 6.10 |
| US | 0.939 | 2.38 | 2.52 | 2.85 | 304 | 254 | 7.87 |
| Overall |  |  |  |  |  |  |  |
| European | 0.916 | 2.09 | 2.28 | 1.81 | 336 | 204 | 6.23 |
| US | 0.957 | 2.42 | 2.53 | 3.00 | 304 | 286 | 7.91 |
| Difference | 0.041 | 0.33 | −0.25 | 1.19 | 32 | 82 | 1.68 |
| Difference as % of US | −4.3 | −13.7 | −9.9 | −39.7 | +10.5 | −28.7 | −21.2 |

[a] The data included three European and four US terminal cross genotypes with 28 or 32 pigs per genotype-sex group. Test period was from 27 to 113.5 kg live weight.

The third concept is that the relative emphasis the selection criterion places on the component traits directs the relative genetic changes in each trait. The selection criterion can be either a designed selection index, a function of performance traits or BLUP genetic merit estimates, or a composite trait like lean tissue feed conversion. The selection criteria for the performance traits evaluated in the specified environmental-management conditions determines the direction and magnitude of genetic change for each of the underlying biological traits (Fowler *et al.*, 1976).

The fourth concept is that pigs from the same genetic source and sex must be considered as a genetic population. Breed or breed-cross is not an adequate definition of genotypes as different seedstock sources have implemented vastly different genetic selection programmes. Pigs of the same breed-cross from different seedstock suppliers can have drastically different lean growth rates and carcass composition (Schinckel, 1994a). Differences in the relative protein accretion rates, feed intakes, backfat depths and lean mass estimates are found to vary between the sexes and among genotypes (Table 2.1). For example, on the average, barrows consume 9 to 10% more feed than gilts at the same live weights. The ratio of feed intakes of barrows relative to gilts actually varies from 1.04 to 1.16 for different genetic populations (Schinckel, 1994a). In general, the higher the feed intake of a genotype, the larger the differences are between the sexes in fat-free lean mass and backfat

**Table 2.2.** Differences in fat depth and fat-free lean mass of three genotypes at four live weights.[a]

| Live weight (kg) | Genotype A | | Genotype B | | Genotype C | | SEM |
|---|---|---|---|---|---|---|---|
| | Barrows | Gilts | Barrows | Gilts | Barrows | Gilts | |
| Fat depth 10 rib (cm) | | | | | | | |
| 84 | 3.12 | 2.54 | 1.90 | 1.88 | 2.48 | 2.36 | 0.10 |
| 100 | 3.38 | 2.67 | 2.18 | 2.10 | 3.04 | 2.81 | 0.11 |
| 114 | 4.62 | 3.38 | 2.38 | 2.12 | 3.68 | 2.90 | 0.12 |
| 128 | 4.93 | 3.96 | 2.79 | 2.54 | 3.93 | 3.37 | 0.13 |
| Fat-free lean (kg) | | | | | | | |
| 84 | 26.8 | 27.7 | 29.2 | 30.3 | 26.2 | 26.7 | 0.41 |
| 100 | 28.2 | 34.0 | 35.1 | 35.2 | 30.4 | 30.9 | 0.52 |
| 114 | 32.7 | 37.7 | 43.8 | 44.9 | 32.3 | 30.7 | 0.61 |
| 128 | 35.5 | 41.3 | 45.7 | 47.9 | 37.7 | 41.5 | 0.65 |

[a] Adjusted for live weight. Genotype A had the highest feed intakes. Genotype B had the lowest feed intake of the seven genotypes (Thompson *et al.*, 1996).

thickness (Table 2.2). Because genotype by sex interactions exist, pigs of each sex should be evaluated for each genetic source.

The use of different selection criteria can cause large genotype differences in feed intake. Divergent selection for increased or decreased growth rate caused 40% differences in feed intake (Woltman *et al.*, 1995). The differences in feed intake were primarily accounted for by differences in fat accretion. Selection for lean growth rate with equal emphasis on backfat and growth rate resulted in very little (1.0%) change in feed intake (Cleveland *et al.*, 1983). Increased emphasis on reducing backfat has resulted in decreased feed intakes (Fowler *et al.*, 1976; Webb, 1989).

Lean efficiency and per cent lean have been the primary traits emphasized in Europe (Mitchell *et al.*, 1982; Webb and Curran, 1986; Webb, 1989). When feed cost is a high percentage of total cost and carcass value programmes are widely implemented, lean efficiency and carcass merit become relatively much more important than growth rate. A number of selection experiments and genetic trend estimates (Ellis *et al.*, 1983; Brandt, 1987; Cameron and Curran, 1994) have shown that feed intakes have decreased substantially in pigs primarily selected for a combination of leanness and lean efficiency. The biological changes likely to have occurred as the result of selection for lean efficiency include the following: increased protein accretion rates, decreased fat accretion, decreased feed intake, increased energy partitioning from lipid to protein accretion (steeper slopes at similar live weights) and increased mature lean masses (Mitchell *et al.*, 1982; Webb, 1989).

Different selection criteria and testing procedures for lean growth or lean efficiency will also produce different correlated responses in postpubertal growth and reproductive performance (Kerr and Cameron, 1995, 1996). Differential correlated responses may differ for several reasons including the fact that reproductive traits including age at puberty, weaning to breeding intervals and ovulation rates may be affected by threshold levels of live weight, body composition (i.e. lipid to protein ratios) and protein or lipid catabolism at crucial times (Kirkwood and Aherne, 1985).

In order to understand fully the effects of selection for growth or leanness on reproduction, lean growth and efficiency should be measured as directly as possible as well as a wide range of components of reproduction (i.e. age and composition at puberty, postpubertal oestrus at conception, expression of oestrus, litter sizes, breeding performance, and sow weight changes during lactation). Relationships between the biological changes produced as a result of selection for lean production and components of reproduction may differ by parity. Thus, multiple parities should be evaluated to determine the effects of selection.

In the past three years a number of European, Canadian and Australian sources of breeding stock have entered the United States. This importation of seedstock has added genetic variation for lean growth, fat accretion and lean efficiency. Differences in feed intake of 30% exist between different geno-types (Schinckel, 1994a). The differences for digestible energy consumed above maintenance approached 60%. In a recent trial, three European termi-nal cross-genotypes and four US genotypes (two Hampshire-Duroc [HD] × Yorkshire-Landrace [YL], one H × YL, one D × YL) were evaluated (Schinckel, unpublished data, 1995). The European pigs grew 4.3% slower, had 10.5% higher lean growth rates, consumed 13.7% less feed, and had substantially lower daily carcass fat growth rates (Table 2.1). This resulted in the European genotypes having higher ratios of carcass lean gain to fat gain (1.65 vs 1.06), and 21.2% better lean feed conversion than US genotypes.

In the future it is likely that swine growth models will be used to refine performance testing and genetic selection procedures. One approach is to consider a biological model of the relationship between protein deposition and food intake (Fowler *et al.*, 1976; De Vries and Kanis, 1992; Kanis and De Vries, 1992). The logical approach would be to select for increased protein accretion, increased partitioning of energy to protein and optimal level of daily feed intake, the minimum level needed to achieve maximum protein accretion (De Vries and Kanis, 1992; Schinckel, 1994a).

Using this approach, economic weights can be calculated for parameters describing the partitioning of energy between protein and lipid accretion, maximum protein accretion and feed intake. The optimum change in feed intake capacity depends upon both the current levels and expected changes in energy partitioning and protein accretion (De Vries and Kanis, 1992). A selection index based on expected future genetic gains can result in lower production costs over a number of generations of selection (Kanis and De Vries, 1992).

The second approach would be to utilize a deterministic model to simulate biological and economic inputs for life cycle pork production (Tess *et al.*, 1983). Independent changes of the biological traits can be used to simulate changes in either feed (Mcal kg$^{-1}$ lean or body weight) or economic efficiency (cost per 100 kg lean or live weight). The economic impact of the direct effects and interactions between reproductive and pig growth traits can be evaluated.

The third approach is to utilize a swine growth model to produce both genetic and environmental variation in the underlying biological traits (Knap, 1995). The biological traits simulated can include parameters which describe protein accretion rates, feed intakes, energy partitioning and maintenance requirements including variation in the fractions of protein present in different tissues and protein turnover rates (Knap, 1995, 1996; Knap and Schrama, 1996). Stochastic growth models allow researchers to examine alternative testing procedures, i.e. diets utilized, method of feeding, on and off test weights, performance measurements, and selection criteria. Stochastic models can also be used to evaluate the use of economic vs biological indexes and alternative genetic merit estimation programmes.

Stochastic models can reproduce known changes on the relationships among performance traits (Knap, 1995). For example, at low feed levels or restricted feeding, the phenotypic and genetic correlations between daily gain and lean percentage are positive. When pigs are fed *ad libitum* and achieve higher feed intakes, the correlations become negative.

The genetic and phenotypic parameters of conventionally measured performance traits also differ between the three genders: boars, gilts and castrates. In some countries, selection is primarily based upon the performance of boars and gilts, while commercial production includes castrates and gilts. This situation makes it difficult to direct the testing and selection procedures to result in optimal genetic changes in the underlying biological traits.

In the United States, both production-management and marketing systems are constantly changing. Currently, producers are implementing management strategies that improve overall herd health status including segregated early weaning and three-site production procedures. Improved health status through these procedures results in 20 to 30% increase in feed intake and increased backfat depths at 100 kg live weight in barrows. Marketing systems are increasing premiums paid for predicted carcass lean percentage and heavier carcasses. Some marketing systems require lean pigs at 120 kg live weight to obtain maximum carcass premiums. Genetic selection must be targeted to increase protein accretion potentials and result in optimal feed intakes under high-health status conditions. In the United States where castrates are fed as market animals, selection must result in castrates that are able to maintain high lean growth rates up to 125 kg live weight. This is not easily accomplished because the protein accretion rates of castrates after 80 kg live weight drop more rapidly than those of boars or gilts. Stochastic models should be able to produce genetic (or sex) by environmental interactions and thus evaluate the optimal genetic selection strategy for different production-marketing systems.

Differences between genetic populations makes it difficult to predict precisely pig growth, composition and carcass value. Genetic selection can alter water : protein ratios at the same live weight and distribution of lean and fat mass within the lean cuts. Genotype-specific relationships between body components (i.e. water : protein, visceral vs carcass growth) should be considered for each widely used genetic population.

## Conclusions

Vastly different performance testing and selection programmes have resulted in substantially different genetic populations of pigs. To understand the biological changes made as a result of selection, genetic evaluation must be directed towards quantifying the underlying biological traits.

The biological parameters needed to define pig growth are protein accretion potentials, lipid accretion potentials, partitioning of energy between protein and lipid growth, energy requirements for maintenance and energy intake potentials. Estimates of these growth parameters are needed as inputs for swine growth models. Alternative methods to estimate these growth parameters have been developed, and they vary in cost, accuracy, practicality, and in their underlying assumptions.

The primary reproductive characteristics include underlying biological traits describing pubertal development, prolificacy and longevity. The evaluation of reproductive characteristics must carefully consider all non-genetic factors that influence their performance. Whenever possible, the underlying biological characteristics of composite traits such as number born alive should be evaluated.

## References

Agricultural Research Council (ARC) (1981) *The Nutrient Requirements of Pigs.* Commonwealth Agricultural Bureaux, Slough, UK.

Aherne, F.X. and Kirkwood, R.N. (1985) Nutrition and sow prolificacy. *Journal of Reproduction and Fertility*, Suppl 33, 169–183.

Akridge, J.T., Brorsen, B.W., Whipker, L.D., Forrest, J.C., Kuei, C.H. and Schinckel, A.P. (1992) Evaluation of alternative techniques to determine pork carcass value. *Journal of Animal Science* 70, 18–28.

Allrich, R.D., Christenson, R.K., Ford, J.J. and Zimmerman, D.R. (1982) Pubertal development of the boar: testosterone, estradiol-17β, cortisol and LH concentrations before and after castration at various ages. *Journal of Animal Science* 55, 1139–1146.

Allrich, R.D., Christenson, R.K., Ford, J.J. and Zimmerman, D.R. (1983) Pubertal development of the boar: age-related changes in testicular morphology and in vitro production of testosterone and estradiol-17β. *Biology of Reproduction* 28, 902–909.

Beltranena, E., Aherne, F.X., Foxcroft, G.R. and Kirkwood, R.N. (1991) Effects of pre- and postpubertal feeding on production traits at first and second estrus in gilts. *Journal of Animal Science* 69, 886–893.

Bennett, G.L. and Leymaster, K.A. (1989) Integration of ovulation rate, potential embryonic viability and uterine capacity into a model of litter size in swine. *Journal of Animal Science* 67, 1230–1241.

Bennett, G.L. and Leymaster, K.A. (1990) Genetic implications of litter size in swine based on ovulation rate, potential embryonic viability and uterine capacity. II. Simulated selection. *Journal of Animal Science* 68, 980–986.

Bikker, P. (1994) Protein and lipid accretion in body components of growing pigs: effects of body weight and nutrient intake. PhD thesis, Wageningen Agricultural University, Wageningen, The Netherlands.

Bikker, P., Karabinas, V., Verstegen, M.W.A. and Campbell, R.B. (1995) Protein and lipid accretion in body components of growing gilts (20–45 kg) as affected by energy intake. *Journal of Animal Science* 73, 2355–2363.

Black, J.L., Davies, G.T., Bray, H.R. and Chapple, R.P. (1995) Modeling the effect of genotype, environment and health on nutrient utilization. In: Danfaer, A. and Lescoat, P. (eds), *Proceeding 4th International Workshop on Modeling Nutrient Utilization in Farm Animals*. National Institute of Animal Science, Tjele, Denmark, pp. 85–105.

Bolet, G., Botte, F.M., Locatelli, A., Gruand, J., Terqui, M. and Berthelot, F. (1986) Components of prolificacy in hyperprolific Large White sows compared with the Meishan and Large White breeds. *Genetique, Selection, Evolution* 18, 333–342.

Brandt, H. (1987) Development and genetic aspects of feed intake in three breeds of pigs at German test stations and measures to prevent further deterioration. *Pig News and Information* 8(1), 29–33.

Branscheid, W. and Sack, E. (1988) Comparison of objective grading devices among extremely different breeds. *Pig News and Information* 9, 129–135.

Brendemuhl, J.H., Lewis, A.J. and Peo, E.R. Jr (1987) Effect of protein and energy intake by primiparous sows during lactation on sow and litter performance and sow serum thyroxine and urea concentrations. *Journal of Animal Science* 64, 1060–1069.

Bridges, T.C., Turner, U.W., Smith, E.M., Stahly, T.S. and Loewer, O.J. (1986) A mathematical procedure for estimating animal growth and body composition. *Transactions of the American Society of Agricultural Engineers* 29, 1342–1347.

Brody, S. (1945) *Bioenergetics and Growth*. Reinhold, New York, USA.

Brooks, P.H. and Smith, D.A. (1980) The effect of mating age on the reproductive performance, food utilization and liveweight change of the female pig. *Livestock Production Science* 7, 67–78.

Brown, J.E., Fitzburgh, H.A., Jr and Cartwright, T.C. (1976) A comparison of nonlinear models for describing weight–age relationships in cattle. *Journal of Animal Science* 42, 810–818.

Burnett, P.J., Walker, N. and Kilpatrick, D.J. (1988) The effect of age and growth traits on puberty and reproductive performance in the gilt. *Animal Production* 46, 427–436.

Cameron, N.D. and Curran, M.K. (1994) Selection for components of efficient lean growth in pigs. 4. Genetic and phenotypic parameter estimates and correlated responses in performance traits with ad-libitum feeding. *Animal Production* 59, 281-291.

Cameron, N.D. and Curran, M.K. (1995) Genotype with feeding regime interaction in pigs divergently selected for components of efficient lean growth rate. *Animal Science* 61, 123-132.

Cameron, R.D.A. (1987) Sexual development and semen production in boars. *Pig News and Information* 8(4), 389-396.

Campbell, R.G. and Taverner, M.R. (1988) Genotype and sex effects on the relationship between energy intake and protein deposition in growing pigs. *Journal of Animal Science* 66, 676-686.

Campbell, R.G., Taverner, M.R. and Curic, D.M. (1983). The influence of feeding level from 20 to 45 kg live weight on the performance and body composition of female and entire male pigs. *Animal Production* 36, 193-199.

Christenson, R.K. (1981) Influence of confinement and season of the year on puberty and estrous activity of gilts. *Journal of Animal Science* 52, 821-830.

Christenson, R.K. (1993) Ovulation rate and embryonic survival in Chinese Meishan and white crossbred pigs. *Journal of Animal Science* 71, 3060-3066.

Christenson, R.K., Ford, J.J. and Redmer, D.A. (1984) Estradiol and testosterone metabolic clearance and production rates during pubertal development in boars. *Biology of Reproduction* 31, 905-912.

Christenson, R.K., Leymaster, K.A. and Young, L.D. (1987) Justification of unilateral hysterectomy-ovariectomy as a model to evaluate uterine capacity in swine. *Journal of Animal Science* 65, 738-744.

Clark, L.K. and Leman, A.D. (1986) Factors that influence litter size in parity two females. In: *Proceeding 9th Congress International Pig Veterinary Society*. Barcelona, Spain, p. 14.

Cleveland, E.R., Johnson, R.K., Mandigo, R.W. and Peo, E.R., Jr (1983) Index selection and feed intake restriction in swine. II. Effect on energy utilization. *Journal of Animal Science* 56, 570-578.

Conniffe, D. and Moran, M.A. (1972) Double sampling with regression in comparative studies of carcass composition. *Biometrics* 28, 1011-1023.

Curtis, S.E., Heidenreich, C.J. and Martin, T.G. (1967) Relationship between body weight and chemical composition of pigs at birth. *Journal of Animal Science* 26, 749-751.

David, P.J. and Johnson, R.K. (1982) Analysis of Nebraska swine field records: Selection practices and genetic trends. *Journal of Animal Science* 55(Suppl. 1), 143.

De Greef, K.H. (1992) Prediction of production; nutrition induced tissue partitioning in growing pigs. PhD thesis, Wageningen Agricultural University, Wageningen, The Netherlands.

De Haer, L.C.M. (1992) Relevance of eating pattern for selection of growing pigs. PhD thesis, Research Institute For Animal Production (IVO-DLO), Schnoord, Zeist, The Netherlands.

De Lange, C.F.M. and Schreurs, H.W.E. (1995) Principles of model application. In: Moughan, P.J., Verstegen, M.W.A. and Visser-Reyneveld M.I. (eds), *Modeling Growth in the Pig*. Wageningen Pers, Wageningen, The Netherlands, pp. 187-208.

De Vries, A.G. and Kanis, E. (1992) A growth model to estimate economic values for food intake capacity in pigs. *Animal Production* 55, 241–246.

De Vries, A.G. and Kanis, E. (1994) Selection for efficiency of lean tissue deposition in pigs. In: Cole, D.J.A., Wiseman, J. and Varley, M.A. (eds), *Principles of Pig Science*. Nottingham University Press, Nottingham, pp. 23–41.

Dunkin, A.C. and Black, J.L. (1987) The relationship between energy intake and nitrogen balance in the growing pig. In: Moe, P.W., Tyrrell, H.F. and Reynolds, P.J. (eds), *Energy Metabolism of Farm Animals*. Rowman & Littlefield, Beltsville. European Association for Animal Production 32, 110–113.

Dyck, G.W. (1974) Effects of a cold environment and growth rate on reproductive efficiency in gilts. *Canadian Journal of Animal Science* 54, 287–292.

Dyck, G.W. (1988) Factors influencing sexual maturation, puberty and reproductive efficiency in the gilt. *Canadian Journal of Animal Science* 68, 1–13.

Dyck, G.W. and Cole, D.J.A. (1986) The effect of restricted energy and nutrient intake after mating on reproductive performance of multiparous sows. *Animal Production* 42, 127–132.

Dyck, G.W. and Strain, J.H. (1983) Postmating feeding level effects on conception rate and embryonic survival in gilts. *Canadian Journal of Animal Science* 63, 579–585.

Efron, B. (1982) The jackknife, the bootstrap and other resampling plans. CBMS-NSF Conference Series in Applied Mathematics. *Society Industrial and Applied Mathematics*. Philadelphia, Pennsylvania, USA.

Ellis, M., Smith, W.O., Henderson, R., Whittemore, C.T. and Laird, R. (1983) Comparative performance and body composition of control and selection lines of large white pig. 2. Feed to appetite for a fixed time. *Animal Production* 36, 407–413.

Emmans, G.C. (1995a) Energy systems and the prediction of energy and feed intakes. In Moughan, P.J., Verstegen, M.W.A. and Visser-Reyneveld, M.I. (eds), *Modelling Growth in the Pig*. Wageningen Pers, Wageningen, The Netherlands, 1995, pp. 115–122.

Emmans, G.C. (1995b) Ways of describing pig growth and food intake using equations. *Pig News and Information* 16(4), 113–116.

Engel, B. and Walstra, P. (1991) Increasing precision or reducing expense in regression experiments by using information from a concomitant variable. *Biometrics* 47, 13–20.

Etienne, M., Camous, S. and Cuvillier, A. (1983) Effect of feed restrictions during growth on puberty and reproductive performance in gilts. *Reproductive and Nutritional Development* 23, 309–319.

Evans, D.G. and Kempster, A.J. (1979) The effects of genotype, sex and feeding regimen on pig carcass development. *Journal of Agricultural Science, Cambridge* 93, 339–347.

Ferguson, N.S. and Gous, R.M. (1993) Evaluation of pig genotypes. 1. Theoretical aspects of measuring genetic parameters. *Animal Production* 56, 233–243.

Ferrell, C.L. (1988) Contribution of visceral organs to animal energy expenditure. *Journal of Animal Science* 66 (Suppl. 3), 23–34.

Ferrell, C.L. and Cornelius, S.G. (1984) Estimation of body composition of pigs. *Journal of Animal Science* 58, 903–912.

Fitzburgh, H.A., Jr (1976) Analysis of growth curves and strategies for altering their shape. *Journal of Animal Science* 42, 1036–1051.

Ford, J.J. (1990) Differentiation of sexual behavior in pigs. *Journal of Reproductive*

*Fertilization* (Suppl. 40), 311–321.

Forrest, J.C., Kuei, C.H., Orcutt, M.W., Schinckel, A.P., Stouffer, J.R. and Judge, M.D. (1989) Potential new methods of on-line pork carcass evaluation. *Journal of Animal Science* 67, 2164–2170.

Fowler, V.R. and Ensminger, M.E. (1960) Interactions between genotype and plane of nutrition in selection for rate of gain in swine. *Journal of Animal Science* 19, 439–449.

Fowler, V.R., Bichard, M. and Pease, A. (1976) Objectives in pig breeding. *Animal Production* 23, 365–387.

Friesen, K.G., Nelssen, J.C., Goodband, R.D., Tokach, M.D., Schinckel, A.P. and Einstein, M.E. (1996) The use of growth curves for assessing the response to dietary lysine by high-lean gain gilts. *Animal Science* 62, 159–169.

Gama, L.T. and Johnson, R.K. (1993) Changes in ovulation rate, uterine capacity, uterine dimensions and parity effects with selection for litter size in swine. *Journal of Animal Science* 71, 608–617.

Gu, Y., Schinckel, A.P., Martin, T.G., Forrest, J.C., Kuei, C.H. and Watkins, L.E. (1992) Genotype and treatment biases in lean estimation of swine carcass. *Journal of Animal Science* 70, 1708–1718.

Gueblez, R., Gestin, J.M. and Le Henaff, G. (1985) Incidence de l'age et de l'epaisseur de lar dorsal a 100 kg sur la carriere reproductive des truies large white. *Journal Recherche Porcine en France* 17, 113–120.

Hammond, J.A. (1952) Physiological limits to intensive production in animals. *British Agricultural Bulletin* 4, 222–233.

Henken, A.M., Brandsma, H.A., Van der hel, W., and Verstegen, M.W.A. (1991) Difference in energy metabolism and protein retention of limit-fed growing pigs of several breeds. *Journal of Animal Science* 69, 1443–1453.

Hughes, P.E. and Verley, M.A. (1980) *Reproduction in the Pig*. Butterworth, London.

Hunter, M.G., Biggs, C., Foxcroft, G.R., McNeilly, A.S. and Tilton, J.E. (1993) Comparisons of endocrinology and behavioral events during the perovulatory period in Meishan and Large White hybrid gilts. *Journal of Reproduction and Fertility* 97, 475–480.

Hutchens, L.K., Hintz, R.L. and Johnson, R.K. (1981) Genetic and phenotypic relationships between puberty and growth characteristics in gilts. *Journal of Animal Science* 53, 946–951.

Hutchens, L.K., Hintz, R.L. and Johnson, R.K. (1982) Breed comparisons for age and weight at puberty in gilts. *Journal of Animal Science* 55, 60–66.

Joubert, D.M. (1963) Puberty in female farm animals. *Animal Breeding Abstracts* 31(3), 95–306.

Kanis, E. (1990) Effect of food intake capacity on genotype by feeding regimen interactions in growing pigs. *Animal Production* 50, 343–351.

Kanis, E, and De Vries A.G. (1992) Optimization of selection for food intake capacity in pigs. *Animal Production* 55, 247–255.

Kennedy, B.W., van der Werf, J.H.F. and Meuwissen, T.H.E. (1993) Genetic and statistical properties of residual feed intake. *Journal of Animal Science* 71, 3239–3250.

Kerr, J.C. and Cameron, N.D. (1995) Reproductive performance of pigs selected for components of efficient lean growth. *Animal Science* 60, 281–290.

Kerr, J.C. and Cameron, N.D. (1996) Responses in gilt traits measured during performance test, at mating and at farrowing with selection for components of efficient lean growth rate. *Animal Science* 63, 235–241.

King, R.H. and Dunkin, A.C. (1986) The effect of nutrition on the reproductive performance of first-litter sows. *Animal Production* 42, 119–125.

King, R.H., Cleary, G.V., Maughan, N. and Power, C. (1984) The effect of initial fat reserves of gilts on their subsequent reproductive performance. *Proceedings of the Australian Society of Animal Production* 15, 702–712.

Kirkwood, R.M. and Aherne, F.X. (1985) Energy intake, body composition and reproductive performance of the gilt. *Journal of Animal Science* 60, 1581–1589.

Knap, P.W. (1995) Aspects of stochasticity: variation between animals. In: Moughan, P.J., Verstegen, M.W.A. and Visser-Reynevald, M.I. (eds), *Modeling Growth in the Pig*. Wageningen Pers. Wageningen, The Netherlands, pp. 165–172.

Knap, P.W. (1996) Stochastic simulation of growth in pigs: protein turnover-dependent relations between body composition and maintenance requirements. *Animal Science* 63, 549–561.

Knap, P.W. and Schrama, J.W. (1996) Simulation of growth in pigs: approximation of protein turnover parameters. *Animal Science* 63, 533–547.

Koong, L.J., Nienaber, J.A. and Mersmann, H.J. (1983) Effects of plane of nutrition on organ size and fasting heat production in genetically lean and obese pigs. *Journal of Nutrition* 113, 1626–1631.

Lamberson, W.R., Johnson, R.K., and Zimmerman, D.R. and Long, T.E. (1991) Direct responses to selection for increased litter size, decreased age at puberty, or random selection following selection for ovulation rate in swine. *Journal of Animal Science* 69, 3129–3143.

Legault, C. (1985) Selection of breeds, strains and individual pigs for prolificacy. Control of Pig Reproduction II. *Journal of Reproduction and Fertility* (Suppl. 33), 151–166.

Legault, C. and Caritez, J.C. (1982) Premier bilan de l'expérimentation sur le porc Chinois en France 1. Performances de reproduction en race pure et en croisement. *Journal Recherche Porcine en France* 14, 127–136.

Lunstra, D.D., Borg, K.E. and Klindt, J. (1993) Changes in porcine testicular structure during development in Meishan boars. *Biology of Reproduction* 48 (Suppl. 1), 126.

Lutz, J.B., Rampacek, G.B. and Kraeling, R.R. (1985) Induction of ovulation in the prepuberal gilt by pulsatile administration of gonadotropin releasing hormone. *Domestic Animal Endocrinology* 2, 61–65.

McPhee, C.P., Williams, K.C. and Daniels, L.J. (1991) The effect for rapid lean growth on the dietary energy requirements of pigs fed to scale. *Livestock Production Science* 27, 185–198.

Mitchell, G., Smith, C., Makowes, M. and Bird, P.J.W.N. (1982) An economic appraisal of pig improvement in Great Britain. *Animal Production* 35, 215–224.

Moughan, P.J., Smith, W.C. and Stevens, E.V.J. (1990) Allometric growth of chemical body components and several organs in the pig (20–90 kg live weight). *New Zealand Journal of Agricultural Research* 33, 77–84.

Mrode, R.A. and Kennedy, B.W. (1993) Genetic variation in measures of food efficiency and their genetic relationships with growth rate and backfat. *Animal Production* 56, 225–232.

Mullan, B.P. and Williams, I.H. (1989) The effect of body reserves at farrowing on the reproductive performance of first-litter sows. *Animal Production* 48, 449–457.

National Research Council (NRC) (1988) *Nutrient Requirements of Swine*. National Academic Press, Washington, DC, USA.

Newton, E.A. and Mahan, D.C. (1992) Effect of feed intake during late development on pubertal onset and resulting body composition in crossbred gilts. *Journal of Animal Science* 70, 3773–3780.

Newton, E.A. and Mahan, D.C. (1993) Effect of initial breeding weight and management system using a high-producing sow genotype on resulting reproductive performance over three parities. *Journal of Animal Science* 71, 1177–1186.

Noblet, J. and Henry, Y. (1991) Energy evaluation systems in pig diets. In: Batterham, E.S. (ed.), *Manipulating Pig Production III*. Australian Pig Science Association, Attwood, Victoria, Australia, pp. 87–110.

Palmer, G.Y., Schinckel, A.P. and Einstein, M.E. (1993) Evaluation of lean growth of seven genotypes of swine. *Journal of Animal Science* 71(Suppl. 1), 244.

Parks, J.R. (1982) *A Theory of Feeding and Growth in Animals*. Springer-Verlag, New York.

Planella, J. and Cook, G.L. (1991) Accuracy and consistency of prediction of pig carcass lean concentration from P2 fat thickness and sample joint dissection. *Animal Production* 53, 45–352.

Quiniou, N., Noblet, J., van Milgen, J. and Dourmad, J.Y. (1995) Effect of energy intake on performance, nutrient and tissue gain and protein and energy utilization in growing boars. *Animal Science* 61, 133–143.

Quiniou, N., Dourmad, J.Y. and Noblet, J. (1996) Effect of energy intake on the performance of different types of pig from 45 to 10 kg body weight. 1. Protein and lipid deposition. *Animal Science* 63, 277–288.

Rae, W., Jones, S.D.M. and Kennedy, B.W. (1985) Pork carcass grading: the magnitude of breed and sex biases in the prediction of meat yield from a single fat measurement. *Canadian Journal of Animal Science* 65, 619–625.

Rozeboom, D.W. (1996) Gilt pool management to maximize longevity and lifetime performance. *Proceedings 1996 Allen D. Leman Swine Conference*. University of Minnesota, Minneapolis, Minnesota, USA, pp. 34–38.

Rydmer, L., Eliasson-Selling, L., Johanson, K., Stern, S. and Andersson, K. (1994) A genetic study of estrus symptoms at puberty and their relationship to growth and leanness in gilts. *Journal of Animal Science* 72, 1964–1970.

Schinckel, A.P. (1994a) Nutrient requirements of modern pig genotypes. In: Garnsworthy, P.C. and Cole, D.J.A. (eds), *Recent Advances in Animal Nutrition*. University of Nottingham Press, Nottingham, UK, pp. 133–169.

Schinckel, A.P. (1994b) The minimum number of traits for an accurate description of growth. In: *Proceedings 5th World Congress on Genetics Applied to Livestock Production*, vol. University of Guelph, Guelph, Canada, pp. 177–184.

Schinckel, A.P. and De Lange, C.F.M. (1996) Characterization of growth parameters needed as inputs for pig growth models. *Journal of Animal Science* 74, 2021–2036.

Schinckel, A.P., Johnson, R.K., Pumfrey, R.A. and Zimmerman, D.R. (1983) Testicular growth in boars of different genetic lines and its relationship to reproductive performance. *Journal of Animal Science* 56, 1065–1076.

Schinckel, A.P., Johnson, R.K. and Kittok, R.J. (1984) Testicular development and endocrine characteristics of boars selected for either high or low testis size. *Journal of Animal Science* 58, 675–685.

Schinckel, A.P., Harris, D.L. and Stewart, T.S. (1985) Swine genetic evaluation and selection programs. *The Professional Animal Scientist* 1, 8–14.

Schinckel, A.P., Berg, E.P., Wagner, J.R., Chen, W.F. and Forrest, J.C. (1995a) A method to reduce genotype and sex bias in prediction equations estimating pork carcass value. *Journal of Animal Science* 73 (Suppl. 1), 162.

Schinckel, A.P., Williams, N.H. and Einstein, M.E. (1995b) Concepts of pig growth. *Agricultural Research Bulletin No. 992*. Purdue University, West Lafayette, Indiana, USA.

Schinckel, A.P., Preckel, P.V. and Einstein, M.E. (1996) Prediction of daily protein accretion rates of pigs from estimates of fat-free lean gain between 20 and 120 kilograms live weight. *Journal of Animal Science* 74, 498–503.

Shields, R.G., Jr, Mahan, D.C. and Graham P.L. (1983) Changes in swine composition from birth to 145 kg. *Journal of Animal Science* 57, 43–54.

Smith, C. and Fowler, V.R. (1978) The importance of selection criteria and feeding regimes in the selection and improvement in pigs. *Livestock Production Science* 5, 415–423.

Stant, E.G., Martin, T.G., Judge, M.D. and Harrington, R.B. (1968) Physical separation and chemical analysis of the porcine carcass at 23, 46, 68 and 91 kilograms liveweight. *Journal of Animal Science* 27, 636–644.

Stein, T.E., Dijkhuizen, A., D'Allaire, S. and Morris, R.S. (1990) Sow culling and mortality in commercial swine breeding herds. *Preventative Veterinary Medicine* 9, 85–94.

Strang, G.S. (1970) Litter productivity in Large White pigs. I. The relative importance of some sources of variation. *Animal Production* 2, 225–231.

Tess, M.W., Bennett, G.L. and Dickerson, G.E. (1983) Simulation of genetic changes in life cycle efficiency of pork production. I. A bioeconomic model. *Journal of Animal Science* 56, 336–353.

Thompson, J.M., Sun, F., Kuczek, T., Schinckel, A.P. and Stewart, T.S. (1996) The effect of genotype and sex on the patterns of protein accretion in pigs. *Animal Science* 63, 265–276.

Wagner, J.R., Schinckel, A.P. and Forrest, J.C. (1993) Genotype and sex biases in the estimation of pork carcass composition. *Proceedings of the National Swine Improvement Federation*, St Louis, Missouri, USA, pp. 47–53.

Wagner, J.R., Schinckel, A.P., Chen, W.F., Forrest, J.C. and Coe, B.L. (1995) Body composition of swine during growth and development. *Journal of Animal Science* 73 (Suppl. 1), 141.

Walstra, P. (1980) Growth and carcass composition from birth to maturity in relation to feeding level and sex in Dutch Landrace pigs. PhD thesis, Agricultural University Wageningen, The Netherlands.

Webb, A.J. (1989) Genetics of feed intake in the pigs. *The Voluntary Feed Intake of Pig*. Pub. No. 13. British Society of Animal Production. Edinburgh, Scotland, pp. 41–50.

Webb, A.J. and Curran, M.K. (1986) Selection regime by production system interaction in pig improvement: a review of possible causes and solutions. *Livestock Production Science* 14, 41–56.

White, B.R., Lan, Y.H., McKeith, F.K., Novakofski, J., Wheeler, M.B. and McLaren, D.G. (1995) Growth and body composition of Meishan and Yorkshire barrows and gilts. *Journal of Animal Science* 73, 738–749.

Whittemore, C.T. (1983) Development of recommended energy and protein allowances for growing pigs. *Agricultural Systems* 11, 159-186.

Whittemore, C.T. (1994) Growth and the simulation of animal responses. In: Cole, D.J.A., Wiseman, J., Varley, M.A. (eds), *Principles of Pig Science*. Nottingham University Press, Nottingham, pp. 55-74.

Whittemore, C.T. and Fawcett, R.H. (1976) Theoretical aspects of a flexible model to simulate protein and lipid growth in pigs. *Animal Production* 22, 87-96.

Whittemore, C.T., Franklin, M.F. and Pearce, B.S. (1980) Fat changes in breeding sows. *Animal Production* 31, 183-190.

Whittemore, C.T., Tullis, J.B. and Emmans, G.C. (1988) Protein growth in pigs. *Animal Production* 46, 437-445.

Whittemore, C.T., Kerr, J.C. and Cameron, N.D. (1995) An approach to prediction of feed intake in growing pigs using simple body measurements. *Agricultural Systems* 47, 235-244.

Williams, N.H., Cline, T.R., Schinckel, A.P. and Jones, D.J. (1994) The impact of ractopamine, energy intake and dietary fat on finisher pig growth performance and carcass merit. *Journal of Animal Science* 72, 3152-3162.

Wilson, E.R., Johnson, R.K. and Wetteman, R.P. (1977) Reproductive and testicular characteristics of purebred and crossbred boars. *Journal of Animal Science* 44, 939-947.

Wise, T., Lunstra, D.D. and Ford, J.J. (1996) Differential pituitary and gonadal function of Chinese Meishan and European white composite boars: Effects of gonadotrophin-releasing hormone stimulation, castration, and steroidal feedback. *Biology of Reproduction* 54, 146-153.

Woltman, M.D., Clutter, A.C. and Buchanan, D.S. (1995) Effect of divergent selection for postweaning average daily gain on front-end structural soundness of market weight pigs. *Journal of Animal Science* 73, 1940-1947.

Yang, H., Eastham, P.R., Phillips, P. and Whittemore, C.T. (1989) Reproductive performance, body weight and body condition of breeding sows with differing body fatness at parturition, differing nutrition during lactation and differing litter size. *Animal Production* 48, 181-201.

Young, L.D. (1995) Reproduction of $F_1$ Meishan, Fengjing, Minzhu and Duroc gilts and sows. *Journal of Animal Science* 73, 711-721.

Young, L.D., Pumfrey, R.A., Cunningham, P.J. and Zimmerman, D.R. (1978) Heritabilities and genetic and phenotypic correlations for prebreeding traits, reproductive traits and principal components. *Journal of Animal Science* 46, 937-949.

Young, L.G., King, G.J., Shaw, J., Quinton, M., Walton, J.S. and McMillan, I. (1991) Interrelationships among age, body weight, backfat and lactation feed intake with reproductive performance and longevity of sows. *Canadian Journal of Animal Science* 71, 567-575.

# The Chemical and Physical Description of Feedstuffs

**3**

## P.J. Moughan,[1] G. Annison,[2] S.M. Rutherfurd[1] and J. Wiseman[3]

[1]Institute of Food, Nutrition and Human Health, Massey University, Palmerston North, New Zealand; [2]Australian Food Council, Barton, ACT 2600, Australia; [3]Faculty of Agricultural and Food Sciences, University of Nottingham, Sutton Bonington Campus, Loughborough, Leicestershire LE12 5RD, UK

## Introduction

The food eaten by an animal not only provides the nutrients which fuel metabolism and growth, but also directly influences numerous physiological processes. There is a reasonably well-developed quantitative understanding concerning the dependence of tissue deposition on levels of absorbed nutrients, with the latter being directly related to the chemical composition of the food. However, there is a much poorer quantitative, though reasonably good qualitative understanding of the influence of physical and chemical properties of a food on physiological function. If we are to realize a more complete quantitative understanding of digestion and metabolism in the pig, then it is critical that a fuller understanding not only of the chemical composition of feedstuffs but also of their physical and indeed biological properties be developed. The latter subject, in particular, is a rapidly emerging area of scientific investigation. It is now well established that, in addition to their recognized nutritional role, breakdown products of digestion, such as volatile fatty acids and small peptides, may have important physiological effects at gut level and systemically.

The objective of this chapter is to provide an overview of the chemical and physical characteristics of feedstuffs and how a description of these may assist in developing a quantitative understanding of pig biology. The contribution addresses the proteinaceous, carbohydrate, fat and mineral components of the feed separately.

# The Proteinaceous Component of Feedstuffs

To examine critically alternative hypotheses concerning whole body protein metabolism and thus to develop quantitative models of metabolism requires that protein and amino acid levels in diverse biological materials be known with accuracy. The amino acid composition of proteins is the main consideration, rather than secondary, tertiary and quaternary structures, although an understanding of the ability of dietary proteins to yield biologically active peptides during digestion is assuming a greater importance.

## Protein content

Protein can be determined by a variety of methods which are reviewed in detail by Pomeranz and Meloan (1980). Historically, crude protein has been used as a benchmark for comparing feedstuffs and is calculated based on a determination of total nitrogen (N). Total N is commonly determined by either the Kjeldahl or Dumas method (Pomeranz and Meloan, 1980; AFRC, 1987). However, crude protein may be inaccurate, since some feedstuffs contain appreciable amounts of non-protein N (AFRC, 1987). Furthermore, it is generally assumed that protein contains 16% N and consequently a conversion factor of 6.25 is used to convert N content to crude protein content. However, the N content can vary from as little as 4.2% up to 30% (Pomeranz and Meloan, 1980) depending on the amino acid composition, rendering the generalized conversion factor of 6.25 unsuitable in many circumstances. In particular, proteins containing high amounts of histidine, tryptophan and basic amino acids such as arginine and lysine will have higher N contents. Consequently, grain millers often use 5.7 and the dairy industry 6.38, as conversion factors (AFRC, 1987). Recently, correction factors all lower than the traditional factor of 6.25 have been reported (Salo-Väänänen and Koivistoinen, 1996). It is ironic that so much attention has gone into improving the accuracy of N determination, only to be negated if inaccurate conversion factors are used.

Rather than protein *per se*, however, it is the amino acids that limit rate of growth and lean tissue deposition. It is thus critical to a quantitative description of pig biology to be able to describe accurately the amino acid contents of feeds, tissues and physiological fluids. There are numerous amino acids and amino acid analysis is not straightforward.

## Amino acids

### Sample hydrolysis

The current standard method for protein hydrolysis consists of incubation of the material in 6 M HCl in an oxygen-free environment for 24 h at 110°C.

While this procedure will allow accurate determination of several amino acids, there are notable exceptions. Isoleucine and valine are not completely released from the protein under these conditions, while serine and threonine, methionine, cysteine, tyrosine, asparagine, glutamine and tryptophan are destroyed to varying degrees.

If complete removal of oxygen is not achieved then methionine and, in particular, cysteine are oxidized during hydrolysis. Consequently these amino acids are usually quantitatively oxidized to cysteic acid and methionine sulphone, respectively, prior to hydrolysis. While this oxidation step does improve the recoveries of both sulphur amino acids, cysteic acid is still partially destroyed during hydrolysis and indeed as much as 30% can be lost (Darragh *et al.*, 1996).

Tyrosine is susceptible to halogenation during hydrolysis, although the addition of phenol largely prevents this reaction from occurring (Nissen, 1992). The presence of iron and copper ions can also reduce tyrosine recoveries, so using redistilled hydrochloric acid is recommended (Finley, 1985). If samples contain more than around 5% fat, tyrosine and lysine can be modified or destroyed, and defatting is recommended (Finley, 1985). Significant losses of tryptophan will occur during acid hydrolysis, particularly in the presence of cysteine or carbohydrates (Finley, 1985). However, in pure proteins the addition of thiol compounds such as β-mercaptoethanol will reduce the loss of tryptophan, though in feedstuffs and even with the addition of thiol compounds the recoveries of tryptophan are low (Ng *et al.*, 1987). Steinhart (1984) has extensively reviewed methods for determining tryptophan and, traditionally, base hydrolysis has been most commonly used. More recently, reduction to dihydrotryptophan using pyridine borane prior to acid hydrolysis has been attempted with some success (Wong *et al.*, 1984). Despite more work having been conducted with tryptophan than with any other amino acid it remains the most difficult amino acid to quantitate.

Asparagine and glutamine are converted to aspartic acid and glutamic acid, respectively, during hydrolysis. Consequently, both aspartic acid and glutamic acid represent the sum of their acid and amide derivatives. Glutamine and asparagine can be determined directly by a variety of methods including esterification–reduction of carboxylic groups (Wilcox, 1967), carbodiimide modification of free carboxylic acid groups (Carraway and Koshland, 1972), after enzymatic hydrolysis (Tower, 1967), and after converting the amide to its corresponding amine (Soby and Johnson, 1981). While some of these methods show promise, few laboratories routinely determine asparagine and glutamine, as such.

Sequential hydrolysis times are necessary for the accurate determination of serine, threonine, isoleucine and valine (Robel and Crane, 1972). Serine and threonine can be more accurately determined in hydrolysates that have undergone hydrolysis for less than 24 h, while for isoleucine and valine an extended hydrolysis is required. Robel and Crane (1972) derived relationships between the amino acid release from a protein and amino acid

**Table 3.1.** The amino acid composition[a] of egg-white lysozyme determined by non-linear least-squares regression of different sequences of hydrolysis intervals ($A_o$) compared with a conventional 24-h hydrolysis value and the composition based on sequencing.[b] (From Darragh et al., 1996, reproduced with permission of the publisher.)

| Amino acid | $A_o$ | 24 h | Actual |
|---|---|---|---|
| Glutamine | 5.8 | 5.3 | 5 |
| Glycine | 12.1 | 10.4 | 12 |
| Serine | 9.8 | 8.4 | 10 |
| Threonine | 6.9 | 6.3 | 7 |
| Aspartic acid | 21.0 | 20.5 | 21 |
| Alanine | 12.0 | 11.3 | 12 |
| Valine | 5.8 | 5.8 | 6 |
| Isoleucine | 6.0 | 5.4 | 6 |
| Leucine | 8.0 | 7.5 | 8 |
| Tyrosine | 3.2 | 2.8 | 3 |
| Phenylalanine | 3.4 | 3.2 | 3 |
| Histidine | 1.1 | 1.1 | 1 |
| Lysine | 6.1 | 5.6 | 6 |
| Arginine | 11.0 | 10.3 | 11 |
| Methionine[c] | 1.9 | 1.9 | 2 |
| Cysteine[d] | 8.0 | 6.5 | 8 |

[a] Expressed as moles of the amino acid per mole of lysozyme protein.
[b] Mean value of duplicate analyses.
[c] Detected as methionine sulphone.
[d] Detected as cysteic acid.

destruction during hydrolysis in an attempt to more accurately determine amino acid contents. More recently, and based on the latter principles, algorithms have been developed to allow accurate prediction of amino acid content in lysozyme (Darragh et al., 1996), milk (Darragh, 1995) and hair (Hendriks et al., 1998). These models take into account the rate of release of amino acids from proteins and their subsequent rate of destruction during hydrolysis to predict accurately the amino acid content from the standard 24 h hydrolysis values. Such an approach enhances analytical accuracy (Table 3.1). While these models can be costly and time consuming to develop, as, ideally, separate models are required for individual protein sources, they do offer considerable promise for the routine correction of 24 h hydrolysis values. The development of a simple method to determine accurately all amino acids in a single assay will probably never be fulfilled, since the chemistry of the amino acids is so diverse. The current method of acid hydrolysis has changed very little since it was developed by Stein, Moore and co-workers in the early 1950s, and despite hundreds of papers being published on the subject in the interim, the hydrolysis step of amino acid analysis still remains the main source of variability (Finley, 1985).

Attempts have been made to reduce amino acid analysis time by reducing both the hydrolysis time and the chromatography run times. Reverse-phase chromatography has reduced the run time for the chromatography step, while utilizing higher temperatures has been reported to be a successful means of reducing hydrolysis times (Gehrke *et al.*, 1985). The subject of hydrolysis times has been reviewed extensively (Blackburn, 1978; Gehrke *et al.*, 1985; Rayner, 1985). Microwave hydrolysis has been examined as a means of reducing the hydrolysis time down to a few minutes and appears to offer promise (Chiou and Wang, 1989).

### The chromatography step

Quantitation of the liberated amino acids is most commonly carried out by either ion exchange or reverse phase chromatography. Gas–liquid chromatography is utilized to a lesser extent. Comprehensive reviews of this topic have been made by several authors including Hare *et al.* (1985) and Williams (1988, 1994).

### Presentation of results

In presenting amino acid compositions, it is usual practice to express them in units of g/100 g. This requires application of the molecular mass for each amino acid to convert moles of amino acid, obtained from analysis, to grams of amino acid. The choice of molecular mass is important since each amino acid has a molecular mass for the free molecule (including the water molecule added during hydrolysis) and a corresponding molecular mass for the amino acid as it is found bound within a protein. It is important when presenting amino acid data that the type of molecular mass (free or bound) used in correction is reported (Rutherford and Moughan, 1993).

### Pitfalls

Amino acid analysis is a complex and difficult method to perform and there are many reasons for variability in amino acid compositional data (Finley, 1985). One major source of error is the sampling of heterogeneous materials. Thorough mixing of samples will usually remedy this problem although extra replicates or larger sample sizes are often required for acceptable precision. Subsampling, pooling and remixing before taking the analytical sample is also recommended to minimize variation in analysis (Finley, 1985).

It is also important that analysts use a range of standards for quality control. Internal standards should be used routinely to account for physical losses of sample during sample preparation, malfunctioning of analytical equipment (for example leaking fittings or faulty injectors), or for incomplete derivatization. The internal standard can be any compound that: (i) is not present in the sample being analysed; (ii) will separate from all other detected compounds during chromatography; and (iii) will be detected using the method employed. Some common examples include norleucine, amino isobutyric acid and norvaline. There is no consensus as to the point of inclusion of the internal standard. Some researchers add the internal standard prior to hydrolysis (van der Meer, 1990); however, since the extent of hydrolytic losses varies with each amino acid, addition of the internal standard prior to hydrolysis may not be appropriate (Robel, 1973). An alternative is to add the internal standard immediately after hydrolysis (Davies and Thomas, 1973; Sarwar *et al.*, 1988; Malmer and Schroeder, 1990), in which case the internal standard acts as a control for the chromatography step of the procedure rather than the hydrolysis step.

It is also good practice to include a standard protein with every batch of samples being hydrolysed and analysed. The standard protein should be of known amino acid content (sequence analysis), and glycoproteins should be avoided. By checking the recovery of amino acids from standard proteins, the overall accuracy of the analytical procedure can be monitored.

Participation in amino acid analysis proficiency programmes is also important for reducing variability between laboratories. This not only gives the analyst confidence that his or her laboratory is performing well in comparison with that of their peers, but also allows assessment of the inter-laboratory variation that can be expected with amino acid analysis. Variation between laboratories depends on several factors including sample type, methods employed, and the amino acids being determined. In recent collaborative studies coefficients of variation were typically between 4 and 20%, with the amino acids tryptophan, cysteine, methionine, histidine and proline being the most variable (van der Meer, 1990; NATA, 1993, 1994; Williams, 1994; Friedman, 1996).

Inappropriate storage of samples prior to analysis can also increase the variability of results. Photo-induced oxidation of cysteine has been discussed by Forbes and Savige (1962), while Finley (1985) has reported losses of methionine and tyrosine, respectively, from samples exposed to ultraviolet light. Furthermore, prolonged storage at ambient temperature, and certain forms of processing can lead to reduced levels of arginine and lysine, which undergo reactions with a variety of compounds found in feedstuffs (Hurrell and Carpenter, 1981). Some of the lysine derivatives can revert back to lysine during acid hydrolysis leading to overestimates of lysine content. This is a particularly serious problem, since lysine is often the first limiting amino acid in diets for the pig. For such samples other methods for detecting the unreacted lysine are required. These include the fluorodinitrobenzene (FDNB), trinitrobenzene sulphonic acid (TNBS) and homoarginine methods (Hurrell and Carpenter, 1981). Recently, a new method for determining absorbed available lysine has been developed (Moughan and Rutherfurd, 1996).

### Physiological samples

Prior to the analysis of free amino acids in physiological samples it is necessary to remove intact protein. Numerous methods have been used to remove protein from such samples and include picric acid or sulphosalicylic acid precipitation, high speed centrifugation, ultrafiltration and ion-exchange resins (Williams, 1988). Use of an internal standard is also recommended here and is most suitably added prior to the ultrafiltration step (Davey and Ersser, 1990). Analysis of physiological samples is often more difficult than for protein hydrolysates since there are usually many more compounds present.

### Summary

Despite considerable effort having been expended over the years on improving analytical methods, amino acid analysis remains a difficult procedure requiring meticulous attention to detail if reliable results are to be achieved. Those seeking a quantitative description of amino acid ingestion, uptake and metabolism must assess sources of data carefully and need to appreciate the inherent limitations. Providing there is a strict adherence to protocol, modern amino acid analysis can provide data with a within-laboratory repeatability of about 5% and a reproducibility between laboratories of around 10%.

# Dietary Carbohydrate

The carbohydrate fraction of a feedstuff gives rise after digestion in the animal to simple sugars or volatile fatty acids and organic acids, all important sources of energy. Traditionally, in pig nutrition, the energy contribution from dietary carbohydrates has been represented indirectly in the digestible energy (DE) measure. However, if a more quantitative understanding of metabolism is sought, it may be necessary to distinguish between the different compounds entering the portal blood, arising from the carbohydrate fraction.

Characterization of the carbohydrate fraction of foods has been the subject of a recent review (Englyst and Hudson, 1996). In addition to discussing the simple sugars, sugar alcohols and oligosaccharides, the latter authors make the valuable distinction between rapidly digestible starch, slowly digestible starch and resistant starch. The latter material is starch and starch degradation products escaping digestion in the small bowel and passing into the large intestine. Part of the resistant starch will be fermented in the large intestine. The nutritional significance of different starch fractions, as far as livestock nutrition is concerned, is yet to be investigated. Certainly, there may be value in assessing the level of 'resistant' starch in feedstuffs for some animals, particularly when starch digestibility is known to be less than complete. Some starches, such as raw potato starch, are poorly digested which relates to the particularly crystalline structure (Annison and Topping, 1994) of tuber starches. Cereal starches are generally well digested in the absence of antinutritive factors while legume starches are intermediate in their digestibility. It should be remembered, however, that processing and subsequent storage conditions can have a profound influence on starch digestibility.

Of major physiological, as opposed to nutritional significance, however, is the non-starch plant polysaccharide (NSP) fraction. The NSPs have major effects on physiological functions such as voluntary food intake, stomach emptying, digesta flow, rate and extent of food digestion and gut endogenous protein loss. NSPs can have a significant influence on the weight and volume of the various digestive organs and on gut cellular proliferation. Description of the NSP (fibre) component is discussed in the following section.

## The NSPs

Until quite recently the fibre fraction of feeds and feed ingredients for monogastrics was considered relatively unimportant. This is because chemical methods to measure it detected fractions (crude fibre, acid detergent fibre and neutral detergent fibre) which were largely biologically inert. Nutritional science has progressed considerably in recent years in response to more sophisticated measures of that part of plant feed ingredients known as 'fibre'. This has led to the realization that fibre fractions can have important nutritional and physiological effects. Firstly, they may contribute significantly to energy supply through fermentation in the hindgut, which is of particular importance to pigs, and secondly, some specific fibre components display antinutritive activity by directly inhibiting digestive processes (Annison, 1993).

Before detailing methods to determine the NSPs, the chemistry and structure of the components are briefly described. This is important for understanding why more traditional methods of fibre determination fail to provide physiologically useful information. Understanding the chemistry also gives an appreciation of the shortcomings of the newer methods and highlights the need for even more advanced techniques.

## NSP structure

Polysaccharides are macromolecular polymers of simple sugars or monosaccharides. The sugars are linked together by a specific type of linkage called a glycosidic bond which is formed between the hemiacetal (or hemiketal) group of one sugar and the hydroxyl group of another. The number of possible glycosidic bonds between monosaccharides is very high. The importance of the bonds in determining the susceptibility to digestive enzymes is illustrated by the resistance of cellulose (a $\beta$-(1–4) glucan) to starch-degrading enzymes. The great majority of NSPs in monogastric diets are of plant origin and in modern diet formulations the variety present may be large, which reflects the diverse ingredients currently available. Many different types of sugars may be present.

The sugars combine to form polysaccharides. The polysaccharides may be relatively simple such as the $\beta$-D-glucans which are linear polymers of glucose with $\beta$-(1–3),(1–4) glycosidic links (Fincher and Stone, 1986). The other major cereal polysaccharides, the arabinoxylans, are more complex being composed of two sugars, arabinose and xylose, in a branched structure (Annison *et al.*, 1992). Even more complex polysaccharides may be present if legumes are used in the ration. The main NSP of lupins is the highly complex branched structure containing long $\beta$-(1–4)-D-galactose side chains attached to a pectin-like main chain of rhamnose and galacturonic acid linked by $\beta$-(1–4) and $\alpha$-(1–2) bonds, respectively. There are also side chains of $\alpha$-(1–5)-L-arabinose (Evans *et al.*, 1993).

## Plant cell wall structure

The great range in the chemical structures of the NSPs means that as isolated materials they have quite different physical properties and nutritional activities. It is important to remember, however, that NSPs do not exist as completely separate materials in feedstuffs. Many NSPs are cell wall components and are closely associated with other polysaccharides or non-carbohydrate material such as protein and lignin (Fincher and Stone, 1986; Selvendran *et al.*, 1987). Cell walls are highly ordered, consisting of different polysaccharides, polyphenols and glycoproteins and glycolipids arranged in three main patterns to give fibrillar polysaccharides (mainly cellulose), matrix polysaccharides (mainly hemicellulose and pectin) and encrusting substances (mainly lignin). A second level of association is the bonding between molecules of the cell wall components. There may be relatively weak hydrogen bonding or stronger ionic cross-linking through cations or covalent bonds. These associations must be considered because they will influence the manner in which the NSPs behave when ingested. Solubility is an important factor in the antinutritive activity of NSPs (Annison, 1993). The solubility of an NSP is determined not only by its primary structure, but also by how it is bound to other cell wall components. The digestion and hindgut fermentation of the polysaccharides may be inhibited due to links to other components.

It is important that these relationships are understood as they impinge directly on the requirements for analysis. It is essential that analysts realize that routine 'off the shelf' procedures may not be appropriate for some feed ingredients – a degree of tailoring will probably be required to provide a procedure which gives meaningful results.

## Analysis of fibre and NSP

The relationship between plant fibre components and the various analytical methods is shown in Table 3.2. This serves as an overall scheme for describing the methods in greater detail.

### Crude fibre (CF)

The determination of crude fibre involves defatting the sample followed by two digestion (boiling) steps. These steps will solubilize considerable portions of the NSP, both NSP which is likely to be soluble when ingested and a fraction which may remain insoluble during *in vivo* digestion. The solid residue is washed, dried and weighed and is termed 'crude fibre'. This material will be unaffected by passage through the small intestine of monogastrics and will probably have little nutritive effect with the exception of possibly having some influence on the bulking of stools. The material may also influence fermentation processes in the hindgut of monogastrics but the nutritive consequences will be minimal.

### Acid detergent fibre (ADF)

The determination of ADF involves a single digestion of the feed sample in boiling acetyl trimethyl ammonium bromide in 0.5 M sulphuric acid. The washed, dried residue is ADF. Once again the digestion step will solubilize and degrade large portions of NSP and so the residue will consist largely of insoluble NSP and lignin associated with little, if any, nutritional activity.

### Neutral detergent fibre (NDF)

In the determination of NDF the sample is treated with a solution containing sodium dodecyl sulphate and ethylene diamine tetra-acetic acid in phosphate buffer at pH 7. This solution will remove readily soluble NSP as well as some less soluble material such as the pectic substances. The residue may well have some NSP remaining which has nutritive activities but it is likely that in most samples, the amount will be low.

The determinations of CF, ADF and NDF do not account for the soluble NSP in feed samples. Thus materials such as soluble β-glucans, arabinoxylans and the polyuronates are not measured. It is precisely these polysaccharides which are known to have antinutritional effects in monogastrics.

## Modern methods for the determination of NSP

NSP determination is difficult because of the diverse structure of NSPs. This has hindered the development of a single routine method. Additionally, NSPs are usually present at low absolute levels and the levels of starch and sometimes simple sugars are relatively high. Both starch and simple carbohydrates must be removed before direct determination of the NSP. The principles behind most methods are shown in Table 3.3. For samples with

**Table 3.2.** Model for definitions and methods for the determination of fibre non-starch polysaccharides. (After Mugford 1993.)

| Resistant to digestion in the small intestine | Total dietary fibre (gravimetric) | Non-starch polysaccharides (NSP) | Non-cellulosic polysaccharides (NCP) | | | Plant cell wall material |
|---|---|---|---|---|---|---|
| | | | Neutral polysaccharides | Soluble fibre | Other sugars | |
| | | | Acidic polysaccharides | | Uronic acids | |
| | | | | | Rhamnose | |
| | | | | | Xylose | |
| | | | | | Arabinose | |
| | | | | | Mannose | |
| | | | | | Galactose | |
| | | | Cellulose | Insoluble fibre | Glucose | |
| | | | | Crude fibre | Lignin | |
| | | | | | Protein residues | |
| | | | | | Minerals | |
| | Resistant starch[a] | | | | | |
| | Oligosaccharides | | | | | |

[a] Resistant starch will be measured as NSP unless specific measures are taken to quantify it separately.

**Table 3.3.** Procedures involved in the determination of non-starch polysaccharides in feeds and feed ingredients. (After Mugford, 1993.)

| Method | Process | | |
|---|---|---|---|
| Sample preparation | homogenize – improves extractions and reproducibility | | |
| Pre-treatment | de-fat – fat interferes with analysis, remove glucose, sucrose, oligosaccharides | | |
| Pre-hydrolysis | autoclave, DMSO or KOH – solubilizes some protein, NSP and starch | | |
| Amylolysis | hydrolysis of starch → Supernatant aliquot – determine glucose colorimetrically, calculate total starch. | | |
| Precipitate-soluble NSP | add ethanol (precipitation of soluble NSP to combine with insoluble NSP) | Add ethanol (precipitation of soluble NSP) | |
| (separation) | (Filter and weigh – total fibre; includes protein, ash) | centrifugation of sample → Supernatant / Residue | |
| Hydrolysis | Hydrolyse (acid) | Hydrolyse (acid) | Hydrolyse (acid) |
| Quantify | Measure neutral sugars by HPLC or GLC (following derivatizations). Measure uronic acids by spectrophotometry | Measure neutral sugars by HPLC or GLC (following derivatizations) Measure uronic acids by spectrophotometry | Measure neutral sugars by HPLC or GLC (following derivatizations) Measure uronic acids by spectrophotometry |
| Calculate | Total neutral non-starch polysaccharide and uronates | Soluble neutral NSP and uronates | Insoluble neutral NSP and uronates |
| | Total fibre NSP | Total soluble NSP | Total insoluble NSP |

high levels of fat (5% +) a defatting procedure is required. The simple sugars (glucose, sucrose and oligosaccharides) are removed by extraction into ethanol or methanol. The next step is the removal of starch by digestion with enzymes (amylases and amyloglucosidase). It is crucial that the digestion is complete as any remaining starch will be recorded as NSP in the next steps. During the digestion of starch some NSP is also solubilized. This provides opportunity to split the sample to allow the determination of soluble and insoluble NSP. If total NSP is required, four volumes of ethanol should be added to precipitate the soluble NSP. The mixture is then centrifuged and the residue is washed and dried. If both soluble and insoluble NSP are to be determined, following starch digestion the reaction mixture is centrifuged. Four volumes of ethanol are added to the supernatant to precipitate the soluble NSP. The centrifuged residue which contains the insoluble NSP is then washed and dried. The residues can be weighed to give measures of soluble, insoluble and total fibre. The fibre residues may also contain appreciable amounts of ash and protein and these must be accounted for (i.e. protease treatment/ash determination) to increase the accuracy of the method.

Alternatively, the NSP can be determined specifically, using procedures which can be applied to the total NSP residue or to the soluble and insoluble NSP residues. To quantify the NSP fractions the polysaccharides must be hydrolysed to their constituent sugars which are then quantified. Several hydrolytic procedures have been developed and some procedures are better suited to some samples. For the hydrolysis of total and insoluble NSP, pretreatment with 12 M sulphuric acid is essential. This dissolves the cellulose in the sample and renders it vulnerable to acid hydrolysis. The subsequent hydrolysis then employs sulphuric acid of varying strengths and different reaction times but the temperature of hydrolysis is usually 100°C. For the soluble NSP, pretreatment is usually unnecessary and the acid is often trifluoroacetic acid (2 M) at a temperature of 125°C. Following hydrolysis, the sugars are quantified. In simple systems this may employ a colorimetric agent (i.e. 3,5-dinitrosalicylic acid) which is specific for the reducing (aldehyde) groups of the free sugars. A problem with this approach is that the presence of protein in the sample may interfere with determination. In addition, it gives no information on the type of polysaccharide present. More specific approaches involve identification and quantification of the monosaccharides either by high performance liquid chromatography (HPLC) or by gas–liquid chromatography (GLC). The advantage of HPLC is that no further sample work-up is required. With GLC, volatile derivatives must be prepared. Most commonly, alditol acetates are used and the procedures for their preparation are well established. The advantage of both the HPLC and GLC procedures is that the sugar profile obtained can often give some indication of the types of polysaccharides present.

Another drawback of the GLC and uronic acid methods is that the acidic sugars (uronic acids) do not behave as the neutral sugars and a further determination must be used. Often a colorimetric procedure is used to determine the uronic acids.

The above discussion describes how the analysis of feeds has been developed to give values for total, soluble and insoluble NSP. While these procedures are useful in helping to define the quality of diets, there is not a direct relationship between NSP level and nutritive value and physiological effects. Not all the NSP has antinutritive activities and little is known about the fermentability of specific NSPs and hence their true energy value. Methods are required which measure specific NSPs directly, rather than quantifying their sugars after hydrolysis.

### Enzymatic methods

Enzymes which cleave polysaccharides are called glycanases. Glycanases with high specificity can be used to cleave specific polysaccharides in feed samples. The release of the sugars is then monitored, usually by a colorimetric procedure. The routine determination of β-glucan in feeds is now possible using specific β-glucanases and glucosidases. The determination of arabinoxylans and galactomannans by specific glycanases is also possible (McCleary, 1993). Detailed methods have still to be developed, but one can imagine a situation in the future where specific enzyme procedures are available to determine specific NSPs with known antinutritive activities.

## Physical properties of 'fibre'

As well as being able to characterize the NSPs chemically, it is equally important to be able to describe their physical properties and thus their physiological effects, which are becoming increasingly understood (Low, 1993), at least from a qualitative perspective. Although related to chemical composition, such effects cannot be predicted based on a knowledge of chemical composition alone. It will be important to be able to characterize NSPs with respect to their potential fermentability in the gut, to their contribution to digesta viscosity and to properties such as water holding capacity or 'bulk'. Such characteristics are known to have physiological influences. For example, degree of fermentability will affect microbial metabolism and thus influence the degradation of materials in the large intestine and through this, the passage of compounds across the colonic mucosa. Volatile fatty acids, the products of microbial fermentation, are known to elicit effects on gut cellular proliferation and overall gut function. The bulkiness of a diet will influence voluntary food intake and may lead to changes in the weight and volume of different parts of the gut (Low, 1993). Fibre-induced increases in weight of the digestive tract as a proportion of total body weight may be quantitatively important. Digesta viscosity affects rate of passage, the extent of digestion of food macromolecules and may lead to increased endogenous protein loss from the small bowel. Much work has been done to show that higher dietary fibre levels are associated with an overall lowering of dietary nutrient digestibility, but this effect has not been related to identifiable properties of fibre, which is required if this important effect is to be predicted more generally. The inclusion of a fibre term in multiple regression equations does assist in the accuracy of prediction of dietary digestible energy and crude protein (Noblet and Perez, 1993). Progress is being made at characterizing NSPs in ways that better correlate with their physiological properties (Smits and Annison, 1996), but much more work needs to be done in this area. The distinction between soluble and insoluble fibre fractions is useful as are *in vitro* measures of viscosity. The recent work of Monro (1993), which describes a procedure for determining dietary soluble fibre that mimics conditions in the gut and correlates well with *in vivo* solubilization, is interesting.

## Viscosity

A majority of polysaccharides, when dissolved in water, give viscous solutions. Viscosity is dependent on a number of factors, including the size of the molecule, whether it is branched or linear, the presence of charged groups and, of course, the concentration. The polysaccharides increase the viscosity of water at low concentrations by directly interacting with the water molecules. The relationship between concentration and viscosity is not linear, but more akin to an exponential function, as the prime determinant is the volume 'swept' as the polysaccharide molecule tumbles through space.

This relationship breaks down as the concentration increases and the molecules of the polysaccharide interact and become entangled in a network. This process can cause great increases in the viscosity and is dependent on the formation of junction zones between the polysaccharide molecules. Gel formation can occur when the interactions of the polysaccharide molecules become great. Increases in digesta viscosity associated with wheat arabinoxylan ingestion have been noted in broiler chickens (Choct *et al.*, 1996) but similar experiments have not been performed in pigs. Given that relationships between dietary NSPs and digesta viscosity postabsorption can be developed, it may be possible to predict certain physiological outcomes. For example, in the work of Larsen *et al.* (1993), viscosity, a specific property of fibre, was shown to be related to increases in small intestinal endogenous protein loss (Fig. 3.1).

## Ion binding

Some NSPs such as pectins may have a high charge density at some pH values due to the presence of acidic groups. Apart from the association of cations with the negatively charged groups, in some NSPs the stereo-structure of the polysaccharide allows the chelating of ions to occur. Indeed, cations can form ionic bridges between NSP molecules, profoundly influencing their viscosity and gel-forming properties.

Although there is little supporting evidence, the possibility of cations

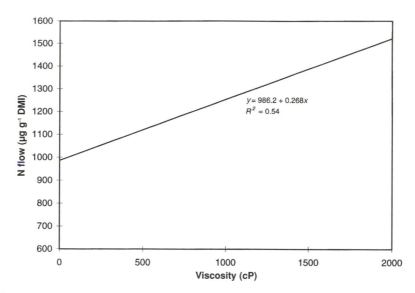

**Fig 3.1.** Endogenous N flow in ileal digesta related to dietary fibre viscosity for growing rats fed a protein-free diet.

acting as ion bridges between polysaccharides and small charged molecules, should not be discounted.

### Surface activity

Polysaccharides can present charged (negative and less commonly positive) as well as weakly hydrophobic and hydrophilic surfaces. When in solution they tend, therefore, to associate with other such surfaces. When ingested these may be the surfaces of food particles, the surface of lipid micelles or the glycocalyx surface of the gut. Once again, molecular mass and chain length play a major role as allosteric effects are almost certain. This effect may influence nutrient digestibility.

### Water-holding activity

Both soluble and insoluble NSPs have water-holding activities. The insoluble forms such as cellulose and xylans behave like sponges whereas the soluble forms can entrap water molecules through the formation of networks. This becomes particularly marked at higher concentrations where gels may form. It has been shown (Decuypere *et al.*, 1994) that the water-holding capacity (WHC) of fibrous feed ingredients is associated with increased endogenous gut nitrogen losses of both secretory and bacterial origin in the pig. Further, the work of Kyriazakis and Emmans (1995) clearly demonstrates an effect of the WHC of a feedstuff on *ad libitum* food intake in pigs. Scaled daily food intake (g kg$^{-1}$ live weight) was found to be directly proportional to the reciprocal of the WHC of the feeds. These results encourage further investigation of WHC as a predictor of maximum feed intake capacity of pigs receiving high-fibre bulky feeds. This has important implications to enabling robust quantitative description of the ingestion and utilization of food by pigs and is a much more satisfying approach than earlier quantitative predictions of voluntary food intake based on dietary crude fibre contents (Agricultural Research Council, 1967). Not only is the amount and type of dietary fibre important in influencing voluntary feed intake in the pig, but the interaction of factors such as live weight, animal adaptation, feeding regimen and climate also need to be taken into account (Black, 1995).

## Summary

The technologies for assessing fibre are slowly improving and as momentum builds to utilize a wider variety of raw materials for animal production further advancements are assured. The wide range of 'fibre' components will challenge the ingenuity of the analyst, particularly in terms of getting meaningful results. It will be necessary to look further than simply estimating to what extent carbohydrates contribute to energy either directly following digestion or as hindgut fermentation products. The physicochemical properties of fibres and how these impact quantitatively on animal function will need to be defined. This may require analytical methods to be developed which are sensitive to quite subtle structural features in the fibre macromolecules.

# The Lipid Component

Fats and oils are important dietary components. Their calorific value is at least twice that of other ingredients, they are sources of essential fatty acids (and, to a lesser extent, fat soluble vitamins), they contribute to palatability particularly if other dietary ingredients are somewhat dry and dusty and act as lubricants during the milling and mixing process. They also have specific physiological influences such as the effect on rate of stomach emptying. They are, however, extremely variable commodities. It is the purpose of this section to review their chemical composition and subsequent dietary energy value.

## Chemical structure of fats and oils

Fats and oils found widely in nature are present invariably as triacylglycerides. These molecules consist of the trihydric alcohol glycerol to which three individual fatty acids are esterified giving neutral lipids. Their primary function within organisms is as a means of storing energy. Another class of lipid, phospholipid, consists of glycerol, two fatty acids and a phosphorylated molecule, for example, choline giving phosphatidyl choline (lecithin). These are present in only very low concentrations in dietary lipid. The concentration of free fatty acids in naturally-occurring fats and oils is usually very low.

Fatty acids consist of a carbon chain with terminal methyl and carboxyl groups. The vast majority of fatty acids employed in pig nutrition have chain lengths of between 14 and 20 carbon atoms, although fatty acids with longer chain lengths (22, 24 and 26) are found in fish oils. The carbon to carbon bonds are either saturated or unsaturated (usually *cis* although *trans* isomers are found). This particular characteristic has a profound effect upon the physical properties of the fatty acid in question. Thus the melting point of *n*-octadecanoic acid (18 atoms of carbon with no double bonds – C18:0, common name stearic acid) is 70°C, whereas that of *cis*-9-octadecanoic (one double bond – C18:1, oleic acid) is 13°C, that of *cis*-9,12-octadecadienoic acid (two double bonds – C18:2, alpha-linoleic acid) is $-5$°C and that of *cis*-9,12,15-octadecatrienoic acid (three double bonds – C18:3, alpha-linolenic acid) is $-11$°C.

The position of the double bond (i.e. precisely which carbon atoms are involved) is of some significance in metabolism. While desaturation (i.e. inserting double bonds) is chemically possible, higher animals including pigs are unable to insert such bonds between the last double bond and the terminal methyl group. Thus the length of the carbon chain between these two is fixed and defines the fatty acid family. Thus oleic acid is referred to as omega-9 (or n-9); *cis*-9-hexadecanoic (one double bond – C16:1, palmitoleic acid) as omega-7 (or n-7); linoleic acid as omega-6 (or n-6) and linolenic as omega-3 (or n-3). The omega-6 and omega-3 fatty acids are metabolized *in vivo* to higher order molecules (such as prostaglandins, prostacyclins, thromboxones), all of which have a fundamental role as controllers of cellular function. In addition, alpha-linoleic acid is metabolized into *cis*-5,8,11,14-eicosatetraenoic acid (four double bonds – C20:4, arachidonic acid which is a key component of cell membranes) which, it should be noted, is still an omega-6 fatty acid. The omega-6 and, to a lesser extent, omega-3 fatty acids are, therefore, of major metabolic significance. Because they cannot be synthesized by the animal from other families, they are termed essential fatty acids.

Triacylglycerides usually consist of more than one type of fatty acid but, in general, they are of a similar nature. Thus plant oils tend to consist predominantly of unsaturated fatty acids whereas animal fats are more saturated. There are, however, specific types of plant and animal lipids characterized by the predominant fatty acid. Because of the influence of degree of saturation of a fatty acid on melting point, the plant oils and animal fats described have similar properties to their constituent fatty acids. Generally, plant oils (with the notable exception of palm oil) are liquid at room and biological temperatures whereas animal fats are not. Fish oils consist predominantly of long carbon chain highly unsaturated fatty acids and are liquid at these temperatures.

## Chemical modifications to fats and oils during processing

Fats and oils destined for inclusion into diets for pigs are invariably subjected to a series of chemical processes. These range from degumming, refining and filtering of vegetable oils (designed to purify the commodity for human consumption) through to the rendering of animal wastes. A further category of so-called 'feed grade' fats are the recovered vegetable oils which have outlived their usefulness as commodities involved in cooking human food items. It is important to appreciate, therefore, that fats and oils employed in animal nutrition are secondary products from another industry. The simplest chemical modification arising from these processes is an increase in the proportion of free fatty acids.

Fats and oils are relatively unstable molecules, the more so the greater the degree of unsaturation, and the heating processes to which they are subjected can alter their chemical structure significantly, particularly if such processing is in the presence of oxygen (Artman, 1969; Wiseman, 1986).

Lipid oxidation is a sequence of particularly complex chemical reactions. However, the major result is that degradation products are generated from the fatty acids and triacylglycerides originally present. These products range from small molecular mass volatile components (e.g. aldehydes and ketones; Chang *et al.*, 1978) through to fatty acids which have additional functional groups present (Artman and Alexander, 1968). More severe conditions will lead to the production of cyclic and dimeric/polymeric fatty acids and triglycerides (McInnes *et al.*, 1961; Barrett and Henry, 1966; Ohfuji and Kaneda, 1973).

## Nutritional value of fats and oils

Fats and oils are important sources of energy and essential fatty acids. To realize this potential, however, the material must be digested. Digestion of fats and oils proceeds through hydrolysis of the triacylglyceride, following emulsification in conjunction with bile salts, under the action of pancreatic lipase, and entry of the hydrolytic products (free fatty acids and 2-monoacylglycerides) into micelles prior to absorption (Freeman, 1984). The variability in chemical composition of fats and oils described above is of fundamental importance to the overall digestive process. Increasing unsaturation of constituent fatty acids leads to greater absorption and, ultimately, higher digestible energy (DE) values whereas elevated levels of free fatty acids in the fat or oil will lead to reduced DE content. Chain length is of some significance, but, in practice, is probably of minor importance.

Complex dimeric and polymeric fatty acid structures arising from oxidation and heat damage are probably completely indigestible. Indeed, they may even have a negative nutritional value as the physicochemical conditions within the gut associated with their presence (excessively viscous digesta) may interfere with the digestion and absorption of other dietary components (Hsieh and Perkins, 1976; Perkins and Taubold, 1978; Combe *et al.*, 1981).

Minor modifications to fatty acids (for example presence of functional groups) may not interfere with absorption but such molecules may not have the same nutritional value as intact fatty acids and may in fact be damaging to the system because of their oxidized nature. Feeding of so-called oxidized fats can lead to gross physiological changes, including liver enlargement, which are symptomatic of the presence of toxins. Hydroperoxides are extremely toxic molecules, and although in all probability only sparingly absorbed, may be damaging (Andrews *et al.*, 1960; Poling *et al.*, 1962). As with polymeric structures, oxidized fats may adversely influence the activity of other nutrients in the diet, particularly fat-soluble vitamins and essential amino acids (Alexander, 1966; Lea *et al.*, 1966).

## The prediction of the DE of fats

Fats and oils as raw materials in pig nutrition are frequently blends of more than one commodity which underlines the importance of being able to predict the DE value. For some considerable time there has been an understanding of the physiological importance of degree of saturation and free fatty acid content of fats and oils on the overall digestive process. However, quantifying the responses has not been extensively considered. A series of comprehensive metabolism experiments (Powles *et al.*, 1993a, b, 1995) were designed with the objective of quantifying the influences of degree of saturation and free fatty acid content on the DE values of fat blends. The approach adopted was structured, with incremental increases in free fatty acid (FFA) and degree of saturation (expressed as the ratio of unsaturated to saturated fatty acids – U/S – in the blend) occurring both separately and together.

Inspection of the response of DE to U/S revealed an exponential relationship, whereas that of DE to FFA was linear. These responses led to the derivation of prediction equations which related both FFA and U/S (acting as separate independent variables with no interaction term) to DE (Table 3.4). The data confirm the marginal superiority of the growing/finishing pig compared to the postweaning pig in terms of digestibility of feed grade fats. Solutions of the functions given in Table 3.4 are presented in Fig. 3.2 for two FFA levels. Although the range of chain lengths for fatty acids within feed grade fats is comparatively limited, it is nevertheless possible that the shorter chain length ones, if saturated (*n*-dodecanoic, C12 : 0, lauric acid), might behave like unsaturated ones in terms of dietary energy value which would alter, albeit only marginally, the prediction based on the U/S ratio.

**Table 3.4.** Prediction equations relating digestible energy (DE, MJ kg$^{-1}$) to the ratio of unsaturated to saturated fatty acids (U/S) and free fatty acid content (FFA, g kg$^{-1}$ fat) of fats and oils (Ages 1 and 2 refer to 10–20 kg and 35–85 kg live weight pigs, respectively).

$$DE\ (MJ\ kg^{-1}\ fat) = A + B \times FFA + C \times e^{(D \times U/S)}$$

| | Age 1 | | Age 2 | |
|---|---|---|---|---|
| | Constant | PV[a] | Constant | PV[a] |
| A | 36.90 ±0.501 | 0.80 | 37.89 ±1.690 | 0.77 |
| B | −0.005 ±0.0010 | | −0.005 ±0.0020 | |
| C | −7.33 ±2.700 | | −8.20 ±1.750 | |
| D | −0.91 ±0.452 | | −0.52 ±0.376 | |

[a] Proportion of variance in the dependent variables accounted for by the function.

## The chemical analysis of fats

Estimation of the total fat content of a compound is of particular importance as fat levels are employed to predict DE values of diets. There are two approaches, being extraction with light petroleum ether and the same extraction preceded by boiling in 3 M HCl. The former is sometimes referred to as ether extract (EE) and the latter as acid ether extract (AEE). It is important to establish, in any discussion of lipid content of raw materials and compound diets, which of the two methods has been employed. The concept of 'net' fat (Hyvönen, 1996) where all the quantitated fatty acids are determined and then converted to triacylglycerols by calculation, is less ambiguous nutritionally than the traditional estimate of extracted total fat.

Estimation of total free fatty acid content, which is of some considerable nutritional significance, is achieved with sodium hydroxide titration with oleic acid as a standard (or, if chain lengths are shorter, with another appropriate fatty acid). Specific fatty acids are usually analysed by use of gas–liquid chromatography after conversion of the fatty acids to their methyl esters.

Feed grade fats and oils may contain a number of impurities and it is conventional to adopt procedures for their determination by measuring the amounts of moisture and unsaponifiable material.

Also, procedures have been developed to monitor processing damage. Free fatty acid content is used in the human food industry as FFA levels will

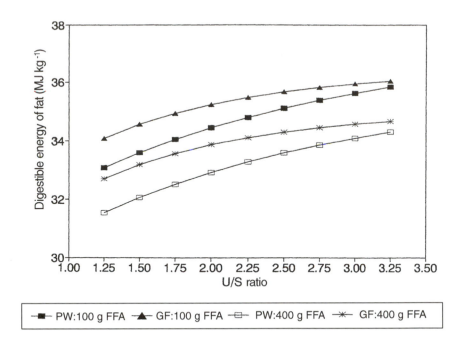

**Fig 3.2.** Prediction of the digestible energy (of fats) for pigs as influenced by age, ratio of unsaturated (U) to saturated (S) fatty acids (U/S) and free fatty acid (FAA) content (employing two levels – 100 and 400 g kg$^{-1}$ fat) for postweaning (PW) and growing/finishing (GF) animals.

increase in purified oils following oxidation. However, this test is inappropriate for feed grade fats, as FFA themselves are acceptable ingredients in fat blends. Peroxide value (PV) has been employed but is not valid as in fact it may decrease with time even though the fat or oil is further oxidizing (Poling *et al.*, 1962). Total non-elutable material (NEM) has become established as a standard procedure for estimating all damaged material within a fat blend although it does not seek to estimate individual categories (Waltking *et al.*, 1975).

## Summary

Diets are usually described in terms of their crude fat (ether extract) contents, but such a measure gives no insight as to the types of fatty acids present. More detailed information on the lipid constituents will lead to a better quantitative understanding of fat digestion and subsequent fatty acid utilization. Fatty acid uptake from the digestive tract is a complex process with synergistic effects (among types of fatty acids) having been observed. Degree of saturation and chain length are also important considerations. Detailed information on the dietary fatty acids does allow prediction, albeit crude, of fatty acid absorption in the pig.

## The Minerals

More than 30 minerals are considered to be essential for growth in the pig. These have diverse functions ranging from major constituents of tissue (e.g. calcium and phosphorus in bone) through to elements which are components of macromolecules (iron in haemoglobin) or co-factors in metabolic pathways (selenium with the enzyme glutathione peroxidase). Minerals are somewhat unusual in so far as they may frequently become toxic at levels of inclusion not greatly above the requirement. Furthermore, interactions between individual minerals, and between minerals and other dietary components, may be important both in terms of absorption and metabolic activity.

Mineral homeostasis is often under endocrine control (for example calcium) which means that even simple estimates of uptake are variable and dependent upon overall dietary provision and metabolic needs. The availability of minerals in feedstuffs may be compromised due to their being incorporated within complex molecules that are not digested by the pig. Perhaps the best known example of this is phosphorus bound within phytic acid (hexainositol phosphate).

Analytical methods (see review by Thompson and Fowler, 1990) for determining the gross mineral contents of feedstuffs are diverse but it is generally accepted that atomic absorption spectroscopy is a particularly appropriate methodology although inductively coupled plasma emission spectroscopy (ICP) has merit. The role of minerals in growth and skeletal development of the pig has been the subject of review (Agricultural Research Council, 1981) as has the physiological effects of dietary electrolytes (Patience, 1989).

## Prediction of the Digestible Energy (DE) Value of Mixed Diets

The ability to predict dietary DE, which is affected by all the chemical components of the diet, and is key to a quantitative understanding of pig metabolism, has been considered in depth (Wiseman and Cole, 1983; Morgan *et al.*, 1987; Batterham, 1990).

The approach is comparatively straightforward and involves the conduct of a series of metabolism trials to establish DE values for a range of compound diets which are then analysed for various chemical constituents. Multiple regression analysis is used to derive relationships between the former and the latter. It is the choice of the independent variables that has been the subject of much debate. Conventionally, starch, fat, protein and fibre have been used. The problem has been approached theoretically by Whittemore (1987) who generated the following equation based upon the mean digestibility coefficients of gross energy for four chemical components:

$$DE \ (MJ \ kg^{-1}) \ = \ 0.016 \ starch \ + \ 0.035 \ fat \ + \ 0.019 \ protein \ + \ 0.001$$
$$fibre \ (with \ all \ constituents \ expressed \ as \ g \ kg^{-1} \ diet).$$

While the chemical constituents selected account for most of the energy-yielding components of a diet, there has been much discussion on their validity and the appropriateness of constant coefficients. As discussed earlier, fibre is an imprecise term. Furthermore, the use of a constant coefficient for fat is inappropriate because of the variability between different fats and mixtures of fats in terms of their DE values (Powles *et al.*, 1995).

The approach adopted by Wiseman and Cole (1983) was to start with the premiss that gross energy (GE) itself is an important determinant of DE. Secondarily, the digestibility of the GE is influenced by chemical composition, e.g.:

$$DE \ (MJ \ kg^{-1}) \ = \ -2.293 \ + \ 1.064 \ GE \ (MJ \ kg^{-1}) \ + \ 0.035\% \ crude$$
$$protein \ - \ 0.053\% \ ether \ extract \ - \ 0.442\% \ crude \ fibre$$
$$- \ 0.0191\% \ ash$$

Addition of non-linear terms for ether extract and crude fibre led to a marginal improvement in accuracy which is evidence of interactive effects involving these two components.

It is important to note that, although the chemical constituents do have some biological validity, these prediction equations are essentially empirical in nature. It is, therefore, particularly important that equations are not applied to diets the chemical composition of which is outside the original data set employed in their derivation or that contain unusual raw materials. Also, such equations predict the amount of energy disappearing from the gut, but give no information concerning the nature of the energy loss nor the amounts of energy absorbed in the form of various chemical compounds. The ratio of dietary net energy to dietary digestible energy is not constant (Noblet, 1996).

A further problem with prediction equations is that they do not account for physical differences between feedstuffs which, while having no influence on chemical composition, may impact upon digestibility and subsequent DE values. The use of *in vitro* digestibility assays in conjunction with chemical composition may enhance the accuracy of DE prediction (van der Meer and Perez, 1992; Jaguelin *et al.*, 1994). Further, in the future emphasis should shift towards predicting the uptake of specific dietary chemical components rather than DE, which then allows a simulation of metabolism and estimation of metabolizable and net energy.

# Other Factors

In addition to the mixture of chemical compounds which makes up a feedstuff and the physiological effects that this may elicit, there is a host of physical attributes, often a reflection of the three-dimensional chemical structures and their associations within the feed matrix, that is important. Some of these properties (e.g. water-holding capacity, bulkiness) have been mentioned already but other factors such as texture, flavour, the presence of mycotoxins and the presence of impurities can have major effects on palatability and thus food intake and on subsequent animal performance (Robb, 1990; Lawrence, 1990a; Moughan and Smith, 1996). It is difficult to describe the effect of these factors quantitatively. Of particular significance is the presence in feeds of non-nutrient compounds that may have direct antinutritional effects. Compounds such as tannins, lectins, phaseolins and trypsin inhibitors as well as certain toxic amino acids can significantly affect digestion and metabolism (Huisman *et al.*, 1990). The data reported by Schulze (1994) demonstrate the quantitatively significant influence of dietary antinutritional factors on gut endogenous amino acid losses in the pig.

Also, any processing to which a feedstuff may be subjected (e.g. heating, grinding) or the physical form in which the feed is presented to the animal (e.g. wet vs dry, meal vs pellets) may influence food intake, digestion, absorption and subsequent nutrient metabolism. This has been the subject of review (Lawrence, 1990b; Moughan and Smith, 1996).

# Overall Conclusion

It should be apparent to the reader that in spite of intensive research over many decades, it is still not possible to characterize completely and accurately the chemical compounds present in a feedstuff on a routine basis. Analysts and biologists need to recognize the inherent limitations of datasets. In the future, greater analytical accuracy will ensue and rapid inexpensive methods for predicting chemical composition will be developed. Of particular promise here is the technology of near-infrared spectroscopy (Osbourne *et al.*, 1993).

Particularly in the case of the plant fibre component of feeds, a better characterization of the physical properties, which are related to the chemical composition, is required. Also, more information is required concerning the effects of physical feed processing methods and specific antinutritional factors on voluntary food intake, digestion and absorption. Feed ingredients need to be described by measures that allow an accurate quantitative prediction of the ability of a mixed feedstuff to deliver specific chemical compounds to the portal and lymphatic circulations and, moreover, provide information which can be used to predict certain physiological outcomes. As a description of feedstuffs becomes less 'chemically' and more 'physiologically' and 'metabolically' orientated, our ability to quantitatively describe and predict pig growth and development will be enhanced.

# References

AFRC (Agricultural and Food Research Council) (1987) Technical committee on responses to nutrients No. 2, Characterisation of feedstuffs: nitrogen. *Nutrition Abstracts and Reviews, Series B: Livestock Feeds and Feeding* 57, 713-736.

Agricultural Research Council (1967) *The Nutrient Requirements of Farm Livestock.* Agricultural Research Council, London, UK.

Agricultural Research Council (1981) *The Nutrient Requirements of Pigs.* Agricultural Research Council, London, UK.

Alexander, J.C. (1966) Effect of diet handling on nutritional studies with used frying fats. *Lipid* 1, 254-257.

Andrews, J.C., Griffith, W.H., Mead, J.F. and Stein, R.A. (1960) Toxicity of air-oxidised soybean oil. *Journal of Nutrition* 70, 199-210.

Annison, G. (1993) The role of wheat non-starch polysaccharides in broiler nutrition. *Australian Journal of Agricultural Research* 44, 405-422.

Annison, G. and Topping, D.L. (1994) Nutritional role of resistant starch: Chemical structure vs. physiological function. *Annual Review of Nutrition* 14, 297-320.

Annison, G., Choct, M. and Cheetham, N.W.H. (1992) Analysis of wheat pentosans from a large-scale isolation. *Carbohydrate Polymers* 19, 151-159.

Artman, N.R. (1969) The chemical and biological properties of heated and oxidised fats. *Advances in Lipid Research* 7, 245-330.

Artman, N.R. and Alexander, J.C. (1968) Characteristics of some heated fat components. *Journal of the American Oil Chemists Society* 45, 643-648.

Barrett, C.B. and Henry, C.M. (1966) Heat induced changes during processing and use of edible fats. *Proceedings of the Nutrition Society* 25, 4-9.

Batterham, E.S. (1990) Prediction of the dietary energy value of diets and raw materials for pigs. In: Wiseman, J. and Cole D.J.A. (eds) *Feedstuff Evaluation.* Nottingham University Press, Nottingham, UK, pp. 267-281.

Black, J.L. (1995) Modelling energy metabolism in the pig – critical evaluation of a simple reference model. In: Moughan, P.J., Verstegen, M.W.A. and Visser-Reyneveld, M.I. (eds), *Modelling Growth in the Pig.* Wageningen Pers, Wageningen, The Netherlands, pp. 87-102.

Blackburn, S. (1978) *Amino Acid Determination: Methods and Techniques,* 2nd edn. Marcel Dekker, New York, pp. 7-37.

Carraway, K.L. and Koshland, D.E. (1972) Carbodiimide modification of proteins. In: Hirs, C.H.W. (ed.), *Methods in Enzymology,* vol. 25. Academic Press, New York, pp. 616-623.

Chang, S.S., Petersen, R.J. and Ho, C.T. (1978) Chemical reactions involved in the deep-fat frying of food. *Journal of the American Oil Chemists Society* 55, 718-727.

Chiou, S-H. and Wang, K-T. (1989) Peptide and protein hydrolysis by microwave irradiation. *Journal of Chromatography* 491, 424-431.

Choct, M., Hughes, R.J., Wang, J., Bedford, M.R., Morgan, A.J. and Annison, G. (1996) Increased small intestinal fermentation is partly responsible for the antinutritive activity of non-starch polysaccharides in chickens. *British Poultry Science* 37, 609–621.

Combe, N., Constatin, M.J. and Entressangles, B. (1981) Lymphatic absorption of non-volatile oxidation products of heated oils in the rat. *Lipids* 16, 8–14.

Darragh, A.J. (1995) The amino acid composition of human milk – towards determining the amino acid requirements of the human infant. PhD Thesis, Massey University, New Zealand.

Darragh, A.J., Garrick, D.J., Moughan, P.J. and Hendriks, W.H. (1996) Correction for amino acid loss during hydrolysis of a purified protein. *Analytical Biochemistry* 236, 199–207a.

Davey, J.F. and Ersser, R.S. (1990) Amino acid analysis of physiological fluids by high-performance liquid chromatography with phenylisocyanate derivatisation and comparison with ion-exchange chromatography. *Journal of Chromatography* 528, 9–23.

Davies, M.G. and Thomas, A.J. (1973) An investigation of hydrolytic techniques for the amino acid analysis of feedstuffs. *Journal of the Science of Food and Agriculture* 24, 1525–1540.

Decuypere, J.A., Spriet, S.M. and van Gils, L.G. (1994) Influence of the water-holding-capacity (WHC) of the feed on the precaecal and faecal apparent digestibility in pigs. In: Souffrant, W.-B. and Hagemeister, H. (eds), *Digestive Physiology in Pigs*. Bad Doberan, Germany, pp. 125–128.

Englyst, H.N. and Hudson, G.J. (1996) The classification and measurement of dietary carbohydrates. *Food Chemistry* 57, 15–21.

Evans, A.J., Cheung, P.C.K. and Cheetham, N.W.H. (1993) The carbohydrate composition of cotyledons and hulls of cultivars of *Lupinus angustifolius* from Western Australia. *Journal of the Science of Food and Agriculture* 61, 189–194.

Fincher, G.B. and Stone, B.A. (1986) Cell walls and their components in cereal technology. In: Pomeranz, Y. (ed.), *Advances in Cereal Science and Technology*, Vol. 8. AACC, Minnesota, pp. 207–295.

Finley, J.W. (1985) Reducing variability in amino acid analysis. In: Finley, J.W. and Hopkins, D.T. (eds), *Digestibility and Amino Acid Availability in Cereals and Oilseeds*. American Association of Cereal Chemists, St Paul, Minnesota, pp. 15–30.

Forbes, W.F. and Savige, W.E. (1962) Photolysis and photooxidation of amino acids and peptides. I. Irradiation of aqueous solutions of cystine with sunlight and other ultraviolet lights sources. *Photochemistry and Photobiology* 1, 1–13.

Freeman, C.P. (1984) The digestion, absorption and transport of fat – Non-ruminants. In: Wiseman, J. (ed.), *Fats in Animal Nutrition*. Butterworths, London, pp. 105–122.

Friedman, M. (1996) Nutritional value of protein from different food sources. A review. *Journal of Agricultural and Food Chemistry* 44, 6–29.

Gehrke, C.W., Wall, L.L., Sr, Absheer, J.S., Kaiser, F.E. and Zumwalt, R.W. (1985) Sample preparation for chromatography of amino acids. Acid hydrolysis of proteins. *Journal of the Association of Analytical Chemistry* 68, 811–821.

Hare, P.E., St John, P.A. and Engel, M.H. (1985) Ion-exchange separation of amino acids. In: Barrett, G.C. (ed.), *Chemistry and Biochemistry of Amino Acids*. Chapman and Hall, London, pp. 415–425.

Hendriks, W.H., Moughan, P.J. and Tartellin, M.F. (1998) The amino acid composition of cat (*Felis catus*) hair. *Animal Science* (in press).

Hsieh, A. and Perkins, E.G. (1976) Nutrition and metabolic studies of methyl esters of dimeric fatty acids in the rat. *Lipids* 11, 763-768.

Huisman, J., van der Poel, A.F.B., Verstegen, M.W.A. and van Weerden, E.J. (1990) Antinutritional factors (ANF) in pig production. *World Review of Animal Production*, vol. XXV, No. 2, pp. 77-82.

Hurrell, R.F. and Carpenter, K.J. (1981) The estimation of available lysine in foodstuffs. In: Eriksson, C. (ed.), *Progress in Food and Nutritional Science, vol. 5. Maillard Reaction in Foods*. Pergamon Press, Oxford, pp. 156-176.

Hyvönen, L. (1996) Approach to fat analysis in foods. *Food Chemistry* 57, 23-26.

Jaguelin, Y., Février, C., Sève, B., Lechevestrier, Y., Mariscal-Landin, G., Leroux, P. and Lebreton, Y. (1994) Assessment of the apparent and true N digestibility in pigs for several class of feedstuffs through an *in vitro* determination. In: Souffrant, W.-B. and Hagemeister, H. (eds), *Digestive Physiology in Pigs*. Bad Doberan, Germany, pp. 114-117.

Kyriazakis, I. and Emmans, G.C. (1995) The voluntary feed intake of pigs given feeds based on wheat bran, dried citrus pulp and grass meal, in relation to measurements of feed bulk. *British Journal of Nutrition* 73, 191-207.

Larsen, F.M., Moughan, P.J. and Wilson, M.N. (1993) Dietary fibre viscosity and endogenous protein excretion at the terminal ileum of growing rats. *Journal of Nutrition* 123, 1898-1904.

Lawrence, T.L.J. (1990a) Concepts of palatability in pig and poultry foods. In: Smith, W.C. (ed.), *Proceedings of the Inaugural Massey Pig and Poultry Symposium*. Massey University, Palmerston North, New Zealand, pp. 112-150.

Lawrence, T.L.J. (1990b) Effects of dietary physical form on health and performance of the growing pig. In: Smith, W.C. (ed.), *Proceedings of the Inaugural Massey Pig and Poultry Symposium*. Massey University, Palmerston North, New Zealand, pp. 112-150.

Lea, C.H., Parr, L.J., L'Estrange, J.L. and Carpenter, K.J. (1966) Nutritional effects of auto-oxidised fats in animal diets. 3. The growth of turkeys on diets containing oxidised fish oil. *British Journal of Nutrition* 20, 123-132.

Low, A.G. (1993) The role of dietary fibre in pig feeds. In: Cole, D.J.A., Haresign, W. and Garnsworthy P.C. (eds), *Recent Developments in Pig Nutrition*, vol. 2. Nottingham University Press, pp. 137-162.

Malmer M.F. and Schroeder, L.A. (1990) Amino acid analysis by high performance liquid chromatography with methanesulphonic acid hydrolysis and 9-fluorenylmethylchloroformate derivatisation. *Journal of Chromatography* 514, 227-239.

McCleary, B.V. (1993) New enzymic methods in determining dietary fibre components. In: Samman, S. and Annison, G. (eds), *Dietary Fibre and Beyond - Australian Perspectives*, vol. 1. Nutrition Society of Australia, Occasional Publication, Sydney, pp. 54-74.

McInnes, A.G., Cooper, F.P. and McDonald, J.A. (1961) Further evidence for cyclic monomers in heated linseed oil. *Canadian Journal of Chemistry* 39, 1906-1914.

Monro, J.A. (1993) A nutritionally valid procedure for measuring soluble dietary fibre. *Food Chemistry* 47, 187-193.

Morgan, C.A., Whittemore, C.T., Phillips, C.T. and Crooks, P. (1987) The prediction of the energy value of compound pig foods from chemical analysis. *Animal Feed Science and Technology* 17, 81-107.

Moughan, P.J. and Rutherfurd, S.M. (1996) A new method for determining digestible reactive lysine in foods. *Journal of Agricultural and Food Chemistry* 44, 2202–2209.

Moughan, P.J. and Smith, W.C. (1996) Principles of pig nutrition. In: Taverner, M.R., Dunkin, A.C. (eds), *Pig Production, World Animal Science, C10*. Elsevier, Amsterdam, pp. 141–167.

Mugford, D.C. (1993) Current methods for measurements of dietary fibre: choices and suitability. In: Samman, S. and Annison, G. (eds), *Dietary Fibre and Beyond – Australian Perspectives*, vol. 1. Nutrition Society of Australia, Occasional Publications, Sydney. pp. 19–36.

NATA (National Association of Testing Authorities) (1993) NATA proficiency testing report.

NATA (National Association of Testing Authorities) (1994) NATA proficiency testing report.

Ng, L.T., Pascaud, A. and Pascaud, M. (1987) Hydrochloric acid hydrolysis of protein and determination of tryptophan by reverse-phase high performance liquid chromatography. *Analytical Biochemistry* 167, 47–52.

Nissen, S. (1992) Amino acid analysis in food and physiological samples. In: Nissen, S. (ed.), *Modern Methods in Protein Nutrition and Metabolism*. Academic Press, San Diego, USA, pp. 1–8.

Noblet, J. (1996) Digestive and metabolic utilisation of dietary energy in pig feeds: Comparison of net energy systems. In: Garnsworthy, P.C., Wiseman, J. and Haresign, W. (eds), *Recent Advances in Animal Nutrition – 1996*. Butterworths, Oxford, pp. 207–231.

Noblet, J. and Perez, J.M. (1993) Prediction of digestibility of nutrients and energy values of pig diets from chemical analysis. *Journal of Animal Science* 71, 3389–3398.

Ohfuji, T. and Kaneda, T. (1973) Characterisation of toxic compounds in thermally oxidised oil. *Lipids* 8, 353–359.

Osbourne, B.G., Fearn, T. and Hindle, P.H. (1993) Applications of near infrared spectroscopy in food and beverage analysis. In: *Practical NIR Spectroscopy with Applications in Food and Beverage Analysis*. Longman, Harlow, UK, pp. 145–200.

Patience, J.F. (1989). The physiological basis of electrolytes in animal nutrition. In: Haresign, W. and Cole, D.J.A. eds, *Recent Advances in Animal Nutrition*. Butterworths, London, pp. 211–228.

Perkins, E.G. and Taubold, R. (1978) Nutritional and metabolic studies of non-cyclic dimeric fatty acid methyl esters in the rat. *Journal of the American Oil Chemists Society* 55, 632–634.

Poling, C.E., Warner, W.D., Mone, P.E. and Rice, E.E. (1962) The influence of temperature, heating time and aeration upon the nutritive value of fats. *Journal of the American Oil Chemists Society* 39, 315–320.

Pomeranz, Y. and Meloan, C.E. (1980) Nitrogenous compound. In: Pomeranz, Y. and Meloan, C.E. (eds), *Food Analysis: Theory and Practice*, revised edition. AVI Publishing Company, Westport, Connecticut, pp. 668–691.

Powles, J., Wiseman, J., Cole, D.J.A. and Hardy B. (1993a) Effect of chemical structure of fats upon their apparent digestible energy value when fed to growing/finishing pigs. *Animal Production* 57, 137–146.

Powles, J., Wiseman, J., Cole, D.J.A. and Hardy, B. (1993b) Effect of chemical structure of fats upon their apparent digestible energy value when given to young pigs. *Animal Production* 58, 411–417.

Powles, J., Wiseman, J., Cole, D.J.A. and Jagger, S. (1995) Prediction of the apparent digestible energy value of fats given to pigs. *Animal Science* 61, 149–154.

Rayner, C.J. (1985) Protein hydrolysis of animal feeds for amino acid content. *Journal of Agricultural Food Chemistry* 33, 217–244.

Robb, J. (1990) Effects of mycotoxins on animal performance. In: Haresign, W. and Cole, D.J.A. (eds), *Recent Advances in Animal Nutrition*. Butterworths, London, pp. 61–76.

Robel, E.J. (1973) Elimination of amino acid losses with protein acid hydrolyzates due to adsorption. *Analytical Biochemistry* 51, 137–145.

Robel, E.J. and Crane, A.B. (1972) An accurate method for correcting unknown amino acid losses from protein hydrolyzates. *Analytical Biochemistry* 48, 233–246.

Rutherfurd, S.M. and Moughan, P.J. (1993) Use of free or bound amino acid molecular weight in the determination of amino acid compositions. In: Batterham, E.S. (ed.), *Manipulating Pig Production IV*. Australasian Pig Science Association, Attwood, Australia, 229 pp.

Salo-Väänänen, P.P. and Koivistoinen, P.E. (1996) Determination of protein in foods: comparison of net protein and crude protein (N $\times$ 6.25) values. *Food Chemistry* 57, 27–31.

Sarwar, G., Botting, H.G. and Peace, R.W. (1988) Complete amino acid analysis in hydrolysates of foods and feces by liquid chromatography of precolumn phenyl-isothiocyanate derivatives. *Journal of the Association of Official Analytical Chemists* 71, 1172–1175.

Schulze, H. (1994) Endogenous ileal nitrogen losses in pigs, dietary factors. PhD thesis, Wageningen University, The Netherlands, 147 pp.

Selvendran, R.R., Stevens, B.J.H. and Du Pont, M.S. (1987) Dietary fibre: Chemistry, analysis and properties. *Advances in Food Research* 31, 117–178.

Smits, C.H.M and Annison, G. (1996) Non-starch plant polysaccharides in broiler nutrition – towards a physiologically valid approach to their determination. *World Poultry Science Journal* 52, 203–221.

Soby, L.M. and Johnson, P. (1981) Determination of asparagine and glutamine in polypeptides using bis(1,1-trifluoroacetoxy)iodobenzene. *Analytical Biochemistry* 113, 149–153.

Steinhart, H. (1984) Summary of the workshop on tryptophan analysis. In: Zebrowska, T., Buraczewska, L., Buraczewski, S., Kowalczyk, J. and Pastuszewska, B. (eds), *Proceedings of the VI International Symposium on Amino Acids*. Polish Scientific Publishers, Warsaw, pp. 434–447.

Thompson, J.K. and Fowler, V.R. (1990) The evaluation of minerals in the diets of farm animals. In: Wiseman, J. and Cole, D.J.A. (eds), *Feedstuff Evaluation*. Nottingham University Press, Nottingham, UK, pp. 235–259.

Tower, D.B. (1967) Enzymatic determination of glutamine and asparagine. In: Hirs, C.H.W. (ed.), *Methods in Enzymology*, vol. 11. Academic Press, New York, pp. 76–93.

van der Meer, J.M. (1990) Amino acid analysis of feeds in the Netherlands: Four-year proficiency study. *Journal of the Association of Official Analytical Chemists* 73, 394–398.

van der Meer, J.M. and Perez, J.M. (1992) *In-vitro* evaluation of European diets for pigs. Prediction of the organic matter digestibility by an enzymic method or by chemical analysis. *Journal of the Science of Food and Agriculture* 59, 359–363.

Waltking, A.E., Seery, W.E. and Bleffert, G.W. (1975) Chemical analysis of polymerisation products in abused fats and oils. *Journal of the American Oil Chemists Society* 52, 96–100.

Whittemore, C.T. (1987) *Elements of Pig Science*. Longman, Harlow, UK.

Wilcox, P.E. (1967) Determination of amide residues by chemical analysis. In: Hirs, C.H.W. (ed.), *Methods in Enzymology* vol. 11. Academic Press, New York, pp. 63–76.

Williams, A.P. (1988) Determination of amino acids. In: MaCrae, R. (ed.), *HPLC in Food Analysis*. Academic Press, London, pp. 441–470.

Williams, A.P. (1994) Recent developments in amino acid analysis. In: D'Mello, J.P.F. (ed.), *Amino Acids in Farm Animal Nutrition*. CAB International Wallingford, UK, pp. 11–36.

Wiseman, J. (1986) Antinutritional factors associated with dietary fats and oils. In: Haresign, W. and Cole, D.J.A. (eds), *Recent Advances in Animal Nutrition – 1986*. Butterworths, London, pp. 47–75.

Wiseman, J. and Cole, D.J.A. (1983) Predicting the energy content of pig feeds. In: Haresign, W. (ed.), *Recent Advances in Animal Nutrition – 1983*. Butterworths, London, pp. 59–70.

Wong, W.S.D., Osuga, D.T., Burcham, T.S. and Feeney, R.E. (1984) Determination of tryptophan as the reduced derivative by acid hydrolysis and chromatography. *Analytical Biochemistry* 143, 62–70.

# The Thermal and Infectious Environment

**4**

## J.L. Black,[1] H.J. Bray[2] and L.R. Giles[3]

[1]John L. Black Consulting, Locked Bag 21, Warrimoo, NSW 2774, Australia; [2]South Australian Research and Development Institute, Pig and Poultry Production Institute, University of Adelaide, Roseworthy SA 5371, Australia; [3]NSW Agriculture, Elizabeth Macarthur Agricultural Institute, PMB 8, Camden, NSW 2570, Australia

## Introduction

The rate of growth and efficiency of feed use for pigs raised commercially are well below the potential of the animals and the levels achieved under experimental conditions (Black and Carr, 1993). This reduced performance appears to be associated primarily with a depression in feed intake, but there is evidence also that these animals tend to be fatter than their counterparts given the same amount of feed under experimental conditions (Chapple, 1993). The depressed intake and changed pathways of nutrient use may be caused by many factors associated with a commercial pig-raising environment, but climatic conditions outside the range of thermal comfort for the pig and infectious disease are likely to contribute.

The aim of this chapter is to review the effects of the thermal and infectious environments on the intake and utilization of nutrients by pigs. The effects of the social environment on pig performance are reviewed in Chapter 5. Although discussed separately within this book, there is clearly considerable interaction between the social, thermal and infectious environments. For example, comparisons of the performance of pigs housed individually with pigs in groups generally, but not always, show greater feed intakes and growth rates for the individually penned animals (Bustamante et al., 1996). One explanation for the apparently contradictory observations could be a change in the thermal environment of pigs in some experiments because of differences in the rates of heat loss between single and group-penned pigs. Similarly, there are interactions between the social and infectious environments where the mixing and transporting of pigs have been demonstrated to

© CAB INTERNATIONAL 1999. *A Quantitative Biology of the Pig*
(ed. I. Kyriazakis)

decrease performance and cellular immunity (Ekkel *et al.*, 1995) and to increase signs of clinical disease (Hessing and Tielen, 1994) compared with pigs maintained in one group. Pigs exposed to draught and low temperatures also show increased signs of clinical disease such as diarrhoea, coughing, sneezing and haemorrhagic ear lesions compared with pigs housed under favourable climatic conditions (Hessing and Tielen, 1994).

## The Thermal Environment

### Heat exchange between the pig and its environment

The factors controlling heat exchange between an animal and its environment have been described previously (Mount, 1974; Curtis, 1983; Bruce, 1993). The pig is able to regulate heat exchange over a wide range of climatic conditions, but there are limits to this regulatory capacity which, when exceeded, result in death from either hypo- or hyperthermia. Thermoregulation, or the control of body temperature, is achieved through various mechanisms including shivering and other thermogenic biochemical reactions, altering feed intake and activity, changing the rate of blood flow to the skin and skin contact with the floor, huddling, seeking radiant heat sources, regulating evaporative heat loss from the lungs through changing respiration rate and from the skin by altering the proportion of skin that is wet. Pigs also allow deep body temperature to change over a narrow range as part of their thermoregulation strategy (Morrison and Mount, 1971; Giles, 1992).

Total heat exchange between a pig and its environment ($Q$, $W$) can be calculated from radiant exchange with the building and other pigs ($Q_r$), heat absorption from a radiant heater ($Q_b$), convection loss ($Q_c$), conduction loss through the floor ($Q_f$) and evaporative loss from the respiratory tract ($Q_l$) and skin ($Q_s$).

$$Q = Q_r - Q_b + Q_c + Q_f + Q_l + Q_s \tag{4.1}$$

Each of these mechanisms can be described quantitatively and are presented in detail by Bruce and Clark (1979) and Black *et al.* (1986). The quantitative relationships described by these authors have been incorporated into computer models used to simulate the growth and body composition of pigs.

### *Critical temperatures*

The zone of thermal comfort which is termed also the thermoneutral zone or the zone of least thermoregulatory effort (Mount, 1974) is defined as the ambient temperature range over which a pig can maintain body temperature within a narrow limit from approximately 38.8 to 39.2°C through regulation of blood flow to the skin and through behavioural and postural changes including huddling. These responses require little expenditure of energy and do not involve a change in voluntary feed intake. As ambient temperature falls within the zone of thermal comfort, the pig will first vasoconstrict skin blood vessels to reduce heat loss from the skin, start seeking sources of radiant heat, huddle with other pigs and change its posture to reduce heat loss to either the air or floor, whichever is the colder.

When a pig has done all it can through vasoconstriction and behavioural changes to minimize heat loss, a further fall in ambient temperature below the zone of thermal comfort will result in a net loss of heat to the environment. The pig must then increase heat production to maintain body temperature within the acceptable range. This is achieved initially through shivering and other thermogenic reactions and is followed, often after several days, by an increase in feed intake (Verhagen *et al.*, 1987). The ambient temperature at which heat production is increased to maintain body temperature is termed the lower critical temperature ($t_l$, °C). A continuing fall in ambient temperature below $t_l$ results in increasing amounts of metabolic heat being produced until the animal is metabolizing at its maximum rate. This maximum metabolic rate is termed summit metabolism (Curtis, 1983) and in several animal species appears to be about 10 times fasting heat production (Black *et al.*, 1994). Furthermore, animals cannot maintain summit metabolism indefinitely and become fatigued over time. Although there appear to be no studies with pigs, sheep are capable of maintaining metabolic rates up to half summit metabolism indefinitely. However, as the animal cools further and metabolic rate approaches summit, it becomes progressively exhausted. Sheep can maintain summit metabolism for only 1 to 2 h before body temperature starts to decline precipitously and death occurs.

As ambient temperature increases towards the upper end of the zone of thermal comfort, vasoconstriction changes to vasodilation to increase heat loss to the air and the pig undertakes behaviours and postures to maximize heat loss to the floor and surrounding environment. With a further increase in ambient temperature above the zone of thermal comfort, thermoregulation can be achieved only by increasing evaporative heat loss from the lungs and skin. The temperature at which evaporative heat loss increases markedly has been termed (Black *et al.*, 1986) the evaporative critical temperature ($t_e$, °C).

Unlike many other mammals, pigs have few sweat glands and evaporative heat loss is achieved through an increase in respiration rate and wetting of the skin. In their natural environment, pigs wallow and water evaporation from mud on the skin is an effective avenue for heat loss. Pigs raised commercially on either slatted or wire floors in hot weather are usually able to wet some skin with drinking water, saliva or urine, but the proportion of skin that becomes wet under different management conditions has not been defined. This information is important for predicting accurately heat loss from commercially raised pigs. Despite being able to wet some skin, it is probable that, for pigs maintained above $t_e$ under many commercial environments, considerable evaporative heat loss is from the lungs where there is a substantial energy cost to increasing respiration rate (Ingram and Legge, 1969).

With increasing ambient temperature above $t_e$, the pig starts to invoke other mechanisms to reduce heat production. Body temperature begins to increase until it reaches about 39.3°C when the pig ceases most activity that is not associated directly with feeding, and voluntary feed intake begins to fall (Giles, 1992). Further increases in ambient temperature result in continuing increases in respiration rate and body temperature, and decreases in feed intake until the animal stops eating and respiration rate is at its maximum. The temperature at which the pig is doing all it can to increase heat loss and reduce heat input is termed the upper critical temperature ($t_u$, °C). Any further increase in ambient temperature above $t_u$ will cause an uncontrollable increase in body temperature and death. In a similar manner to summit metabolism, animals cannot maintain maximum respiration rate indefinitely and become exhausted after a few hours when, in sheep, open-mouth panting ceases (Khogali *et al.*, 1983).

## *Prediction of critical temperatures*

All three critical temperatures can be calculated from the same equation by substituting appropriate values for parameters that influence the rate of heat exchange between the pig and its environment. Parameters for calculation of $t_l$, $t_e$ and $t_u$ are set, respectively, to minimize total heat loss when body temperature is 39°C, to maximize sensible heat loss from the body through convection and conduction when body temperature is 39°C, and to maximize heat loss from the body for the calculated body temperature when respiration rate is maximal.

$$t_l \text{ or } t_e \text{ or } t_u = t_b - (\{[Q_n - Q_l - A.A_f/A(t_b - t_f)/(R_f + R_t)](R_a + R_t) - Q_sR_a - Q_bR_a\}/[A(1 + A_f/A - A_c/A)]) \tag{4.2}$$

where $Q_n$ (W) is heat production under thermoneutral conditions which can be calculated from the sum of energy required for maintenance and metabolic heat associated with the utilization of nutrients by the animal for productive functions (Black, 1995). The latter can be calculated from the stoichiometry of metabolic reactions in relation to the tissues synthesized and nutrients available for metabolism. The pig is assumed to move away from radiant heat when ambient temperature approaches $t_e$ so $Q_b$ would be set to zero for calculation of $t_e$ and $t_u$. Other symbols in equation 4.2 are: $Q_l$ and $Q_s$, heat loss from the lungs and skin; $A$, $A_c$ and $A_f$, total area of the pig and area in contact with other pigs and the floor; $t_b$ and $t_f$, deep body temperature and temperature of the floor; $R_a$, $R_f$ and $R_t$, thermal resistance of the air, floor and body tissue. The algorithms used to establish the variables in the equation for calculating critical temperatures are described by Bruce and Clark (1979) and Black *et al.* (1986).

The external thermal resistance of the skin exposed to air ($R_a$, °C m$^{-2}$ W$^{-1}$)

is calculated in relation to radiant heat exchange and heat loss from convection:

$$R_a = 1/\{0.8\,\sigma\,\varepsilon\,[273 + (t_s + t_r)2]^3 + 2.64(t_s + t_a)^{0.25}/W^{0.082}$$
$$+ 13.4V^{0.6}\,W^{-0.13}\} \qquad (4.3)$$

where $\sigma$ is the Stefan–Boltzmann constant ($5.67 \times 10^{-8}$ W m$^{-2}$ K$^{-4}$), $\varepsilon$ is a constant relating to the emissivity of the pig skin and common building materials, $t_s$ (°C) is skin temperature, $t_r$ (°C) is the temperature of the radiating surroundings, $t_a$ (°C) is air temperature, $V$ (m s$^{-1}$) is air velocity and $W$ (kg) is live weight. Because external thermal resistance ($R_a$) depends, among other factors, on skin temperature and skin temperature is influenced by $R_a$ and $t_a$, it is essential to perform a series of calculations to ensure that equilibrium values for $R_a$ and $t_s$ are obtained when using simulation models to predict heat exchange between a pig and its environment.

$$t_s = \{t_b/R_t + t_a/R_a - Q_s/[A(1 - A_f/A - A_c/A)]\}/(1/R_t + 1/R_a) \qquad (4.4)$$

Tissue thermal resistance ($R_t$) for pigs below the lower critical temperature was assumed by Bruce and Clark (1979) to be $0.02W^{0.33}$. Although tissue resistance for vasoconstricted cold pigs is thought to increase with increasing thickness of subcutaneous fat, there is little information to support this contention. Hovell *et al.* (1977) showed that tissue insulation in 'standard' sows was higher, but not significantly, than in 'thin' sows. Tissue insulation for vasodilated pigs is considerably lower than for vasoconstricted pigs and is unaffected by subcutaneous fat depth; a value for $R_t$, $0.005W^{0.33}$ was assumed (Black *et al.*, 1986).

## Effect of ambient temperature on voluntary feed intake

Voluntary feed intake of pigs housed individually in an ideal stress-free environment within the zone of thermal comfort is related closely to their capacity to utilize nutrients and varies with genotype and physiological state (Black, 1995). Once ambient temperature falls below the lower critical temperature, energy expenditure of the pig increases and voluntary feed intake tends to increase, often after a delay, to compensate for the additional heat loss. Conversely, at temperatures above the evaporative critical temperature, feed intake declines. The magnitude of the rise in feed intake at low temperatures and the fall at high temperatures is influenced by diet composition, stocking arrangements and the weight of pigs. Close (1989) collated information on the effect of ambient temperature on the intake of metabolizable energy for pigs ranging in live weight from 18 to 90 kg and showed that feed intake fell for all animals at high ambient temperatures, whereas the increase in intake at low temperatures was greater in the heavier than in the lighter pigs.

*Below lower critical temperature*

Three factors, the capacity of the digestive tract, stocking arrangements and eating rate, appear capable of limiting the ability of pigs in cold situations to increase feed intake sufficiently to compensate for heat lost to the environment. There is considerable evidence that the capacity of the digestive tract can limit feed intake of pigs, particularly for low-body-weight animals (Black *et al.*, 1986) and this may prevent intake rising sufficiently to maintain energy balance as was seen in the experiments with light-weight pigs reported by Close (1989).

Pigs in groups that are exposed to cold conditions spend more time huddling than in feeding activities compared with pigs in a thermally comfortable environment (Parker *et al.*, 1980). These observations indicate that feed intake of pigs exposed to cold conditions could be influenced by the number of pigs in the pen and be less for pigs housed in a group than for individually penned pigs. Nienaber *et al.* (1990, 1991) subjected pigs housed either separately or in groups of four to ambient temperatures 12°C below the estimated lower critical temperature and observed a 30% increase in feed intake for the cold-exposed single pig compared with only a 11% increase for the cold, group-housed pigs. This difference in feed intake could relate to the rate of heat loss from a pig when huddled in a group compared with the heat loss when it stands to eat. Presumably the pig prefers to remain warm in the group and eats only when the hunger drive is high and then for shorter periods of time compared with pigs housed individually. The length of a meal observed by Nienaber *et al.* (1990) for cold, separately housed pigs at 20 weeks of age was similar to that of the thermally comfortable pigs, whereas the meal length of the cold, group-housed pigs of the same age was only half that of the warm pigs (Nienaber *et al.*, 1990).

It is possible also that eating rate may limit the intake of pigs in extremely cold conditions. Nienaber *et al.* (1990) exposed pigs to a temperature 12°C below their lower critical temperature and found the eating rate was only 0.73 of that observed for pigs exposed either to a temperature 4°C below or 4°C above lower critical temperature. Although the severely cold-exposed group spent a longer time eating and consumed 30% more feed than the control animals, this was insufficient to compensate for the increased heat loss and growth rate was depressed by 19%.

*Above evaporative critical temperature*

Feed intake of pigs exposed to high temperatures has been shown in several studies to be negatively associated with body temperature (Fuller, 1965; Stahly and Cromwell, 1979; Schoenherr *et al.*, 1989; Giles, 1992; Lorschy, 1994). The relationship between feed intake of growing pigs exposed to various high temperature regimes relative to their intake under thermoneutral conditions and body temperature is shown in Fig. 4.1 for several experiments conducted by Giles (1992). The strong negative correlation suggests that feed intake falls by approximately 0.4 for every 1°C increase in

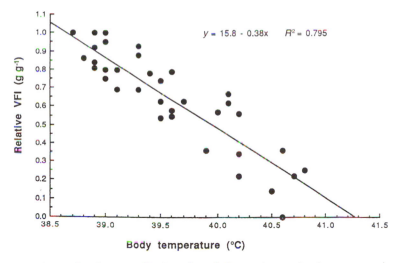

**Fig. 4.1.** Relationship between feed intake relative to that under thermoneutral conditions and deep body temperature for pigs weighing 90 kg. (From Giles, 1992.)

body temperature above about 39.2°C and that animals cease eating when body temperature reaches about 41.3°C. Although almost 0.80 of the variance is described by the relationship, considerable variation remains between animals.

One experiment (Giles and Black, 1991) involved maintaining pigs, which were unable to wet their skin, for 12 days at 31°C. Feed intake declined by 0.60 within two days as body temperature rose to 40.5°C. However, after 9 days body temperature had fallen to 40.0°C while feed intake remained unchanged. A similar observation was made by Morrison and Mount (1971) where feed intake remained unchanged when pigs were exposed to high temperature over an extended period of 30 days. In other experiments, Lorschy *et al.*(1991) exposed pigs weighing 80–90 kg for 9 days to fluctuating daily temperatures with either 16 or 20 h at 31°C and the remainder at 22°C. The mean daily pattern of feed intake and deep body temperature for each treatment (Fig. 4.2) shows that daily feed intake declined as the period of the day at 31°C increased and that more feed was consumed during the period of high temperature as the proportion of the day at 22°C declined.

The experiments of Giles and Black (1991), Giles *et al.* (1991) and Lorschy *et al.* (1991) confirm that, although deep body temperature has a major influence on the feed intake of pigs exposed to heat, other factors appear to be involved. One interpretation of these observations is that the depression in feed intake at any body temperature above normal is influenced by the degree of satiety or the difference between the energy demand and the energy supply to the animal. Such an interpretation would mean that when pigs are hungry, they are prepared to eat at a body temperature they would

(a)

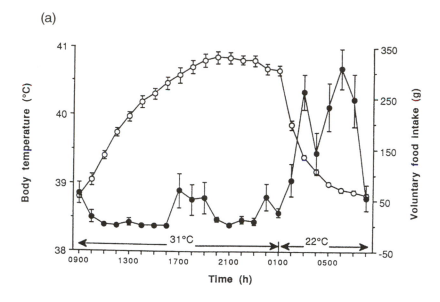

(b)

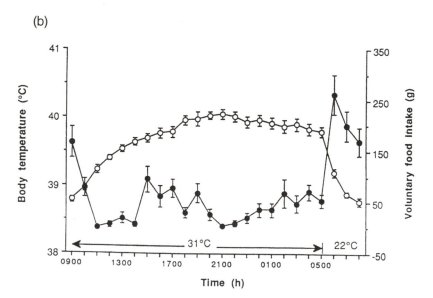

**Fig. 4.2.** Effect of period of day for pigs kept at either 16 h (a) or 20 h (b) at 31°C and the remainder at 22°C for 12 days on deep body temperature (O) and hourly voluntary feed intake (●). (From Lorschy *et al.*, 1991.)

not be prepared to eat at if nearly satiated. This concept has been confirmed in a recent experiment (Giles, unpublished) where pigs housed at 22°C were fed 0.25, 0.5, 0.75 or 1.0 of their *ad libitum* intake in hourly meals for 12 h prior to being exposed to 31°C. Feed was then removed and offered again to the pigs when their body temperature reached 40°C, confirming that intake over the next few hours was inversely related to feed intake during the last 12 h in thermoneutral conditions.

## Effect of ambient temperature on body composition

The effect of the climatic environment on the rates of fat and protein deposition in pigs is influenced by the feeding regime. Under conditions of constant feed intake, the rate of protein deposition is affected little by moderate depressions in temperature below the lower critical temperature and the additional energy required to maintain body temperature comes at the expense of fat deposition (Verstegen *et al.*, 1973; Le Dividich and Noblet, 1982). Consequently, under conditions of restricted feeding, pigs exposed to moderate cold become leaner. Alternatively, when individually housed pigs exposed to cold are given free access to feed and can increase consumption sufficiently to compensate for the additional heat loss, there appear to be no effects of temperature on the rates of either fat or protein deposition (Verstegen *et al.*, 1985; Rinaldo and Le Dividich, 1991). However, there is evidence (Panaretto, 1968) that, under more extreme conditions of cold when metabolic rate approaches half summit, the concentrations of plasma cortisol increase and the rate of protein deposition declines substantially and animals may become fatter than those of the same weight reared under thermoneutral conditions.

When pigs exposed to hot temperatures are given the same feed intake as pigs under thermoneutral conditions, the rates of either fat or protein deposition and body composition are not affected by temperature (Rinaldo and Le Dividich, 1991). However, when feed is freely available, the intake of hot pigs is less than for pigs in a comfortable environment and the rate of fat deposition declines more than the rate of protein deposition with the pigs becoming leaner (Verstegen *et al.*, 1973; Close *et al.*, 1978). This response to a reduction in feed intake is seen typically in pigs housed under ideal environmental conditions (Campbell, 1988) and there is no evidence for a direct effect of temperature on the body composition of pigs other than through a change in energy intake. This appears to be true even under conditions of extreme heat because L.R. Giles (unpublished) observed no change in the rate of protein deposition or in plasma cortisol concentrations in pigs when body temperatures were elevated to 41.0°C.

The weight of visceral organs, particularly the liver, empty digestive tract and heart, declines in pigs exposed for extended periods to hot conditions (Sugahara *et al.*, 1970; Rinaldo and Le Dividich, 1991). This decline in liver and gut weight appears to be directly related to the decline in feed intake, whereas the decline in heart weight is likely to be a response to the reduced demand for oxygen by the tissues resulting from reduced activity of the pig. Both the decline in weight of visceral organs and activity would be expected to lower the fasting heat production of pigs exposed to heat for extended periods (Koong *et al.*, 1983). The reduced maintenance energy requirement with reduced demand for feed may explain why pigs exposed to hot conditions for long periods of time do not increase feed intake despite a fall in body temperature as observed by Giles and Black (1991).

## Predicted effects of the thermal environment on heat exchange and performance

The concepts and algorithms outlined above have been incorporated into computer models such as AUSPIG (Black *et al.*, 1986; Davies *et al.*, 1993) that simulate energy and amino acid utilization in pigs exposed to different environments. The AUSPIG model is used to illustrate quantitatively the predicted effects of changing aspects of the thermal environment on heat exchange between a pig and its surroundings, on critical temperatures and on several animal performance characteristics.

Figure 4.3 shows the predicted effect of ambient temperature from $-5$ to 45°C on the response of a 90 kg entire male pig housed in a separate pen and given free access to a diet containing 14 MJ kg$^{-1}$ of digestible energy (DE). The pig was assumed to be on a slatted concrete floor, with an air speed of 0.2 m s$^{-1}$, an air water content of 10 g kg$^{-1}$ and to have no wet skin when near or below its lower critical temperature, but to be able to wet 0.15 of its skin when near or above its evaporative critical temperature. The pig was predicted to just survive a temperature of 45°C when the air water content was held constant at 10 g kg$^{-1}$ which produced a relative humidity of 15%, but the ambient temperature associated with survival was predicted to be extremely dependent on relative humidity and the pig was predicted to die at 38°C when the air water content was 40 g kg$^{-1}$ which produced a relative humidity of 92%.

The zone of thermal comfort was predicted under the conditions specified to be from 17.2 to 25.7°C with a feed intake of 2.34 kg day$^{-1}$ and a total heat production of 18.73 MJ day$^{-1}$. If the pig had been prevented from wetting any skin the temperature range for thermal comfort would be 17.2 to 22.9. The minimum evaporative heat loss was predicted to be 2.24 MJ day$^{-1}$, with half coming each from the lungs and skin, and the sensible heat loss was 16.49 MJ day$^{-1}$ at the lower end of the zone of thermal comfort. At some temperature close to the evaporative critical temperature, the pig is assumed to wet 0.15 of its skin with a substantial rise in evaporative heat loss from the skin to over 8 MJ day$^{-1}$.

Feed intake was predicted to increase to 3.65 kg day$^{-1}$ when ambient

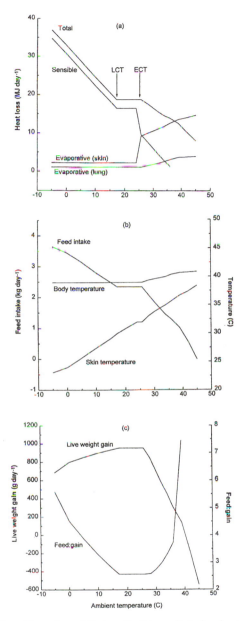

**Fig. 4.3.** (a) Predicted heat loss (MJ day$^{-1}$); (b) predicted voluntary feed intake, deep body temperature and skin temperature; and (c) predicted live weight gain and feed conversion efficiency (feed : gain) for a 90 kg entire male pig in relation to ambient temperature. The pig was penned separately on a slatted concrete floor with 0.2 m s$^{-1}$ air speed, an air water content of 10 g kg$^{-1}$, with zero wet skin at or below its lower critical temperature and 0.15 wet skin at or above its evaporative critical temperature and given free access to a 14 MJ kg$^{-1}$ digestible energy diet. LCT, lower critical temperature; ECT, evaporative critical temperature. Predictions from AUSPIG (Davies *et al.*, 1993).

temperature fell to −5°C and decrease to zero as the temperature approached 45°C (Fig. 4.3b). The fall in feed intake over the temperature range of −5°C to 43°C was equivalent to 76 g or 1.06 MJ DE per 1°C rise in air temperature. The predicted change in feed intake between 5°C and 30°C was 52 g per 1°C or 1.64% per 1°C and compares with the estimate made by the National Research Council (NRC, 1987) of 1.65% per 1°C change in air temperature for pigs weighing 80 kg. The actual change in feed intake with temperature depends on the live weight of the pig and the number of pigs in the group which will determine the extent that intake will increase in the cold, the temperature range of the thermoneutral zone and the maximum and minimum temperatures investigated. The rate of change in intake above the evaporative critical temperature was predicted to be 0.1 kg or 1.4 MJ DE per 1°C rise in air temperature.

Body temperature increased steadily as air temperature increased above the evaporative critical temperature until at about 46°C the pig could no longer control its body temperature and death was predicted to occur within one day. Skin temperature was predicted to range from 23.1°C when air temperature was −5°C to 38.4°C when air temperature was 45°C (Fig. 4.3b). Live weight gain was predicted to be 955 g day$^{-1}$ under thermoneutral conditions and despite the substantial increase in feed intake was predicted to fall to 687 g day$^{-1}$ at −5°C (Fig. 4.3c). The pig was predicted to start losing weight when the ambient temperature reached about 40°C and the feed : gain ratio increased steeply once the ambient temperature moved either below or above the thermoneutral zone.

Heat exchange between a pig and its environment and therefore critical temperatures are affected by several factors including live weight and number of pigs in a pen (Fig. 4.4), feed intake (Fig. 4.5), air speed (Table 4.1), floor type (Table 4.2), and composition of the diet (Table 4.3). The predictions presented are all for the environmental situations specified and will vary as conditions change. Nevertheless, the model illustrates quantitatively the sensitivity of the pig to different aspects of its thermal environment and demonstrates the important role played by computer models, which simulate accurately heat exchange between a pig and its environment, in evaluating the consequences of housing and management strategies on the performance of pigs.

# The Infectious Environment

Disease has been defined as the physiological responses of an animal to an insult or injury to its body (Murray and Phillips, 1995). Diseases can be caused by a range of infectious organisms including viruses, bacteria, protozoa, fungi and prions. Respiratory tract, digestive tract and skin diseases are prevalent in the pig industry but, throughout the world, respiratory diseases are the most important economically. Several associations have been established between

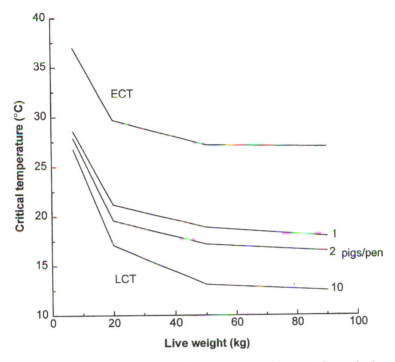

**Fig. 4.4.** Predicted effects of number of pigs per pen and live weight on the lower critical temperature (LCT) and evaporative critical temperature (ECT, independent of number of pigs per pen because no huddling is assumed) for pigs housed in conditions similar to those described in Fig. 4.3.

the severity of respiratory disease, as measured by lung damage at slaughter, and the growth rate of pigs and their efficiency of feed use (Straw *et al.*, 1990; Hill *et al.*, 1992; Pointon *et al.*, 1993). However, these associations do not provide an understanding of the physiological mechanisms responsible for the observed effects on the animal and they do not accommodate interactions between the infectious, climatic or social environments. The aim of this section is to examine quantitatively the effects of one disease, pleuropneumonia, on energy metabolism and other physiological responses and to use it as an example to develop concepts that may allow prediction within animal simulation models of the effects of the infectious environment on animal performance.

## Quantitative effects of pleuropneumonia

Although diseases are known to affect physiological processes such as feed intake, oxygen uptake, the efficiency of energy utilization, body protein

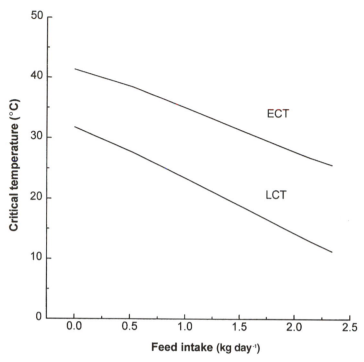

**Fig. 4.5.** Predicted effect of intake of a 14 MJ kg$^{-1}$ digestible energy diet on the lower critical temperature (LCT) and evaporative critical temperature (ECT) of a 90 kg entire male pig penned separately under conditions similar to those described in Fig. 4.3.

**Table 4.1.** Predicted effect of air speed on performance of single penned 90 kg pig at 32°C, relative humidity of 50% and 15% wet skin.

|  |  | \multicolumn{6}{c}{Air speed (m s$^{-1}$)} |
|---|---|---|---|---|---|---|---|
|  |  | 0.1 | 0.2 | 0.4 | 0.8 | 1.6 | 3.2 |
| Feed intake | (kg day$^{-1}$) | 1.39 | 1.53 | 1.75 | 1.99 | 2.25 | 2.34 |
| Live weight gain | (g day$^{-1}$) | 424 | 511 | 641 | 791 | 933 | 954 |
| Body temperature | (°C) | 40.1 | 39.9 | 39.6 | 39.4 | 39.2 | 39.0 |
| Skin temperature | (°C) | 35.6 | 34.7 | 33.9 | 32.9 | 31.8 | 30.8 |
| Respiration rate | (breaths min$^{-1}$) | 147 | 116 | 92 | 65 | 35 | 25 |
| Heat loss from skin | (MJ day$^{-1}$) | 6.59 | 7.97 | 9.86 | 12.32 | 15.59 | 19.66 |

**Table 4.2.** Predicted effect of floor type on critical temperatures of single and group-housed entire male pigs weighing 90 kg and consuming 2.34 kg day$^{-1}$ of a diet containing 14 MJ kg$^{-1}$ digestible energy. The pigs were exposed to an ambient temperature of 32°C, 50% relative humidity, 0.2 m s$^{-1}$ air speed, 15% wet skin for the calculation of the evaporative critical temperature (ECT, °C) and the upper critical temperature (UCT, °C) and no wet skin for the calculation of the lower critical temperature (LCT, °C).

| Floor type | LCT | | ECT | UCT |
|---|---|---|---|---|
| | 1 pig | 10 pigs | | |
| Concrete | 17.2 | 11.3 | 25.7 | 31.9 |
| Straw | 11.6 | 5.7 | 21.0 | 28.4 |
| Timber | 13.2 | 6.7 | 22.1 | 29.2 |
| Wire | 15.3 | 11.6 | 23.7 | 30.5 |

**Table 4.3.** Predicted effect of diet composition on voluntary feed intake and performance of a 90 kg entire male pig exposed to an ambient temperature of 32°C, 50% relative humidity, 0.2 m s$^{-1}$ air speed and 15% wet skin.

| Diet | High fat | High fibre | High protein |
|---|---|---|---|
| *Diet composition* | | | |
| Digestive energy (MJ kg$^{-1}$) | 16.04 | 13.46 | 14.05 |
| Fat (%) | 10.51 | 2.49 | 3.53 |
| ADF (%) | 4.70 | 17.24 | 12.24 |
| Protein (%) | 17.01 | 17.65 | 24.75 |
| | | | |
| *Animal performance* | | | |
| Feed intake (kg day$^{-1}$) | 1.49 | 1.67 | 1.64 |
| Live weight gain (g day$^{-1}$) | 692 | 587 | 558 |
| Heat production (MJ day$^{-1}$) | 14.97 | 15.37 | 15.47 |
| Body temperature (°C) | 39.61 | 39.76 | 39.79 |
| Skin temperature (°C) | 34.29 | 34.37 | 34.39 |
| Respiration rate (breaths min$^{-1}$) | 85 | 103 | 106 |

synthesis and catabolism, body temperature control and tolerance to heat stress (Symons, 1989), little quantitative information exists for these responses in pigs. However, recent studies by Bray (1996) have commenced to provide important results for pigs infected with *Actinobacillus pleuropneumoniae* (App). Pigs weighing approximately 65 kg were prepared surgically to measure oxygen consumption through measurements of cardiac output and arterio-venous differences in oxygen concentration across the lungs. After a period of control measurements, the pigs were intra-bronchially

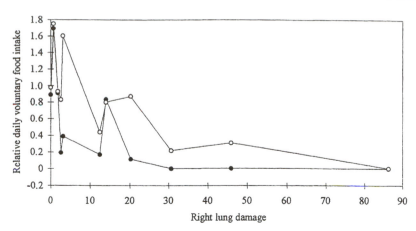

**Fig. 4.6.** Relative (postinoculation/preinoculation) voluntary feed intake on day 2 (●) and day 5 (○) postinoculation with *Actinobacillus pleuropneumoniae* in relation to right lung damage (%) measured 7 days postinoculation. (From Bray, 1996.)

inoculated into the right lung with either saline or one of a range of doses of a pathogenic strain of App serovar 1. Physiological variables were measured for 7 days after inoculation and the severity of pleuropneumonia assessed from measurements of lung damage determined by the image analysis of serial lung slices.

### Voluntary feed intake and body temperature

Voluntary feed intake fell markedly within 24 h of inoculation in all pigs that exhibited lung damage 7 days postinoculation. There was some recovery in feed intake over time, but it did not return to preinoculation values except for three pigs with less than 5% right lung damage. Feed intakes on day 2 and day 5 postinoculation, relative to preinoculation values (Fig. 4.6), were severely reduced in all pigs once more than 20% of the right lung was damaged.

Body temperature in all pigs treated with App rose to a peak of 41°C or higher between 5 and 12 h postinoculation, except for one pig with a peak temperature of 41.8°C which occurred 31 h postinoculation. Body temperature of pigs inoculated with saline did not exceed 39.1°C. Pigs with less than 10% right lung damage had body temperatures near normal by day 2 postinoculation, whereas body temperature remained elevated at day 2 in pigs with more than 10% lung damage. However, by day 5 body temperature had declined to control values in all pigs except for the pig with 87% right lung damage, whose temperature declined to less than the control value.

### Lung physiology and respiration rate

Lung disease affects directly the function of the respiratory tract. Measurements of blood gases and lung mechanics indicated that there was an inequality between ventilation (air reaching the alveolar capillaries) and perfusion (blood reaching the alveolar capillaries). This inequality was particularly evident when pigs stood, suggesting that thickening of the alveolar walls due to bacterial damage may be limiting the diffusion of oxygen through the capillary wall when blood flow through the capillaries was increased due to activity. Pigs with pleuropneumonia also had a reduced ability to generate negative pressures around the lung (Bray, 1996) which resulted in an increase in the work of breathing and in oxygen consumption by the respiratory muscles (Coast and Krause, 1993).

Pigs with 3% or greater lung damage showed an increase in respiration rate with the greatest increase of 2.2 times that measured during the control period occurring on day 2 postinoculation in the pig with 87% lung damage. By day 5 postinoculation, respiration rates for all pigs except the one with greatest lung damage had returned to preinoculation values. The highest respiration rate recorded of 65 breaths per minute was substantially lower than the 150–200 breaths per minute observed in pigs with similar body temperatures but exposed to hot conditions (Giles, 1992). In contrast to heat-stressed animals which are attempting to remove as much heat as possible from the body, febrile animals coordinate thermoregulatory mechanisms to increase body temperature through both the generation and conservation of heat (Moltz, 1993).

### Oxygen consumption

Average oxygen consumption for pigs during the control period was 420 ml min$^{-1}$ and, by day 2 postinoculation, this had increased in pigs with greater than 40% lung damage despite a marked decline in feed intake. By day 5 postinoculation, oxygen consumption for all pigs was between 0.65 and 0.9 of the control values. Previous measurements of oxygen consumption and heat production in pigs with respiratory disease by Intraraksa *et al.* (1984) and Verhagen *et al.* (1987) showed a close association between the decline in feed intake and the decline in oxygen consumption. Figure 4.7 shows that, for pigs inoculated with saline, oxygen consumption relative to the preinoculation values fell with the decline in feed intake. A similar relationship has been observed for pigs with reduced feed intake caused by exposure to high ambient temperatures (Giles and Black, 1991). However, for pigs inoculated with App, relative oxygen consumption did not fall significantly with a decrease in relative feed intake (Fig. 4.7) indicating that maintenance energy requirements of diseased pigs was increased substantially. Similar conclusions have been drawn for humans with sepsis who have increased oxygen consumption (Clevenger, 1993) and a 0.5 rise in heat production (Blaxter, 1989).

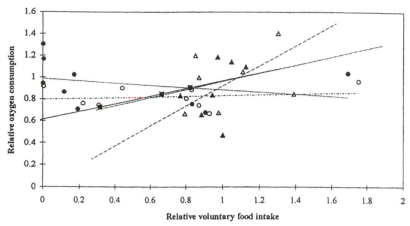

**Fig. 4.7.** Relative (treatment/control period) oxygen consumption (VO₂) in relation to relative (treatment/control period) voluntary feed intake (VFI) for pigs exposed to high ambient temperatures (■—, Giles and Black, 1991), saline inoculated pigs 2 days (▲---) and 5 days (△···, Bray, 1996) postinoculation and *Actinobacillus pleuropneumoniae* inoculated pigs 2 days (●···) and 5 days (○---, Bray, 1996) postinoculation.

$VO_2 = 0.358\ VFI + 0.612$, $R^2 = 0.999$; Giles and Black (1991)
$VO_2 = 0.916\ VFI + 0.007$, $R^2 = 0.155$; saline, day 2
$VO_2 = 0.376\ VFI + 0.586$, $R^2 = 0.104$; saline, day 5
$VO_2 = -0.100\ VFI + 0.987$, $R^2 = 0.076$; *A. pleuropneumoniae*, day 2
$VO_2 = 0.036\ VFI + 0.797$, $R^2 = 0.036$; *A. pleuropneumoniae*, day 5

## Acute and chronic aspects of respiratory disease

Terminology used to describe the severity and stage of disease is subjective and includes the terms peracute, acute, subacute and chronic, where peracute is the most severe. Acute disease may become chronic but still influence an animal's metabolism. For the purpose of this chapter, a simplified definition of the pathological condition is used where acute disease is defined as the period when body temperature is elevated above normal and chronic disease is when the effects of infection are still apparent but body temperature is close to normal.

### Regulation of voluntary feed intake

During acute disease, stimulation of the immune system and release of the cytokines, interleukin-1 (IL-1), IL-6 and tumour necrosis factor (TNF), are likely to be the major reasons for the initial decline in feed intake in pigs with an acute inflammatory disease. IL-1, IL-6 and TNF are directly anorexigenic in mammals (Klasing and Johnstone, 1991), with IL-1 causing a greater response than TNF. Infection of pigs with App has been shown to cause a localized cytokine response within the lung (Baarsh *et al.*, 1995). There is evidence that animals become tolerant to cytokines after a few days because IL-1 infusion was shown to cause an initial depression in the feed intake of rats, but after several days, intake returned to normal despite continued infusion of IL-1 (Mrosovsky *et al.*, 1989).

Although a strong negative relationship between body temperature and feed intake has been observed for pigs exposed to hot conditions (p. 76), no similar strong association exists for pigs infected with pleuropneumonia. Other factors possibly contributing to the reduced feed intake of animals during acute disease include a direct effect on gut function (Mowatt-Larssen and Brown, 1993) through a β-endorphin induced reduction in blood flow to the gut (Navaratnam *et al.*, 1992) and reduced oxygen availability. A reduction in blood oxygen saturation due to pleuropneumonia may also lead to the release from abdominal organs of toxic substances affecting appetite (Lefer, 1987). In addition, an increase in insulin concentration, which has been observed during sepsis (Clowes *et al.*, 1980; Clevenger, 1993), may reduce feed intake (Revell and Williams, 1993). There is little quantitative information on these associations in pigs with disease and further research is required before the mechanisms responsible for reduced feed intake during acute disease can be determined and therefore modelled realistically.

It has been suggested by Bray (1996) that the continued reduction of feed intake in pigs with chronic, severe lung damage is due primarily to a reduction in arterial oxygen saturation caused by a ventilation–perfusion inequality. Figure 4.8 shows arterial and venous blood oxygen saturation, time standing and time feeding on day 6 postinoculation, respectively, for a control pig inoculated with saline (Fig. 4.8a) and a pig inoculated with App and having 31% right lung damage (Fig. 4.8b). Daily feed intake for the pig with pleuropneumonia was only 0.24 of the preinoculation value compared with 0.97 for the control pig. The infected pig had a reduced time standing, smaller meals and a lower venous oxygen saturation than the saline inoculated pig. These observations show that, relative to the control pig, the App infected pig removed more oxygen from the blood during standing and eating. It is speculated that the duration of the meal was shortened due to an inability to meet the oxygen demand for standing during long periods of time. Support for this suggestion comes from observations in humans with chronic, obstructive, pulmonary disease where muscle intracellular substrate concentrations were found at rest to be similar to unaffected people, but impaired muscular oxidative capacity was seen in the forearm of diseased patients following increasing amounts of muscular activity (Wuyam *et al.*, 1992).

Thielscher (1987) suggests that the pig is particularly sensitive to oxygen supply to the heart. The normal pig extracts approximately 0.97 of oxygen from blood as it passes through the heart with the oxygen saturation in coronary blood falling to 0.03. Thielscher (1987) suggests that, compared with other mammals, the pig's heart has approximately 0.22 less oxygen available for oxidative metabolism. Even at rest, pigs have an abnormal ratio of recovery phase to contractile phase of the myocardium. Thus, any interference with oxygen uptake from the lungs could have a substantial effect on the capacity of a pig for activity. Bray (1996) observed that cardiac output increased by up to 0.50 when uninfected pigs stood to eat, whereas cardiac output increased by only 0.075 in the pig with 31% right lung damage. Consequently, arterial oxygen saturation remained virtually stable irrespective of activity in the control pig, whereas the act of standing caused it to fall

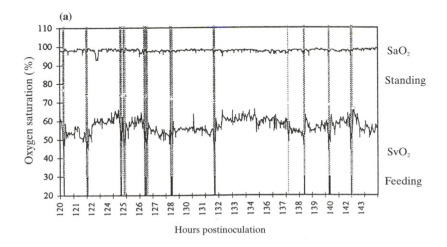

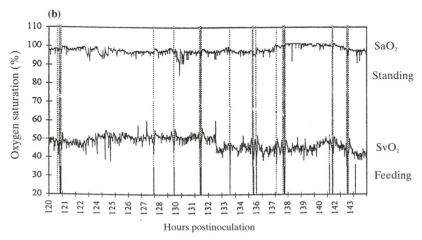

**Fig. 4.8.** Time spent standing and feeding, arterial (SaO$_2$) and venous (SvO$_2$) blood oxygen saturation in (a) a control pig 6 days after inoculation with saline and (b) a pig with 31% right lung damage 6 days after inoculation with *A. pleuropneumoniae*. (From Bray, 1996.)

from 0.97 to 0.86 in the pig with 31% right lung damage. The reduced arterial oxygen saturation in the pig with chronic pleuropneumonia would reduce oxygen supply to the heart, further increasing its dependency on anaerobic metabolism (Thielscher, 1987) and reducing its capacity for activity and depressing feed intake.

### Efficiency of energy utilization and body composition

During acute disease, cytokines cause an increase in oxygen consumption and heat production in order to increase body temperature and generate fever (DeRijk *et al.*, 1993). The administration of an endotoxin to sheep, which resulted in a 1.5°C increase in body temperature, was shown to be associated with an increase of 0.33 in heat production (Baracos *et al.*, 1987). Similarly, Bray (1996) observed an increase in oxygen consumption in pigs with acute infections of App despite a reduction in feed intake. In addition, plasma cortisol concentrations in pigs increased by up to fivefold within hours following infection with App (Bray, 1996). Cytokines, particularly IL-1 and TNF, synergize the effects of cortisol to stimulate proteolysis and reduce protein accretion rates in diseased pigs (Zamir *et al.*, 1992). Thus, during acute disease, feed intake is depressed, maintenance energy requirements are increased and protein accretion rates are reduced, with a resulting substantial reduction in the efficiency of energy utilization for growth.

The evidence presented above for pigs with 20% or greater right lung damage indicates that, during the chronic phase of pleuropneumonia, feed intake remains depressed and oxygen consumption increased relative to preinoculation values. The efficiency of feed use therefore must be depressed by the presence of chronic disease. In addition, plasma cortisol concentrations remained elevated in pigs with chronic disease, further reducing the rate of protein accretion and increasing the proportion of fat in the animal's body (Klasing *et al.*, 1991). There is evidence also that chronic stimulation of the immune system in pigs decreases the rate of protein accretion and requirements for amino acids relative to the intake of digestible energy (Stahly, 1996).

## Predicting the effects of the infectious environment on pig performance

Using the example of pleuropneumonia, disease appears to affect the growth and body composition of pigs mainly through: (i) depressing voluntary feed intake; (ii) increasing maintenance energy requirements; and (iii) decreasing the rate of protein accretion for a given availability of digestible energy. The magnitude of the responses varies between the acute and chronic phases of disease and with its severity. Information is required beyond the 7 days observed by Bray (1996) to establish relationships between the long-term infection and severity of disease on intake, maintenance requirements and protein deposition rates. Nevertheless, the results from Bray (1996) are sufficient to allow the concepts to be incorporated into computer models simulating the growth of pigs.

Disease has been incorporated into the AUSPIG simulation model by increasing maintenance energy requirements by up to 1.3 times the normal predicted value, decreasing the rate of protein deposition by 0.9 times normal and decreasing feed intake down to zero depending on the severity and duration of the disease. This represents an initial step in the full representation of the physiological effects of disease in pigs. Further information is required to quantify better the effects of different severity of different diseases on the physiological determinants of feed intake, growth and nutrient utilization.

It is essential to relate physiological reponses to the severity of disease on farms and a method must be derived to allow prediction on individual pig production units of the likely prevalence and severity of different diseases. The epidemiological studies of Madec and Tillon (1985) and Buddle *et al.* (1992) provide the information and equations that allow these predictions to be made and, if interfaced with the simulation model, the impact of disease on pig performance in commercial units could be assessed quantitatively.

## Conclusions

The performance of pigs raised under commercial conditions is only 0.70 to 0.80 of that observed for pigs raised under ideal conditions because of differences in the social, thermal and infectious environments. There is sufficient information, as summarized in this chapter, to predict accurately the effects on feed intake and nutrient utilization for the thermal environment at temperatures both above and below the zone of thermal comfort. Several examples are given on the way animal performance and critical temperatures are affected by numerous housing and climatic factors. One area requiring additional information is the way housing and stocking conditions affect the proportion of the skin that a pig will wet when it is exposed to both hot and cold conditions. Although less quantitative information is available about the infectious environment, the experiments of Bray (1996) and others provide a sound understanding of the impact of both acute and chronic pleuropneumonia on voluntary feed intake, maintenance energy requirements and body composition of pigs. However, additional information is required for periods of chronic disease beyond 7 days from infection studied by Bray (1996) and for other major diseases of the pig industry. Although the concepts about how the infectious environment affects pig performance can be incorporated into models that simulate pig performance, little evaluation of these concepts has yet been undertaken.

# References

Baarsch, M.J., Scamurra, R.W., Burger, K., Foss, D.L., Maheswaran, S.K. and Murtaugh, M.P. (1995) Inflammatory cytokine expression in swine experimentally infected with *Actinobacillus pleuropneumoniae*. *Infection and Immunity* 63, 3587–3594.

Baracos, V.E., Whitmore, W.T. and Gale, R. (1987) The metabolic cost of fever. *Canadian Journal of Physiology and Pharmacology* 65, 1248.

Black, J.L. (1995) Modelling energy metabolism in the pig – critical evaluation of a simple reference model. In: Moughan, P.J., Verstegen, M.A. and Visser-Reyneveld, M.I. (ed), *Modelling Growth in the Pig*. Wagengin Pers, Wageningen, The Netherlands, pp. 87–102.

Black, J.L. and Carr, J.R. (1993) A symposium – Stocking density and pig performance. In: Batterham, E.S. (ed.), *Manipulating Pig Production IV*. Australasian Pig Science Association, Attwood, Victoria, p. 84.

Black, J.L., Campbell, R.G., Williams, I.H., James, K.J. and Davies, G.T. (1986) Simulation of energy and amino acid utilisation in the pig. *Research and Development in Agriculture* 3, 121–145.

Black, J.L., Davies, G.T., Bray, H.J., Giles, L.R. and Chapple, R.P. (1994) Modelling the effects of genotype, environment and health on nutrient utilisation. In: Danfaer, A. and Lescoat, P. (eds), *Proceedings IVth International Workshop on Modelling Nutrient Utilisation in Farm Animals*. National Institute of Animal Science, Foulum, Denmark, pp. 85–105.

Blaxter, K.L. (1989) *Energy Metabolism in Animals and Man*. Cambridge University Press, London.

Bray, H.J. (1996) The physiological response of growing pigs to pleuropneumonia. PhD thesis, University of Sydney.

Bruce, J.M. (1993) Interactions between the animal and its environment. In: Forbes, J.M. and France, J. (eds), *Quantitative Aspects of Ruminant Digestion and Metabolism*. CAB International, Wallingford, UK, pp. 495–508.

Bruce, J.M. and Clark, J.J. (1979) Models of heat production and critical temperature for growing pigs. *Animal Production* 28, 353–369.

Buddle, J.R., Mercy, A.R., Skirrow, S.Z. Madec, F. and Nicholls, R.R. (1992) *Epidemiological Studies of Pig Diseases. 1. Identification of Risk Factors*. Report to Pig Research and Development Corporation. Murdoch University, Perth, Australia.

Bustamante, M., Jesse, G.W., Becker, B.A. and Krause, G.F. (1996) Effects of individual vs group penning on the performance of weanling pigs. *Journal of Animal Science* 74, 1457–1461.

Campbell, R.G. (1988) Nutritional constraints to lean tissue accretion in farm animals. *Nutrition Research Reviews* 1, 233–253.

Chapple, R.P. (1993) Effect of stocking arrangement on pig performance. In: Batterham, E.S. (ed.), *Manipulating Pig Production IV*. Australasian Pig Science Association, Attwood, Victoria, pp. 87–97.

Clevenger, F.W. (1993) Nutritional support in the patient with the systemic inflammatory response syndrome. *The American Journal of Surgery* 165, 68S–74S.

Close, W.H. (1989) The influence of the thermal environment on the voluntary food intake of pigs. In: Forbes, J.M., Varley, M.A. and Lawrence, T.L.J. (eds), *The Voluntary Food Intake of Pigs*. British Society of Animal Production, Edinburgh, pp. 87–96.

Close, W.H., Mount, L.E. and Brown, D. (1978) The effects of plane of nutrition and environmental temperature on the energy metabolism of the growing pig. II. Growth rate, including protein and fat deposition. *British Journal of Nutrition* 40, 423–431.

Clowes, G.H.A., Jr, Randall, H.T. and Cha, C.-J. (1980) Amino acid and energy metabolism in septic and traumatised patients. *Journal of Parenteral and Enteral Nutrition* 4, 195–205.

Coast, J.R and Krause, K.M. (1993) Relationship of oxygen consumption and cardiac output to work of breathing. *Medicine and Science in Sports and Medicine* 25, 335–340.

Curtis, S.E. (1983) *Environmental Management in Animal Agriculture*. Iowa State University Press, Ames.

Davies, G.T., Black, J.L. and James, K.J. (1993) *AUSPIG - A Decision Support System for Pig Farm Management: User Guide and Reference, Version 3.00*. CSIRO, Division of Animal Production, Sydney.

DeRijk, R.H., Strijbos, P.J.L.M., van Rooijen, N., Rothwell, N.J. and Berkenbosch, F. (1993) Fever and thermogenesis in response to bacterial endotoxin involve macrophage-dependant mechanisms in rats. *American Journal of Physiology* 265, R1179–R1183.

Ekkel, E.D., van Doorn, C.E.A., Hessing, M.J.C. and Tielen, M.J.M. (1995) The specific-stress-free housing system has positive effects on productivity, health and welfare of pigs. *Journal of Animal Science* 73, 1544–1551.

Fuller, M.F. (1965) The effect of environmental temperature on the nitrogen metabolism and growth of the young pig. *British Journal of Nutrition* 19, 531–546.

Giles, L.R. (1992) Energy expenditure of growing pigs at high ambient temperatures. PhD thesis, University of Sydney.

Giles, L.R. and Black, J.L. (1991) Voluntary food intake in growing pigs at ambient temperatures above the zone of thermal comfort. In: Batterham, E.S. (ed.), *Manipulating Pig Production III*. Australasian Pig Science Association, Attwood, Victoria, pp. 162–166.

Giles, L.R., Black, J.L., Gooden, J.M. and Annison, E.F. (1991) Energy expenditure of pigs maintained at high ambient temperature. In: Wenk, C. and Boessinger, M. (eds), *Energy Metabolism in Farm Animals*. European Association of Animal Production, pp. 52–55.

Hessing, M.J.C. and Tielen, M.J.M. (1994) The effect of climatic environment and relocating and mixing on health status and productivity of pigs. *Animal Production* 59, 131–139.

Hill, M.A., Scheidt, A.B., Teclaw, R.F., Clark, L.K., Knox, K.E. and Jordan, M. (1992) Association between growth indicators and volume of lesions in lungs from pigs at slaughter. *American Journal of Veterinary Research* 53, 2221–2223.

Hovell, F.D.DeB, Gordon, J.G. and MacPherson, R.M. (1977) Thin sows. 2. Observations on the energy and nitrogen exchanges of thin and normal sows in environmental temperatures of 20 and 5 degrees C. *Journal of Agricultural Science* 89, 523–533.

Ingram, D.L. and Legge, K.F. (1969) The effect of environmental temperature on respiratory ventilation in the pig. *Respiration Physiology* 8, 1–12.

Intraraksa, Y., Engen, R.L. and Switzer, W.P. (1984) Pulmonary and hematologic changes in swine with *Mycoplasma hyopneumoniae* pneumonia. *American Journal of Veterinary Research* 45, 474–477.

Klasing, K.C. and Johnstone, B.J. (1991) Monokines in growth and development. *Poultry Science* 70, 1781-1789.

Klasing, K.C., Johnstone, B.J. and Benson, B.N. (1991) Implications of an immune response on growth and nutrient requirements of chicks. In: Haresign, W. and Cole, D.J.A. (eds), *Recent Advances in Animal Nutrition*. Nottingham University Press, Nottingham, pp. 135-146.

Khogali, M., Elkhatib, G. Attia, Mustafa, M.K.Y., Gumaa, K., El-Din, A.N. and Ai-Adnani, M.S. (1983) Induced heat stroke: a model in sheep. In: Khogali, M. and Hales, J.R.S. (eds), *Heat Stroke and Temperature Regulation*. Academic Press, Sydney, pp. 253-261.

Koong, L.J., Nienaber, J.A. and Mersmann, H.J. (1983) Effects of plane of nutrition on organ size and fasting heat production of genetically obese and lean pigs. *Journal of Nutrition* 113, 1616-1631.

Le Dividich, J. and Noblet, J. (1982) Growth rate and protein and fat gain in early weaned piglets housed below thermoneutrality. *Livestock Production Science* 9, 731-742.

Lefer, A.M. (1987) Interaction between myocardial depressant factor and vasoactive mediators with ischemia and shock. *American Journal of Physiology* 252, R193-R205.

Lorschy, M.L. (1994) The physiological regulation of heat exchange in the lactating sow exposed to high ambient temperatures. PhD, thesis, University of Sydney.

Lorschy, M.L., Giles, L.R., Bray, H.J., Gooden, J.M. and Black, J.L. (1991) The hourly pattern of voluntary food intake in finisher pigs maintained at fluctuating ambient temperature. In: Batterham, E.S. (ed.), *Manipulating Pig Production III*. Australasian Pig Science Association, Attwood, Victoria, 181 pp.

Madec, F. and Tillon, J.P. (1985) The ecopathological approach in pig veterinary practice. *Pig Veterinary Society Proceedings*, 15, 7-28.

Moltz, H. (1993) Fever: Causes and consequences. *Neuroscience and Behavioral Reviews* 17, 237-269.

Morrison, S.R. and Mount, L.E. (1971) Adaption of growing pigs to changes in environmental temperature. *Animal Production* 13, 51-57.

Mount, L.E. (1974) The concept of thermal neutrality. In: Monteith, J.L. and Mount, L.E. (eds), *Heat Loss from Animals and Man*. Butterworths, London, pp. 425-439.

Mowatt-Larssen, C.A. and Brown, R.O. (1993) Specialised nutritional support in respiratory disease. *Clinical Pharmacy* 12, 276-292.

Mrosovsky, N., Molony, L.A., Conn, C.A. and Kluger, M.J. (1989) Anorexic effects of interleukin 1 in the rat. *American Journal of Physiology* 257, R1315-1321.

Murray, P.G. and Phillips, J.D. (1995) The nature of disease. In: Phillips, J., Murray, P. and Crocker, J. (eds), *The Biology of Disease*. Blackwell Scientific, Oxford, UK, pp. 3-6.

National Research Council (1987) *Predicting Feed Intake of Food Producing Animals*. National Academy of Sciences, Washington, DC.

Navaratnam, N., Herndon, D.N., Woodson, L.C., Linares, H.A., Morris, S. and Traber, D.L. (1992) Endorphin mediation of mesenteric blood flow after endotoxemia in sheep. *Critical Care Medicine* 20, 402-408.

Nienaber, J.A., McDonald, T.P., Hahn, G.L. and Chen, Y.R. (1990) Eating dynamics of the growing-finishing swine. *Transactions of the American Society of Agricultural Engineers* 33, 2011-2018.

Nienaber, J.A., McDonald, T.P., Hahn, G.L. and Chen, Y.R. (1991) Group feeding behaviour of swine. *Transactions of the American Society of Agricultural Engineers* 34, 289-294.

Panaretto, B.A. (1968) Some metabolic effects of cold stress on undernourished non-pregnant ewes. *Australian Journal of Agricultural Science* 19, 273-282.

Parker, R.O., Williams, P.E.V., Aherne, F.X. and Young, B.A. (1980) Serum concentration changes in protein, glucose, urea, thyroxine and triiodothyronine and thermostability of neonatal pigs farrowed at 25 and 10°C. *Canadian Journal of Animal Science* 60, 503-511.

Pointon, A.M., Moore, M. and Cargill, C. (1993) Trends in chronic respiratory disease across Australian piggeries. In: Batterham, E.S. (ed.), *Manipulating Pig Production IV*. Australasian Pig Science Association, Attwood, Victoria, 257 pp.

Revell, D.K. and Williams, I.H. (1993) Physiological control and manipulation of voluntary feed intake. In: Batterham, E.S. (ed.), *Manipulating Pig Production IV*. Australasian Pig Science Association, Attwood, Victoria, pp. 55-80.

Rinaldo, D. and Le Dividich, J. (1991) Assessment of optimal temperature for performance and chemical body composition of growing pigs. *Livestock Production Science* 29, 61-75.

Schoenherr, W.D., Stahly, T.S. and Cromwell, G.L. (1989) The effects of dietary fat and fibre addition on yield and composition of milk from sows housed in a warm or hot environment. *Journal of Animal Science* 67, 482-495.

Stahly, T. (1996) Impact of immune system activation on growth and optimal dietary regimens of pigs. In: Garnsworthy, P.C., Wiseman, J. and Haresign, W. (eds), *Recent Advances in Animal Nutrition*. Nottingham University Press, Nottingham.

Stahly, T.S. and Cromwell, G.L. (1979) Effect of environmental temperature and dietary fat supplement on the performance and carcass characteristics of growing and finishing swine. *Journal of Animal Science* 49, 1478-1488.

Straw, B.E., Shin, S.J. and Yeager, A.E. (1990) Effect of pneumonia on growth rate and feed efficiency of minimal disease pigs exposed to *Actinobacillus pleuropneumoniae* and *Mycoplasma hyopneumoniae*. *Preventative Veterinary Medicine* 9, 287-294.

Sugahara, M., Baker, D.H., Harmon, B.G. and Jensen, A.H. (1970) Effect of ambient temperature on performance and carcass development in young swine. *Journal of Animal Science* 31, 59-62.

Symons, L.E.A. (1989) *Pathophysiology of Endoparasitic Infection Compared with Ectoparacitic Infection and Microbial Infection*. Academic Press, Sydney.

Thielscher, H.-H. (1987) The pig's heart – a problem of pathophysiology. *Pro Veterinario* 3, 12.

Verhagen, J.M.F., Groen, A., Jacobs, J. and Boon, J.H. (1987) The effect of different climatic environment on metabolism and its relation to time of *Haemophilus pleuropneumoniae* infection in pigs. *Livestock Production Science* 17, 365-379.

Verstegen, M.W.A., Close, W.H., Start, I.B. and Mount, L.E. (1973) The effect of environmental temperature and plane of nutrition on heat loss, energy retention and deposition of protein and fat in groups of growing pigs. *British Journal of Nutrition* 30, 21-35.

Verstegen, M.W.A., Brandsma, H.A. and Mateman, G. (1985) Effect of ambient temperature and feeding level on slaughter quality in fattening pigs. *Netherlands*

*Journal of Agricultural Science* 33, 1-15.

Wuyam, B., Payen, J.F., Levy, P., Bessaidne, H., Reutenauer, H., Le Bas, J.F. and Benabid, A.L. (1992) Metabolism and aerobic capacity of skeletal muscle in chronic respiratory failure related to chronic obstructive pulmonary disease. *European Respiratory Journal* 5, 157-162.

Zamir, O. and Hasselgren, P.O., Kunkel, S.L., Frederick, J., Higashiguchi, T. and Fischer, J.E. (1992) Evidence that tumor necrosis factor participates in the regulation of muscle proteolysis during sepsis. *Archives of Surgery* 127, 170-174.

# Describing the Social Environment and its Effects on Food Intake and Growth

**5**

## C.A. Morgan,[1] B.L. Nielsen,[1] A.B. Lawrence[1] and M.T. Mendl[2]

*[1]Animal Biology Division, Scottish Agricultural College, West Mains Road, Edinburgh EH9 3JG, UK; [2]Division of Animal Health and Husbandry, Department of Clinical Veterinary Science, University of Bristol, Langford, Bristol BS18 7DU, UK*

## Introduction

The pig is a social animal by nature but the groupings imposed on pigs in agricultural systems are not those which would apply in the wild. In wild boar and feral pigs, natural groups are composed of up to four females and their offspring (a number of unweaned juveniles and weaned sub-adults) and one or two adult males. Distribution of the groups and their proximity to each other is determined by food and habitat resources (see review by Mendl, 1994). Stolba and Wood-Gush (1989) observed a similar social organization in intensively reared domestic pigs when allowed to live under natural conditions. Growing pigs (or weaned sub-adults) in modern production systems are kept in groups of similar sized (aged) animals, sometimes of one sex only, in contrast to the natural groups of both sexes, young and old. It has been recognized that such groupings can affect the production performance of individuals within the group. Several experiments, which are reviewed here, show that food intake and growth rate of individually penned animals are greater than those of similar individuals kept in a group. However, it is not clear whether these effects are entirely a result of factors associated with group size (e.g. stocking density/space allowance) or if they are a consequence of grouping *per se* (i.e. single pigs vs groups). The purpose of this chapter is to identify the factors responsible for the effects of grouping and, where possible, to quantify the scale of their importance.

# Nutrient Demands

It is generally assumed that, under conditions of no nutrient constraint, growth of the individual is controlled by its genetic potential (see Chapter 10). The pig can therefore be considered to have a potential rate of protein growth and a desired rate of lipid growth to achieve its target mature protein weight and a target lipid to protein ratio at maturity. Desired food intake will be a function of the potential growth and the utilizable nutrient content of the food. Since the processes of metabolism associated with maintenance and growth produce heat, the thermal environment will also affect food intake. If heat cannot be lost to the environment then food intake and growth will be depressed. Conversely, food intake will increase to meet the energy demands of cold environments (see Chapter 4). These matters have been presented and quantified elsewhere (Chapter 10). Feeding behaviour forms the link between other (social) behaviours and food intake and growth. Feeding behaviour may be influenced by competing motivations, such as the preference for remaining close to pen mates, and in this way other motivations may influence food (nutrient) intake.

# Describing Feeding Behaviour

Our ability to measure feeding behaviour of housed animals has been strongly influenced by available technology. In the past feeding behaviour was estimated from measurements of food intake based on refusals (e.g. Gonyou *et al.*, 1992), from feeder visits measured by recording breakage of an infrared beam (e.g. Hsia and Wood-Gush, 1984a) or, more recently, from operant feeders measuring feeder visits and intake by requiring the pig to press panels ('work') to obtain small food rewards (e.g. Bigelow and Houpt, 1988). Clearly none of the above methods provides a complete description of feeding behaviour. In addition the methodology itself, such as requiring animals to work for food, can influence feeding behaviour variables (e.g. Johnson and Collier, 1994; Morrow and Walker, 1994a). These problems have largely been resolved through the development of computerized food intake recording (CFIR) systems, which allow the measurement of unconstrained feeding pattern and food intake.

   When given access to a single homogeneous food source, the feeding behaviour of pigs can be described in terms of three base variables: (i) number of daily feeder visits (NDV); (ii) feed intake per visit (FIV); and (iii) duration of visit (DUV). Other parameters can be derived from these variables: (i) daily food intake (DFI); (ii) feeding rate (FR); and (iii) daily feeder occupation (DFO). No implicit causal relationship should be deduced from the way these six variables are presented. However, it should be noted that for any given set of two derived variables (e.g. DFI and FR) the third derived parameter (DFO) is fixed. The term 'visit' denotes the act of entering a feeder

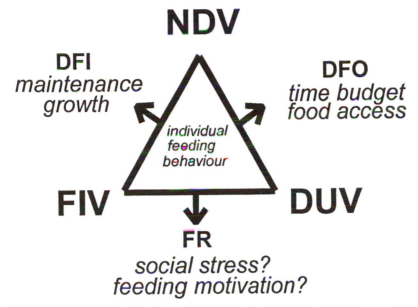

**Fig. 5.1.** Components of feeding behaviour and their interrelations modified from Nielsen (1998). NDV = number of daily feeder visits; FIV = feed intake per visit; DUV = duration of visit; DFI = daily food intake (which is available for maintenance and growth); DFO = daily feeder occupation (which reflects time spent accessing food); and FR = feeding rate (which will reflect body size but perhaps also the degree of social stress facing an individual or its feeding motivation). See text for details.

or feed trough, and has been chosen over the term 'meal' or 'feeding bout' to avoid confusion over choice of meal criterion (see below). The following equations represent the feeding behaviour of any individual animal, and the relationships between these individual components are represented diagramatically in Fig. 5.1.

$$\text{DFI (g day}^{-1}) = \text{NDV (visits day}^{-1}) \times \text{FIV (g visit}^{-1}) \tag{5.1}$$

$$\text{FR (g min}^{-1}) = \text{FIV (g visit}^{-1})/\text{DUV (min visit}^{-1}) \tag{5.2}$$

$$\text{DFO (min day}^{-1}) = \text{NDV (visits day}^{-1}) \times \text{DUV (min visit}^{-1}) \tag{5.3}$$

### Visits, bouts and meals

Various approaches to feeding behaviour parameters are shown in Table 5.1, which compares measures of feeding behaviour of individually housed growing pigs from five experiments.

There is confusion over the use of different terms to describe the basic unit of feeding behaviour. The simplest basic unit is the visit of the pig to the

**Table 5.1.** A comparison of feeding behaviour of individually housed growing pigs measured in six separate locations/experiments.

| | Feeding behaviour parameters[a] | | | | | | | | |
|---|---|---|---|---|---|---|---|---|---|
| | FIV | FIM | DUV | DUM | NDV | NDM | FR | DFI | DFO |
| Paper | | | | | | | | | |
| 1 | — | 1146 | — | 15.4 | — | 3.6 | 74 | 4297 | 55 |
| 2 | — | — | — | — | — | 15.1 | 40 | 2578 | 77 |
| 3 | — | 216 | — | 12.6 | — | 10.7 | 19 | 2069 | 128 |
| 4 | 35 | 104 | 1.38 | 4.0 | 72.8 | 22.9 | 27 | 2075 | 83 |
| 5 | 106 | — | 4.35 | — | 17.0 | — | 25 | 1602 | 66 |

[a] **Description of parameters**: FIV: food intake per visit (g); FIM: food intake per meal (g); DUV: duration of visit (min); DUM: duration of meal (min); NDV: number of daily visits; NDM: number of daily meals; FR: feeding rate (g min$^{-1}$); DFI: daily food intake (g); DFO: daily feeder occupation (min).

**Paper 1: Auffray and Marcilloux (1983):** Weight range 176–275 kg live weight; all entire Large White males (*n*=5); space allowance: 7–10 m$^2$; feeding behaviour monitored using a unique system of infrared beams and a system of rotating food troughs; *ad libitum* access to food; ambient temperature: 18–20°C.

**Paper 2: Hsia and Wood-Gush (1984a):** Weight range 20–80 kg live weight; all Landrace × Large White castrated males (*n*=4); no details of penning; feeding behaviour monitored using infrared light beams and daily weighing of refusals; *ad libitum* access to food (no details of diet composition given); ambient temperature adjusted for body weight: 26°C between 14 and 25 kg to 18°C between 60 and 84 kg live weight.

**Paper 3: Bigelow and Houpt (1988):** Weight range 30–100 kg live weight; all immature Yorkshire females (*n*=6); socially isolated in 2 × 3 m$^2$ pens with concrete floors; feeding behaviour monitored by requiring pigs to press a panel 10 times to receive food rewards; size of food reward adjusted for weight (range 5–10 g per reward); food contained 7.1 MJ net energy kg$^{-1}$ and 20% crude protein; ambient temperature 22–23°C.

**Paper 4: De Haer and de Vries (1993):** Weight range: 25–100 kg live weight: Dutch Landrace males (*n*=12) and females (*n*=8); space allowance: 3.3 m$^2$; feeding behaviour and food intake of individuals and groups of eight measured by a single IVOG feeding station per pen; *ad libitum* access to pelleted: (i) starter feed: 9.4 MJ net energy kg$^{-1}$; 18.2% crude protein; (ii) grower diet: 9.1 MJ net energy kg$^{-1}$; 16.8% crude protein; no mention of ambient temperature.

**Paper 5: Nielsen *et al.* (1996b):** Weight range: 39–50 kg live weight: pigs had previously all been housed in groups of ten in the same building; all entire Landrace × Large White males (*n*=12); space allowance: 13 m$^2$ and straw bedding provided; feeding behaviour monitored for 2 weeks using a FIRE (Hunday Electronics Ltd) feeding station; *ad libitum* access to a pelleted food containing 13.5 MJ DE kg$^{-1}$ and 20% protein; naturally ventilated rooms with insulated kennels provided.

feeder. However, only the two studies that have used CFIR systems (Papers 4 and 5) have reported visits. Most other studies have reported either bouts or meals and only Paper 4 describes both visits and meals. We understand the term meal to mean a bout of visits that are statistically separated (by an inter-bout/meal interval) from other bouts/meals of feeding behaviour (see Martin and Bateson, 1986). There will always be some difficulty in comparing bouts/meals because different statistical procedures can be used to derive the criterion to separate the bouts/meals (e.g. Sibly *et al.*, 1990). The effect of using an inter-meal criterion to 'collapse' visits into meals is illustrated by Paper 4. The visits described in Paper 5 appear similar to the meals reported in Paper 4; this may reflect different software approaches to registering visits between the CFIR systems used. In general, as the weight of the pig and its food intake increase, the frequency and duration of meals decrease, and the size of meals and the rate at which the food is eaten increase (e.g. Paper 1). Bigelow and Houpt (1988) provide detailed information on changes in feeding behaviour over time. The lower number of meals and higher amount eaten per meal reported in Paper 3 compared with Paper 4, with pigs of a similar weight, may well be a consequence of the pigs working for the food rewards, which is known to reduce meal number and increase meal size (e.g. Morrow and Walker, 1994a). These studies with individually housed pigs may be considered to represent the unconstrained feeding pattern. However, whether this can ever be truly observed without interference from the method of measurement is doubtful.

## Relationships between feeding pattern and daily food intake

Considerable variations in the feeding behaviour between individuals can be seen within groups of pigs. Such individual differences in meal patterns have led to speculations about the relationship between meal pattern and daily food intake, and some authors have found significant correlations between feeding frequency and production parameters (Cohn *et al.*, 1962; De Haer, 1992). De Haer *et al.* (1993) found a significant positive relationship between number of daily feeder visits and daily food intake ($r=0.38$; $n=352$) in growing pigs kept in groups of eight with access to a single-space computer-ized feeder. However, Nielsen (1995) found an overall correlation of only 0.19 between number of daily feeder visits and daily food intake for twelve groups of 8–10 pigs across four experiments ($n=113$).

The following example (Fig. 5.2) illustrates why a close relationship between feed intake and feeding frequency is not to be expected. The daily food intake of an animal is achieved through a combination of number of daily feeder visits and feed intake per visit (equation 5.1), and the same level of intake can be obtained by various combinations of these two variables. Pig A and pig B were both kept in the same group of 20, both consumed 1260 g day$^{-1}$, but had substantially different feeding patterns (Fig. 5.2). It is clear

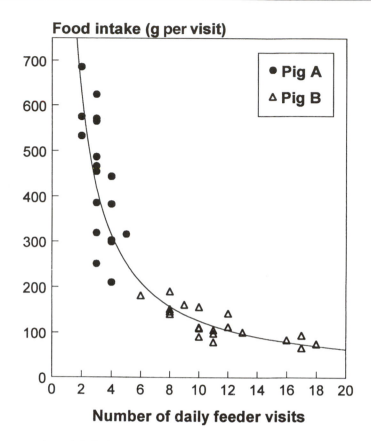

**Fig. 5.2.** Number of daily feeder visits (NDV; recorded over 21 days) plotted against mean feed intake per visit (FIV) for two pigs of similar size housed within a group of twenty. Both pigs had similar levels of daily food intake (1255 and 1264 day$^{-1}$, pigs A and B, respectively), and the pictured isoline represents all combinations of NDV and FIV resulting in a daily intake of 1260 g. (Data from Nielsen *et al.*, 1995a.)

from the spread of the points for each pig, that the pig with a low number of daily feeder visits (pig A) displayed a much higher variation in feed intake per visit than pig B, which visited the feeder much more frequently. Pig B, in turn, had a higher variation in the feeding frequency than pig A. Both pigs varied their feeding pattern along the isoline of equal daily food intake ($x \times y = 1260$ g day$^{-1}$). It can be seen from Fig. 5.2, that even small changes in the number of feeder visits made a large difference to the daily food intake for pig A, whereas similar changes had hardly any consequences for pig B. Thus, pigs with a high number of visits may have more flexible feeding patterns, simply because each visit becomes less crucial for sustaining a given level of food intake. From this example it can be seen that a range of feeding frequencies may lead to similar levels of intake, and that differences in number of feeder visits between individuals are unlikely to explain a significant part of the variation seen between pigs in daily food intake.

## Group Versus Individual Housing

In a recent review, Chapple (1993) has suggested that group-housed growing pigs experience a depression in food intake and growth in comparison to individually housed animals. Table 5.2 provides a summary of the data reviewed by Chapple; an additional experiment from our own work (Nielsen *et al.*, 1996a) is also included. It should be noted that in the latter experiment, the period of individual housing followed the period of group housing; it was not therefore a contemporaneous comparison of individual and group housing.

The degree to which group living is related to a depression in food intake and growth is variable but none the less consistent across all four sets of results. However, the depression in food intake is only reported as significant by Gonyou *et al.* (1992) and Nielsen *et al.* (1996a). All the studies (with the exception of Chapple (1993) who makes no reference to statistical analysis of his data) showed significantly higher growth rates in the individually housed animals. Other studies have shown similar effects: Patterson (1985) found that growth rates of pigs between 37 and 80 kg live weight were depressed by 4% ($P < 0.01$) in groups of five relative to individually housed pigs. Spicer and Aherne (1987) reported that food intake of pigs in pens of four was depressed by 17% and 15% compared to pens of one and two pigs respectively (both comparisons $P < 0.05$); live weight gain was depressed by 10% and 9% for the same comparisons (again $P < 0.05$).

Figure 5.3 shows the daily food intake for pigs in groups of nine or ten at a range of body weights in four separate experiments at SAC Edinburgh. The observed daily food intakes for these grouped animals are contrasted with expected values from two equations based on data from individually housed pigs. One equation is derived from experimental work by Kyriazakis *et al.* (1993), using the same genotype of pig (between 44 and 103 kg live weight) but housed individually with a greater space allowance. The other equation is from Chadd *et al.* (1993) for pigs between 25 and 120 kg live weight. Although this second equation shows some of the group-housed pigs eating the same as, or more than, expected in individually housed animals, the proportion is small (16 out of 104). These data are indicative of grouped pigs having depressed intakes relative to individually housed pigs. It is clear that the considerable variation among the group-housed animals cannot explain the overall lower food intake; no grouped animals reach the mean level of

**Table 5.2.** The effect of group housing on daily food intake (DFI, kg), average daily gain (ADG, kg) and food conversion ratio (FCR, kg food per kg gain) measured in four separate locations/experiments. Percentages in brackets indicate the direction and magnitude of change compared to individual housing.

| | Group size | | | | |
|---|---|---|---|---|---|
| | 1 | 3 | 5 | 8 | 10 |
| **Papers** | | | | | |
| **Chapple**[a] | | | | | |
| – DFI | 2.41 | 2.30 (−5%) | 2.19 (−9%) | – | – |
| – ADG | 0.89 | 0.87 (−2%) | 0.84 (−6%) | – | – |
| – FCR | 2.71 | 2.66 (+2%) | 2.64 (+3%) | – | – |
| **Gonyou**[b] | | | | | |
| – DFI | 2.71 | – | 2.58 (−5%) | | – |
| – ADG | 0.84 | – | 0.81 (−5%) | | – |
| – FCR | 3.21 | – | 3.18 (−1%) | | – |
| **de Haer**[c] | | | | | |
| – DFI | 2.08 | – | – | 1.93 (−7%) | – |
| – ADG | 0.74 | – | – | 0.64 (−16%) | – |
| – FCR | 2.98 | – | – | 3.17 (−6%) | – |
| **Nielsen**[d] | | | | | |
| – DFI | 1.60 | – | – | – | 1.36 (−15%) |
| – ADG | 0.81 | – | – | – | 0.69 (−15%) |
| – FCR | 1.99 | – | – | – | 2.00 (NC) |

NC = no change.

[a] **Chapple (1993):** Weight range: 20–100 kg live weight, space allowances: 1.3, 0.9 and 0.8 $m^2$ $pig^{-1}$ for groups sizes 1, 3 and 5 respectively; no other details available.

[b] **Gonyou *et al.* (1992):** Weight range: 31–90 kg live weight; 50 : 50 males : females (20 pens of five and 60 individuals): all pigs subject to regrouping before start of experimental period; partially slatted pens; space allowances: 1.2 $m^2$ and 0.9 $m^2$ $pig^{-1}$ for individuals and groups of five respectively between 31 and 55 kg; 1.6 $m^2$ and 1.2 $m^2$ $pig^{-1}$ or individuals and groups of five between 55 and 90 kg; individual pigs and groups of five provided with one two-space feeder; *ad libitum* access to a standard diet containing 16% crude protein: pens under constant illumination; no mention of ambient temperature.

[c] **De Haer and de Vries (1993):** See Table 5.1 for full details of methods (20 pens of eight and 20 individuals); space allowances: 3.3 $m^2$ and 0.76 $m^2$ for individuals and groups of eight respectively.

[d] **Nielsen *et al.* (1996a):** See Table 5.1 for full details of methods (three pens of ten and 12 individuals); space allowances: 13.0 $m^2$ and 1.3 $m^2$ for individuals and groups of ten respectively.

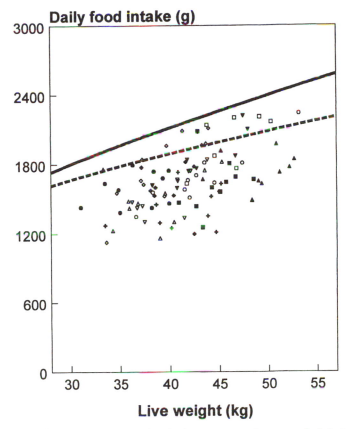

**Fig. 5.3.** Daily food intake (g) for individuals in groups of 10 recorded during four separate experiments (Nielsen *et al.*, 1995a, b, 1996a, b). Pigs within a group are indicated by the same symbol. Individual food intakes were taken at their mid-growth point for the experimental period. These observed values are contrasted to expected values for individually housed pigs using an equation (indicated by the solid line) derived from the work of Kyriazakis *et al.* (1993), who used pigs of the same genotype, and a similar equation (indicated by the broken line) from the work of Chadd *et al.* (1993).

intake achieved by the individually housed animals from the same stock as used by Kyriazakis *et al.* (1993). The grouped pigs in these Edinburgh experiments also grew less quickly but with the same food conversion efficiency compared to the values obtained by Kyriazakis *et al.* (1993).

Group living appears not to influence the efficiency of turning food into growth. Chapple (1993) is alone in suggesting that grouping improves food conversion (although again no statistics are reported). All other studies, including Patterson (1985) and Spicer and Aherne (1987), reported no significant effect of housing on food conversion efficiency.

There is some indication that the effect of grouping on feeding behaviour, food intake and growth is most profound following the initial introduction to the group. Although Gonyou *et al.* (1992) regrouped all pigs (to control for the 'stress' of mixing), their data clearly show that food intake and growth were most depressed between weeks 0 and 4 of the experimental period of 10 weeks.

The CFIR generated data also suggest that reduction in food intake in group-housed pigs is associated with changes in overall feeding behaviour. De Haer and de Vries (1993) reported that grouped pigs ate fewer meals (10.9 vs 22.9 day$^{-1}$) of a larger size (0.22 vs 0.10 kg meal$^{-1}$) which they consumed at a faster rate (32.4 vs 26.9 g min$^{-1}$) relative to individually housed pigs. Individual animals spent more time eating (83.2 vs 62.5 min day$^{-1}$) than grouped pigs. All these effects were significant ($P < 0.01$). Similarly, Nielsen *et al.* (1996a) found that pigs moved from group to individual housing, increased their feeding frequency (14.2 vs 17.0 visits day$^{-1}$) and increased their feeder occupation (52.3 vs 65.9 min day$^{-1}$ both $P < 0.05$). There was, however, no effect of housing on food intake per visit or on feeding rate.

The available evidence therefore supports the view that pigs in groups consume less food and grow less quickly than we would expect if we base our estimates on data from individually housed pigs. This alteration in growth and food intake is apparently achieved without significant alteration to the efficiency of food conversion into live weight gain.

Having so far described the observed effects of grouping on the individual, an explanation and quantification of these effects is now required so that their influence on food intake and growth can be predicted. Grouping may exert its effects by two distinct routes: first, by altering the physical environment (e.g. thermal environment); second, by altering the social environment (e.g. dominance rank). We consider these in turn below.

## Physical Effects of Grouping

There are two physical consequences of grouping which may influence food intake and growth: the level of activity and the thermal environment within the group.

## Activity

One possible explanation for the reduced growth rate lies in the proposition that pigs are more active in a group than when kept individually because they interact with each other (Black, 1995). This increase in activity will require resources in terms of energy and hence some of the energy normally used for growth will be diverted to meet the demands of activity. If this were the case and food intake did not increase to compensate, then growth rate and food conversion would be depressed.

In a study with growing pigs kept individually or in groups of five, Gonyou *et al.* (1992) reported that the grouped animals spent 8.4% of their time standing compared to 7.0% for the individuals (*P*=0.02). The mean weight of the pigs was 60 kg and using the estimate of the energy cost of standing of Noblet *et al.* (1993), this is equivalent to 776 and 647 kJ day$^{-1}$, respectively. The difference of 129 kJ could support 2.9 g of protein growth (10 g lean) or 2.4 g of lipid growth day$^{-1}$ (Close, 1994). The actual difference in growth rate was 32 g LW day$^{-1}$ so the standing component of activity did not account for all of the difference between individuals and groups.

Gonyou *et al.* (1992) did not make any measurements of energy balance and we were unable to find a direct comparison of activity levels and energy balance in growing pigs kept individually or in groups. Such comparisons have, however, been made with pregnant sows. Cronin *et al.* (1986) studied sows for periods of 15 days while tethered and subsequently when kept in a group. When tethered the sows showed slightly less activity (13% vs 17% of observations; NS) and heat production associated with activity increased from 0.20 to 0.24 of the total heat production. However, the increased activity on release from the tethers gradually declined over 10 days to be similar to pre-release levels. The heat production associated with activity was 80.4 and 99.6 kJ LW$^{-0.75}$ (24 h)$^{-1}$ for (tethered) individual and (loose housed) grouped sows. For a 60 kg pig a similar increase in activity energy requirement would correspond to 415 kJ day$^{-1}$, which would lead to a reduction in daily growth of only 9 g protein (approx. 30 g lean) or 8 g lipid.

In contrast, Geuyen *et al.* (1984) found that individually housed sows had slightly higher heat production associated with activity than group-housed sows (83.3 vs 61.5 kJ LW$^{-0.75}$, respectively, at 20°C). The ratio of activity-free heat production to heat production related to activity was similar in groups and individual housing and activity accounted for about 15% of the total heat production in both cases, a value similar to that found by Noblet *et al.* (1993) for individual sows.

Thus, the evidence for increased energy demand associated with the activity in groups causing a reduction in live weight gain is far from being unequivocal. Since, for a wide range of food energy content, growing pigs normally eat to their energy demand (Cole and Chadd, 1989), it would be expected that any increase in energy demand associated with activity would lead to a compensatory increase in food intake but a lowering in food conversion ratio. So, even if the increased activity in groups could explain the reduced growth, it fails to account for the reduction in food intake and it appears likely that the depression in growth is most often a consequence of the latter.

## Thermal environment

The housing of pigs in groups has the potential to affect the thermal environment of the individual. If the pig is unable to lose body heat, food intake will be reduced and the close proximity of other pigs may prevent effective loss of heat. Conversely, if the individually housed pig is cold it will eat more food, whereas group-housed pigs may huddle together to conserve heat and thus require less food increment than the individually penned animal. Therefore, in both cases grouping may increase the effective temperature and reduce food intake. These effects have been reviewed (NRC, 1987) and quantified elsewhere (Chapter 4), and complex models involving humidity, air movement, and floor type (Holmes and Close, 1977), as well as critical temperature, heat production and exchange have been developed (Bruce and Clark, 1979) and extended (Black *et al.*, 1986).

However, it is not clear whether the observed effects on food intake in experiments with individually housed pigs compared with those in groups can be explained by differences in the thermal environment. In the work by De Haer and Merks (1992) individually housed pigs were compared to group-housed conspecifics. The individuals and groups were housed in the same building with automatic ventilation and heating and were therefore maintained in the same thermal environment (J.W.M. Merks, The Netherlands, 1996, personal communication) and the lower food intake of the group-housed animals cannot be explained thus. Even if the effective ambient temperature was higher than the recorded temperatures within the group, the pigs had the opportunity to move apart and thereby reduce the effects of temperature, enabling them to consume sufficient food to achieve their potential growth rate. The observed difference in food intake, therefore, must have been due to some factor(s) other than temperature.

From the above it appears unlikely that the effects of activity and the thermal environment can explain the depression in food intake and growth associated with grouping. In the Introduction to this chapter two possible outcomes of grouping were suggested, i.e. a space/stocking density effect or the effect *per se* of being in a group compared to being a single individual. Both may influence, through social behaviour, the feeding behaviour and food intake of growing pigs.

## Space allowance

The most comprehensive attempt to quantify the effects of grouping on food intake, growth rate and efficiency was made by Kornegay and Notter (1984). Floor space per pig and group size were used as the controlling variables and equations were developed to explain the intake, growth and efficiency of pigs in three weight range classes. Examples of these equations for growing pigs (27–54 kg live weight) are:

| daily gain (kg) | $= 0.489 + 0.520\,S - 0.218\,S^2$ | (5.4) |
|---|---|---|
| | $= 0.6407 - 0.0019\,N$ | (5.5) |
| daily food intake (kg) | $= 1.542 + 0.856\,S - 0.404\,S^2$ | (5.6) |
| | $= 1.5950 - 0.0025\,N$ | (5.7) |
| food/gain | $= 3.3037 - 0.734\,S + 0.406\,S^2$ | (5.8) |
| | $= 2.4974 + 0.0037\,N$ | (5.9) |

where $S$ is the space allowance per pig (m$^2$) and $N$ is the number of pigs per pen.

Whereas these predictions can be applied in practice, there are various shortcomings as described by Chapple (1993). In summary, the purely statistical nature of the equations cannot be interpreted biologically (e.g. positive intercepts predicting growth with no space and no pigs) and the experiments often confounded space allowance with number of pigs in the pen. There are inconsistencies in the predictions by the equations for different weight ranges of pig and they become less accurate as the pig size increases. As Chapple (1993) correctly identified, these problems arise from the strictly empirical approach adopted. In addition, this approach does not distinguish between the effects of group size and the effect of being in a group compared to being individually housed.

In an attempt to achieve a more biological description of space allowance for pigs, Petherick (1983) derived an estimate based on the area covered by the pig when lying on its sternum or side. The allowances were 0.019 and 0.047 m$^2$ LW$^{-0.66}$, respectively, where LW is live weight (kg). Pigs will attempt to position themselves within this range to achieve thermal comfort but the social space requirement cannot be defined so simply and may vary according to the structure of the group.

Edwards *et al.* (1988) proposed that pigs require 0.03 m$^2$ LW$^{-0.67}$ to optimize performance on slats and Black *et al.* (1995) concluded that food intake is depressed once the floor space allowance falls below 0.035–0.039 m$^2$ LW$^{-0.67}$ and that this depression is 20% when the allowance is 0.02$^2$ LW$^{-0.67}$. Using the equation 5.6 above of Kornegay and Notter (1984), for a pig weighing 50 kg, 0.035 m$^2$ LW$^{-0.67}$ would result in a food intake of 1.86 kg whereas at 0.02 m$^2$ LW$^{-0.67}$ the predicted food intake is 1.75 kg, a depression of approximately 6%. Similar calculations using the equation appropriate for a pig weighing 70 kg give a reduction of 11%. Black *et al.* (1995) quoted unpublished work which showed that, for a group of more than five pigs, the depression in food intake was linear as the floor area falls below 0.035 m$^2$ LW$^{-0.67}$. Although the equation of Kornegay and Notter (1984) is a quadratic, the predominant effect is linear. The results of Chapple (1993) showed that when the number of pigs in the pen increased from one to three or one to five then food intake was depressed by 5% and 9%, respectively. Gonyou *et al.* (1992) reported a 5% depression in food intake on going from one to five pigs in a pen. Black *et al.* (1995) concluded that there was little further change as pig numbers increased above five in the pen.

Space allowance and group size interactions will also be affected by the pig : trough ratio present. The relationship between number of animals and number of feeding troughs can be modified through changes to one or both parameters. At a constant space allowance, pigs kept in groups of 20 visited a single-space feeder less often, but ate more per visit than pigs in smaller groups (Nielsen *et al.*, 1995a). However, changes in the group size confound the effects of altering the pig : trough ratio with differences in the social constraints of the animals. Morrow and Walker (1994b) provided groups of 20 pigs with either one or two single-space feeders. The frequency of visits per pig increased significantly when two feeders were present (23.3 vs 30.3 visits day$^{-1}$). This reduction in pig : trough ratio from 20 : 1 to 20 : 2 can be compared to the reduction from 20 : 1 to 10 : 1 obtained by Walker (1991), who used the same experimental set-up, but did not find differences between the two group sizes in the number of visits (in contrast to Nielsen *et al.*, 1995a). Thus, a pig : trough ratio of 20 : 2 differs from a ratio of 10 : 1 in terms of feeding pattern, illustrating the interactive effect of group size and pig : trough ratio.

As yet, no reported experiments have fully controlled for all the major factors that might be expected to affect the performance of pigs in groups. Most notably, in all of the work described above comparing different aspects of performance in individually and group-housed pigs, there has been a confounding of group size and animal density with individual pigs being invariably given more space (e.g. see supplementary notes to Table 5.2). Work comparing food intake and growth in group sizes between 5 and 30 pigs and where stocking density has been kept constant has failed to find any effect of group size on performance (Randolph *et al.*, 1981; McConnell *et al.*, 1987; Petherick *et al.*, 1989; Walker, 1991; Nielsen *et al.*, 1995a). However, as pointed out by Black (1993), group housing will often be accompanied by changes in air quality, disease incidence as well as reductions in access to resources such as food and water, brought about by the presence of pen-mates. In addition, there may be more subtle and less observable influences of the social environment such as either the effects of social 'stress' (Chapple, 1993) or of competing motivations, like the preference for remaining close to pen-mates, which may have direct effects on either nutrient demand or food intake. The extent to which stress or competing motivations are components of or additional to 'stocking density effects' is at present unclear. Quantifying their importance in terms of their effect on food intake, growth and efficiency could potentially allow us to remove the constraints on performance that apparently operate to prevent modern genotypes of pigs achieving their genetic potential for growth under farm conditions. The quantification process is also likely to be of major benefit in the further development of food intake prediction models and their use in the efficient allocation of nutrients to growing pigs (Chapter 10).

# Social Effects of Grouping

For growing pigs kept in groups, the social environment impinges on almost every action performed by the individual animals. Undoubtedly the likely increase in the incidence of disease with grouping will affect performance and this is considered in detail by Black *et al.* (Chapter 4). In the previous section we discussed how the physical aspects of the environment, such as space allowance and number of feeding troughs, interact with the number of animals in a group. We have, however, still not found an explanation for the differences observed in food intake and growth between pigs housed individually and pigs in groups. What are the important aspects of being 'one' compared to 'more than one' in a pen, and how can we quantify them? Is there some form of 'social constraint' responsible for the performance differences observed? The following seeks to pinpoint potential social effects on feeding behaviour and, consequently, feed intake and growth.

## Feeding behaviour

The feeding behaviour of pigs housed in groups has been found to differ significantly from that of individually penned pigs. Within the social environment growing pigs eat substantially fewer, but much larger meals than individually housed animals (De Haer and Merks, 1992; De Haer and de Vries, 1993). It has been assumed that these differences in feeding behaviour were a consequence of the different social environments, in terms of competition and social interactions between group-housed animals. However, these animals had access to only one single-space feeder, and Nielsen *et al.* (1996b) found that the feeding pattern of group-housed pigs given access to four trough spaces was similar to that of individually housed animals with a large number of visits. This increase in frequency of visits did not, however, increase the food intake of these group-housed animals compared to groups with access to only one trough space. Thus, the different feeding patterns of grouped and individual animals is a consequence of the number of trough spaces available, and not a reflection of the social environment. Also the reduced food intake of grouped animals on this evidence does not appear to be caused by the pig : trough ratio.

## Dominance rank

Individual differences in feeding behaviour, such as those described previously, may be a reflection of the social hierarchy. A number of experiments

have reported positive correlations between dominance rank and performance (McBride *et al.*, 1964; Beilharz and Cox, 1967; Hansen *et al.*, 1982)
especially when access to feed has been restricted. However, the causal
relationship between dominance value and production performance (e.g.
live weight) is not clear and some of the reported correlations are weak,
possibly because social rank does not adequately express the role of an animal
within a group (Meese and Ewbank, 1973; Bernstein, 1981). For example,
McBride *et al.* (1964) found that social rank accounted for only 13% of the
total variation in growth, and they note that social behaviour has a minimum
effect on productivity when husbandry conditions are optimal. Nielsen *et al.*
(1995a) found no relationship between either aggression or position in the
hierarchy and any feeding behaviour or performance variables when pigs had
a high space allowance and were provided with straw. Some data suggest that
dominant individuals may spend more time defending their food rather than
eating it (Brouns and Edwards, 1994), and dominance assessed in a non-
feeding situation may show little correlation with that displayed in a feeding
context.

The disagreement in the literature with regard to the effects of dominance hierarchy on performance could be due to our inability to describe
properly the social organization of a group of pigs (Meese and Ewbank, 1973).
Alternatively, any positive correlations found between social rank and production performance may be the result of a third common causal factor, or
rank may be a function of performance.

## Behavioural synchrony

The daily (and nightly) rhythm of behaviours such as feeding may shed some
light on the observed differences in performance between social and non-
social environments. De Haer and Merks (1992) describe the pattern of
feeding behaviour over 24 h for individually and group-housed pigs. In both
housing systems most feeding activity occurred during the day, but the
diurnal pattern of individually housed animals was more equally distributed
across the day compared to the bimodal distribution found in group-housed
pigs. So despite having access to only one feeding trough, the pigs housed in
groups of eight showed more clustering in time of feeding behaviour than
pigs penned individually with unlimited access to food. Group-housed pigs
feeding from a multi-space trough display a high level of simultaneous feeding
and show a preference for troughs directly adjacent to those occupied by
another feeding pig (Nielsen *et al.*, 1996b). The desire to feed at the same
time as conspecifics may interact with the social organization of the group so
that the 'true' value of being dominant could be to have the ability (or choice)
to feed within peak periods of the circadian pattern. It is likely that the
circadian feeding patterns of individually housed pigs are influenced by the

sight or sound of conspecifics in neighbouring pens (e.g. Hsia and Wood-Gush, 1984a), and no behavioural data appear to be available from individual pigs housed in isolation.

The synchrony of behaviour within groups of pigs includes bouts of inactivity as well as bouts of feeding. The importance of behaving in the same way as one's pen-mates may be strong enough to override the desire to feed when the food trough is available if the rest of the group is lying down (Nielsen, 1998). The adaptive value to the individual of synchronous feeding would, in terms of predation, be increased vigilance, a dilution effect, as well as improved defence (Pulliam and Caraco, 1984). Although such evolutionary benefits may appear far removed from our modern farm animals the inherent behavioural mechanisms are still likely to be present in today's pig genotypes (Price, 1984), and little, if any, selection pressure exists against these behaviour patterns.

It should be mentioned here that the importance of allowing simultaneous feeding in pigs has in the past been emphasized as a means to increase food intake. However, no differences in food intake were found by Nielsen *et al.* (1996b) when they compared group-housed pigs given access to single- or multi-space feeders. Indications of enhanced food intake in social environments have previously been reported in growing pigs, but most investigations have been of relatively short duration (Hsia and Wood-Gush, 1983, 1984b). Savory and MacLeod (1980) found that only feeding activity and not food intake was enhanced by the social environment of chickens, a finding recently confirmed by Keeling and Hurnik (1996). Thus the ability, and apparent preference, to feed at the same time as a conspecific may lead to increased feeding activity and occasional and transient increases in food intake, which are not sustained over longer periods.

Thus the observed difference in feeding frequency between group-housed and individually housed pigs is due to physical aspects of the social environment, i.e. differences in the pig : trough ratio, and is not an effect of the social interactions. A high level of simultaneous feeding observed in groups may be part of a more general behavioural synchrony within a group of pigs. Such synchrony of behaviour may lead to pigs choosing to rest instead of eat, even though their daily food intake is below that found in individually housed litter mates (De Haer and Merks, 1992), and despite the feeding trough being available when other group members are resting. If daily feeder occupation is limited by a desire to feed at certain times coinciding with other group members, this may lead to higher feeding rates for some animals in order to obtain their desired daily food intake (see Fig. 5.1). Pigs of high dominance rank may be able to feed at peak times, but this is not necessarily reflected in the level of food intake leading to poor correlations between performance and rank status.

In conclusion, the behavioural pattern of animals within a group, both in terms of feeding and social behaviours, may be affected sufficiently by the

**Table 5.3.** Measured (Obs.) and predicted (Pred.) gain, food intake and backfat thickness (P2) and predicted protein and fat deposition for pigs kept in different group sizes. (From Chapple, 1993.)

| | Group size | | | | | | | |
| | 1 | | 3 | | | 5 | | |
| | Obs. | Pred. | Obs. | Pred1 | Pred2 | Obs | Pred1 | Pred2 |
|---|---|---|---|---|---|---|---|---|
| Gain (kg day$^{-1}$) | 0.89 | 0.89 | 0.87 | 0.86 | 0.86 | 0.84 | 0.83 | 0.83 |
| Food intake (kg day$^{-1}$) | 2.41 | 2.40 | 2.30 | 2.30 | 2.31 | 2.19 | 2.21 | 2.21 |
| Backfat thickness (mm) | 18.7 | | 20.1 | | | 21.7 | | |
| Protein deposition (g day$^{-1}$) | | 132 | | 129 | 125 | | 127 | 119 |
| Fat deposition (g day$^{-1}$) | | 251 | | 234 | 252 | | 217 | 251 |
| Fat tissue (%) | | 26.5 | | 25.7 | 27.3 | | 24.7 | 28.2 |

Pred1 = predicted results of food intake reduction.
Pred2 = predicted results of reduction in protein deposition.

tendency to synchronize behaviour to result in overall lower intakes of group-housed pigs as compared to individually housed conspecifics.

## Stress and Growth

It has been established that grouping has consequences for feeding behaviour, food intake and growth of the individual within the group. Few attempts have been made to derive an understanding of the underlying biological mechanisms responsible for the observed effects.

One possible underlying mechanism may be the involvement of stress in growth regulation. This was recently suggested in a report by Chapple (1993). He measured live weight gain, food intake and backfat thickness of pigs growing from 20 to 100 kg and housed singly or in groups of three or five (Table 5.3: Obs). Using these data and a computer model of pig growth (AUSPIG: Black *et al.*, 1986), he predicted the rate of protein and fat deposition, either according to a straightforward reduction in food intake, or according to a change in the proportions of tissue growth. In the prediction of the effects of a reduction in food intake there was a reduction in both protein and fat deposition (Table 5.3: Pred1). In the experiment, growth rate was reduced whereas the backfat thickness remained constant and, in order to reproduce these observations, the model had to be manipulated such that there was a reduction in protein deposition of 7 g day$^{-1}$ and 13 g day$^{-1}$ on going from one to three or one to five pigs in the group, respectively, with fat deposition remaining constant. Since the observed effects were more consistent with the reduction in protein deposition, Chapple concluded that a simple reduction in food intake was not responsible for the observed reduction in growth. He hypothesized that the protein deposition rate was the governing factor and, since this was reduced, the pigs' demand for protein was also reduced and food intake decreased. He then argued that the reduced protein deposition rate was due to stress on the pig of maintaining its social order within the group. This stress acted via biochemical factors directly on tissue growth (growth hormone, cortisol, insulin-like growth factor and cytokines). A reduction in circulating growth hormone could be expected to reduce protein deposition (MacRae and Lobley, 1991), although the mechanism by which stress should cause such a reduction in growth hormone secretion is not clear. Cortisol increases proteolysis to supply amino acids for gluconeogenesis (Oliverio, 1987). Cytokines also have this effect and in addition reduce food intake (Grimble, 1990) and thus the action of these three factors is consistent with the mechanism for the reduced protein deposition and food intake postulated by Chapple (1993). However, cortisol and cytokines also stimulate lipolysis (Oliverio, 1987; Grimble, 1990) which is in contrast to Chapple's unchanged lipid deposition. Furthermore, frequent social interactions would be expected to increase the episodes of adrenaline release and, although it has been shown to reduce intake (Forbes, 1995), adrenaline also directs nutrients to glucose and the mobilization of fat reserves (Oliverio, 1987). Although corticosteroids are catabolic in laboratory animals, their administration to sheep and cattle has increased fatness, but food intake was also increased (Forbes, 1995). Chronic administration of glucocorticoids results in a redistribution of body fat from the limbs to the trunk (Dickson, 1993) raising the possibility that backfat is maintained at the expense of limb fat.

Von Borell *et al.* (1992) showed that chronic treatment of growing pigs with pharmacological doses of corticosteroid resulted in suppression of growth rate. This demonstrates the possible role of these 'stress hormones' in modulating growth, perhaps through reduction of nutrient demand for protein synthesis. However, the key issues here are whether physiological doses have these effects and, if so, exactly how socially induced stress influences the timing and pattern of the release of these hormones. Cortisol levels may remain elevated for several hours following an environmental or social challenge to the animal (Bradshaw *et al.*, 1996), so repeated challenges could result in chronically higher levels. However, adaptation to perceived stressors may also occur with time such that basal glucocorticoid levels return to near normal (e.g. Ladewig and Smidt, 1989). Paterson and Pearce (1991) reported that pigs kept under crowded conditions (0.025 $m^2$ $LW^{-0.66}$) had higher peak plasma cortisol concentrations than uncrowded pigs (0.048 $m^2$ $LW^{-0.66}$) following ACTH administration (80.7 ± 7.7 vs 60.5 ± 2.7 ng $ml^{-1}$) and had poorer growth rates (0.658 vs 0.735 kg $day^{-1}$). However, the mean cortisol concentration in ten hourly samples was not affected by crowding.

Without detailed data on long-term changes in glucocorticoids, it is difficult to determine the extent to which these may mediate socially induced stress effects on growth. Similar problems exist for other measures used as indicators of a stress response. For example, heart rate may become temporarily elevated during agonistic encounters in pigs (Marchant *et al.*, 1995), but chronic elevations have not been measured in this species. However, chronically elevated heart rate has been observed in subordinate group members in other species (e.g. tree shrews: von Holst *et al.*, 1983), and the associated release of adrenaline could result in changes in the partition of nutrients towards maintenance activity and away from growth.

Another possible route whereby socially induced stress might influence nutrient demand and growth rate is through effects on the animal's immune status. McGlone *et al.* (1993) showed that low ranking pigs suffered a decrease in natural killer cell cytotoxicity following transport in comparison to dominant pigs. Therefore, stressed pigs of low social status might be more susceptible to infections which, in turn, could lead to decreased growth via clinical or subclinical pathogenic routes (see Johnson and von Borell, 1994). Williams *et al.* (1994) reported that pigs with a lower level of activation of the immune system (by medicated early weaning) had a lower ratio of T helper/T suppressor cytotoxic cells, a greater capacity for protein growth and, consequently, higher amino acid requirements than conventionally reared pigs. Socially induced stress may be more apparent in some individuals than others. In some cases, high ranking individuals show lower levels of stress indicators (e.g. McGlone *et al.*, 1993), although there need not necessarily be a linear relationship between measures of stress and measures of social status (cf. Mendl *et al.*, 1992; Creel *et al.*, 1996).

Although there is some circumstantial evidence for the suggestion by Chapple (1993), the full picture is not clear and quantification of these effects is the subject of a current extensive Australian research programme (Black, 1997). Nevertheless, despite the fact that only gain and backfat thickness were actually measured and a computer model was used, Chapple made a first attempt to quantify and explain the underlying mechanisms of the effects of grouping. What is required now is a detailed experiment with pigs in different group sizes where not only food intake and growth rate are measured but also the composition of the growth and physiological parameters relating to stress and the control of growth.

## Conclusions and Future Research Directions

Pigs are considered to eat to satisfy the nutrient requirements for potential protein and lipid growth. However, it is evident that pigs in groups do not consume as much food as those kept singly and thus they fail to meet their potential. Obviously, there are differences in the behavioural constraints related to the social environment: in some way social living affects food intake and growth.

One of the first steps in the quantification of behavioural aspects related to food intake is the requirement for a description of feeding behaviour in terms of visits to the trough. A number of authors have attempted quantification but have used arbitrary and differing methods of analysing meals or bouts of feeding.

There are clear differences between individuals in their feeding behaviour and the way in which they obtain their total daily food intake (feeding frequency and amount eaten). These differences in feeding behaviour may have an effect on growth but the evidence is not conclusive.

It appears that the level of physical activity and the local thermal environment within the group cannot account for the reduced food intake and growth of group-housed pigs.

Although several experiments have been carried out to examine the effect of space allowance and group size (summarized by Kornegay and Notter, 1984), the causes of the effects of these two variables are not known. These experiments have often confounded group size and space allowance and none have controlled completely for the allocation of all resources (number of troughs, drinkers, etc.). Synchronization of feeding behaviour often results in certain members of the group not being able to feed when they wish if there is insufficient trough space and they may not take the opportunity to feed at other times when the trough is vacant, preferring instead to remain with group mates.

Further work should measure feeding behaviour, food intake and growth where pigs are kept singly initially and then mixed into small groups while floor space, number of troughs and number of drinkers per pig are kept constant. Such an experiment would quantify the effects of grouping and indicate if these effects are due to space allocation, etc., or grouping *per se*. To supplement this approach further experiments are required which will provide an understanding of the underlying biological mechanisms. Thus the social effects of grouping could be investigated and quantified by a two pronged approach involving the factors outlined in Fig. 5.4. On the left-hand side of the figure there are the descriptive variables which will provide an empirical quantification of the type initiated by Kornegay and Notter (1984) and further developed by Black (1995) and Black *et al.* (1995). Being strictly empirical in nature, these experiments taken in isolation will still be subject to the same criticism of the previous studies. Therefore, the second line of approach, based on that of Chapple (1993), on the right-hand side of the figure is required to provide a complete understanding of the causation of responses and to contribute to a more generally applicable quantitative model.

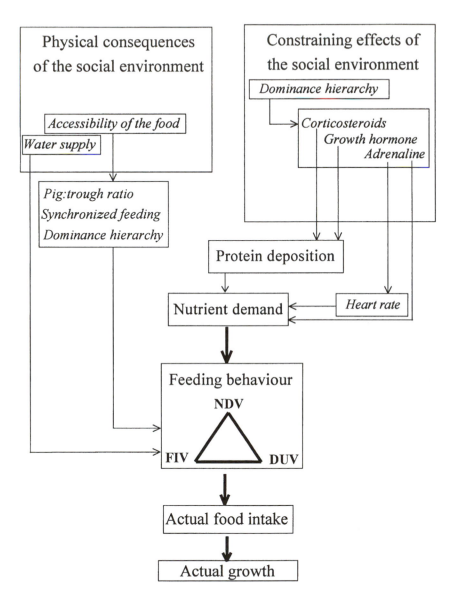

**Fig. 5.4.** A conceptual framework for understanding the potential relationships between social environment, feeding behaviour and growth.

# References

Auffray, P. and Marcilloux, J.C. (1983) Etude de la sequence alimentaire du porc adulte. *Reproduction, Nutrition et Development* 23, 517–524.

Beilharz, R.G. and Cox, D.F. (1967) Social dominance in swine. *Animal Behaviour* 15, 117–122.

Bernstein, I.S. (1981) Dominance: the baby and the bathwater. *Behavior and Brain Science* 4, 419–457.

Bigelow, J.A. and Houpt, T.R. (1988) Feeding and drinking patterns in young pigs. *Physiology and Behavior* 43, 99–109.

Black, J.L. (1993) Symposium conclusions. In: Batterham, E.S. (ed.), *Manipulating Pig Production IV*. Australian Pig Science Association, Victoria, p. 113.

Black, J.L. (1995) Modelling energy metabolism in the pig – critical evaluation of a simple reference model. In: Moughan, P.J., Verstegen, M.W.A. and Visser-Reyneveld, M.I. (eds), *Modelling Growth in the Pig*. Wageningen Pers, The Wageningen, Netherlands, pp. 87–102.

Black, J.L. (1997) The growth gap program. In *Proceedings of the 1997 Pig Research and Development Corporation Meeting*. Melbourne, Australia, pp. 12–14.

Black, J.L., Campbell, R.G., Williams, I.H., James, K.J. and Davies, G.T. (1986) Simulation of energy and amino acid utilisation in the pig. *Research and Development in Agriculture* 3, 121–145.

Black, J.L., Davies, G.T., Bray, H.J., Giles, L.R. and Chapple, R.P. (1995) Modelling the effects of genotype, environment and health on nutrient utilisation. In: Danfaer, A. and Lescoat, P. (eds), *Proceedings of IVth International Workshop on Modelling Nutrient Utilisation in Farm Animals*. National Institute of Animal Science, Denmark, pp. 85–105.

Bradshaw, R.H., Parrott, R.F., Goode, J.A., Rodway, R.G. and Broom, D.M. (1996) Behavioural and hormonal responses of pigs during transport: effect of mixing and duration of journey. *Animal Science* 62, 547–554.

Brouns, F. and Edwards, S.A. (1994) Social rank and feeding behaviour of group-housed sows fed competitively or *ad libitum*. *Applied Animal Behaviour Science* 39, 225–235.

Bruce, J.M. and Clark, J.J. (1979) Models of heat production and critical temperature for growing pigs. *Animal Production* 28, 353–369.

Chadd, S.A., Cole, D.J.A. and Waters, J.R. (1993) The food intake, performance and carcass characteristics of two pig genotypes grown to 120 kg live weight. *Animal Production* 57, 473–481.

Chapple, R.P. (1993) Effect of stocking arrangement on pig performance. In: Batterham, E.S (ed.) *Manipulating Pig Production IV*. Australian Pig Science Association, Victoria, pp. 87–97.

Close, W.H. (1994) Feeding new genotypes: establishing amino acid/energy requirements. In: Cole, D.J.A., Wiseman, J. and Varley, M.A. (eds), *Principles of Pig Science*. Nottingham University Press, Nottingham, pp. 123–140.

Cohn, C., Joseph, D. and Allweiss, M.D. (1962) Nutritional effects of feeding frequency. *American Journal of Clinical Nutrition* 11, 356–361.

Cole, D.J.A. and Chadd, S.A. (1989). Voluntary food intake of growing pigs. In: Forbes, J.M., Varley, M.A. and Lawrence, T.L.J. (eds), *The Voluntary Food Intake of Pigs*. British Society of Animal Production, Edinburgh, pp. 61–70.

Creel, S., Creel, N.M. and Monfort, S.L. (1996) Social stress and dominance. *Nature* 379, 312.

Cronin, G.M., van Tarkwijk, J.M.F.M., van der Hel, W. and Verstegen, M.W.A. (1986) The influence of degree of adaptation to tether – housing by sows in relation to behaviour and energy metabolism. *Animal Production* 42, 257–268.

De Haer, L.C.M. (1992) Relevance of eating pattern for selection of growing pigs. PhD thesis, Wageningen University, The Netherlands, 159 pp.

De Haer, L.C.M. and De Vries, A.G. (1993) Feed intake patterns of and feed digestibility in growing pigs housed individually or in groups. *Livestock Production Science* 33, 277–292.

De Haer, L.C.M. and Merks, J.W.M. (1992) Patterns of daily food intake in growing pigs. *Animal Production* 54, 95–104.

De Haer, L.C.M., Luiting, P. and Aarts, H.L.M. (1993) Relations among individual (residual) feed intake, growth performance and feed intake pattern of growing pigs in group housing. *Livestock Production Science* 36, 233–253.

Dickson, W.H. (1993) Endocrine glands. In: Swenson, M.J. and Reece, W.O. (eds), *Duke's Physiology of Domestic Animals*. Cornell University Press, New York, pp. 629–664.

Edwards, S.A., Armsby, A.W. and Spechter, H.H. (1988) Effects of floor area allowance on performance of growing pigs kept on fully slatted floors. *Animal Production* 46, 453–459.

Forbes, J.M. (1995) *Voluntary Food Intake and Diet Selection in Farm Animals*. CAB International, Wallingford, UK, 532pp.

Geuyen, T.P.A., Verhagen, J.M.F. and Verstegen, M.W.A. (1984) Effect of housing and temperature on metabolic rate of pregnant sows. *Animal Production* 38, 477–485.

Gonyou, H.W., Chapple, R.P. and Frank, G.R. (1992) Productivity, time budgets and social aspects of eating in pigs penned in groups of five or individually. *Applied Animal Behaviour Science* 34, 291–301.

Grimble, R.F. (1990) Nutrition and cytokine action. *Nutrition Research Reviews* 3, 193–210.

Hansen, L.L., Hagelsø, A.M. and Madsen, A. (1982) Behavioural results and performance of bacon pigs fed *ad libitum* from one or several self-feeders. *Applied Animal Ethology* 8, 307–333.

Holmes, C.W. and Close, W.H. (1977) The influence of climatic variables on energy metabolism and associated aspects of productivity in the pig. In: Haresign, W., Swan, H. and Lewis, D. (eds), *Nutrition and the Climatic Environment*. Butterworths, London, pp. 51–73.

Hsia, L.C. and Wood-Gush, D.G.M. (1983) A note on social facilitation and competition in the feeding behaviour of pigs. *Animal Production* 37, 149–152.

Hsia, L.C. and Wood-Gush, D.G.M. (1984a) The temporal patterns of food intake and allelomimetic feeding by pigs of different ages. *Applied Animal Ethology* 11, 271–282.

Hsia, L.C. and Wood-Gush, D.G.M. (1984b) Social facilitation in the feeding behaviour of pigs and the effect of rank. *Applied Animal Ethology* 11, 265–270.

Johnson, D.F. and Collier, G. (1994) Meal patterns of rats encountering variable food procurement cost. *Animal Behaviour* 47, 1279–1287.

Johnson, R.W. and von Borell, E. (1994) Lipopolysaccharide-induced sickness behavior in pigs is inhibited by pretreatment with indomethacin. *Journal of Animal Science* 72, 309–314.

Keeling, L.J. and Hurnik, J.F. (1996) Social facilitation acts more on the appetitive than the consummatory phase of feeding behaviour in domestic fowl. *Animal Behaviour* 52, 11–15.

Kornegay, E.T. and Notter, D.R. (1984) Effects of floor space and number of pigs per pen on performance. *Pigs News and Information* 5, 23-33.

Kyriazakis, I., Emmans, G.C. and Taylor, A.J. (1993) A note on the diets selected by boars given a choice between two foods of different protein concentrations from 44 to 103 kg live weight. *Animal Production* 56, 151-154.

Ladewig, J. and Smidt, D. (1989) Behavior, episodic secretion of cortisol, and adreno-cortical reactivity in bulls subjected to tethering. *Hormones and Behavior* 23, 344-360.

MacRae, J.C. and Lobley, G.E. (1991) Physiological and metabolic implications of conventional and novel methods for the manipulation of growth and production. *Livestock Production Science* 27, 43-59.

Marchant, J.N., Mendl, M.T., Rudd, A.R. and Broom, D.M. (1995) The effects of agonistic interactions on the heart rate of group-housed sows. *Applied Animal Behaviour Science* 46, 49-56.

Martin, P. and Bateson, P. (1986) *Measuring Behaviour*. Cambridge University Press, Cambridge, 200pp.

McBride, G., James, J.W. and Hodgens, N.W. (1964) Social behaviour of domestic animals. IV. Growing pigs. *Animal Production* 6, 129-140.

McConnell, J.C., Eargle, J.C. and Waldorf, R.C. (1987) Effects of weaning weight, co-mingling, group size and room temperature on pig performance. *Journal of Animal Science* 65, 1201-1206.

McGlone, J.J., Salak, J.L., Lumpkin, E.A., Nicholson, R.I., Gibson, M. and Norman, R.L. (1993) Shipping stress and social status effects on pig performance, plasma cortisol, natural killer cell activity, and leukocyte numbers. *Journal of Animal Science* 71, 888-896.

Meese, G.B. and Ewbank, R. (1973) The establishment and nature of the dominance hierarchy in the domesticated pig. *Animal Behaviour* 21, 326-334.

Mendl, M.T. (1994) The social behaviour of non-lactating sows and its implications for managing sow aggression. *The Pig Journal* 34, 9-20.

Mendl, M.T., Zanella, A.J. and Broom, D.M. (1992) Physiological and reproductive correlates of behavioural strategies in female domestic pigs. *Animal Behaviour* 44, 1107-1121.

Morrow, A.T.S. and Walker, N. (1994a) The behavioural and production responses of finishing pigs to increasing workload to obtain food *ad libitum* from hopper feeders. *Animal Production* 59, 125-129.

Morrow, A.T.S. and Walker, N. (1994b) Effects of number and siting of single-space feeders on performance and feeding behaviour of growing pigs. *Journal of Agricultural Science* 122, 465-470.

Nielsen, B.L. (1995) Feeding behaviour of growing pigs: Effects of the social and physical environment. PhD thesis, University of Edinburgh, UK, 123pp.

Nielsen, B.L. (1998) On the interpretation of feeding behaviour measures and the use of feeding rate as an indicator of social restraint. *Livestock Production Science* (in press).

Nielsen, B.L., Lawrence, A.B. and Whittemore, C.T. (1995a) Effect of group size on feeding behaviour, social behaviour, and performance of growing pigs using single-space feeders. *Livestock Production Science* 44, 73-85.

Nielsen, B.L., Lawrence, A.B. and Whittemore, C.T. (1995b) Effects of single-space feeder design on feeding behaviour and performance of growing pigs. *Animal Science* 61, 575-579.

Nielsen, B.L., Lawrence, A.B. and Whittemore, C.T. (1996a) Effect of individual housing on the feeding behaviour of previously group housed growing pigs. *Applied Animal Behaviour Science* 47, 149-161.

Nielsen, B.L., Lawrence, A.B. and Whittemore, C.T. (1996b) Feeding behaviour of growing pigs using single or multi-space feeders. *Applied Animal Behaviour Science* 47, 235-246.

Noblet, J., Shi, X.S. and Dubois, S. (1993) Energy cost of standing activity in sows. *Livestock Production Science* 34, 127-136.

NRC (1987) *Predicting Feed Intake of Food - Producing Animals*. National Academy Press, Washington DC, 85 pp.

Oliverio, A. (1987) Endocrine aspects of stress: central and peripheral mechanisms. In: Wiepkema, P.R. and van Adrichem, P.W.M. (eds), *Biology of Stress in Farm Animals: An Integrative Approach*. Martinus Nijhoff Publishers, Dordrecht, pp. 3-12.

Paterson, A.M. and Pearce, G.P. (1991) The effect of space restriction during rearing on growth and cortisol levels of male pigs. In: Batterham, E.S. (ed.), *Manipulating Pig Production III*. Australian Pig Science Association, Victoria, p. 68.

Patterson, D.C. (1985) A note on the effect of individual penning on the performance of fattening pigs. *Animal Production* 40, 185-188.

Petherick, J.C. (1983) A biological basis for the design of space in livestock housing. In: Baxter, S.H., Baxter, M.R. and MacCormach, J.A.D. (eds), *Farm Animal Housing and Welfare*. Martinus Nijhof, The Hague, pp. 103-120.

Petherick, J.C., Beattie, A.W. and Bodero, D.A.V. (1989) The effect of group size on the performance of growing pigs. *Animal Production* 49, 497-502.

Price, E.O. (1984) Behavioural aspects of animal domestication. *Quarterly Review of Biology* 59, 1-32.

Pulliam, H.R. and Caraco, T. (1984) Living in groups: is there an optimal group size? In: Krebs, J.R. and Davies, N.B. (eds), *Behavioural Ecology. An Evolutionary Approach*, 3rd edn. Blackwell Scientific Publications, Oxford, pp. 122-147.

Randolph, J.H., Cromwell, G.L., Stahly, T.S. and Kratzer, D.D. (1981) Effects of group size and space allowance on performance and behaviour of growing swine. *Journal of Animal Science* 53, 922-927.

Savory, C.J. and MacLeod, M.G. (1980) Effects of grouping and isolation on feeding food conversion and energy expenditure of domestic chicks. *Behavioural Processes* 5, 187-200.

Sibly, R.M., Nott, H.M.R. and Fletcher, D.J. (1990) Splitting behaviour into bouts. *Animal Behaviour* 39, 63-69.

Spicer, H.M. and Aherne, F.X. (1987) The effects of group size/stocking density on weanling pig performance and behaviour. *Applied Animal Behaviour Science* 19, 89-98.

Stolba, A. and Wood-Gush, D.G.M. (1989) The behaviour of pigs in a semi-natural environment. *Animal Production* 48, 419-425.

von Borell, E., Biensen, N.J. and Gatnau, R. (1992) Effects of stress hormones on pig performance and health. *1992 Swine Research Report*. Iowa State University, Iowa, pp. 77-78.

von Holst, D., Fuchs, E. and Stohr, W. (1983) Physiological changes in *Tupaia belangeri* under different types of social stress. In: Dembrowski, T.M., Schmidt, T.H. and Blumchen, G. (eds), *Biobehavioral Bases of Coronary Heart Disease*. Karger, Basel, pp. 382-390.

Walker, N. (1991) The effects on performance and behaviour of number of growing pigs per mono-place feeder. *Animal Feed Science and Technology* 35, 3–13.

Williams, N.H., Stahly, T.S. and Zimmerman, D.R. (1994) Impact of immune system activation on growth and amino acid needs of pigs from 6 to 114 kg body weight. *Journal of Animal Science* 72 (Suppl. 2), 57.

# Processes and Control

# Mating, Pregnancy and Prenatal Growth

6

## J.-Y. Dourmad, J. Noblet, M.C. Père and M. Étienne

*INRA Station de Recherches Porcines, 35590 Saint-Gilles, France*

## Introduction

Mating and pregnancy periods are of great importance in practical pig production, litter size at birth and at weaning being major components of sow productivity. The viability of the young pig around farrowing is highly dependent on its energy reserves, which are related to body weight and composition at birth (Le Dividich *et al.*, 1991), and these reserves are built during prenatal growth. The period of pregnancy appears also to be the main period during which the body reserves of the sow can be restored, and this can affect the long-term reproductive performance. Indeed, most of the effects of nutrient supplies on reproductive performance and longevity are associated with extreme variations of body weight or fat reserves (Dourmad *et al.*, 1994). This can occur on a short-term basis or as the consequence of small deviations on a long-term basis (Close and Cole, 1986). Energy under-feeding during pregnancy is associated with lower body fat reserves at farrowing or at weaning and generally return to oestrus is delayed and conception rate is lowered. With severe energy restriction, culling rate for reproductive disorders increases after the first parity whereas only long-term effects are noticed with moderate restriction. On the other hand, overfeeding during gestation increases weight and condition of the sow at the end of pregnancy and this can cause farrowing difficulties and lactation problems and culling for poor lactation performance or locomotion problems.

Thus, optimal sow performance and longevity require a flexible approach to determination of nutrient requirement during pregnancy, in order to control the evolution of the sow's body reserves. This requires a quantified prediction of the response of pregnant sows and their conceptus to the nutrient supply and to the environmental conditions. Empirical prediction equations, obtained from field or experimental trials, have been proposed to evaluate the effects of protein and energy supply on body weight gain or backfat depth changes, over the whole gestation period (Henry and Etienne, 1978; ARC, 1981; Whittemore and Morgan, 1990). However, the flexibility of these predictions is rather limited and they are appropriate only to specific conditions. For these reasons, more dynamic descriptions of nutrient utilization by the pregnant sow, on a daily basis, have been proposed (Williams *et al.*, 1985; Dourmad, 1987; Pomar *et al.*, 1991; Walker and Young 1992). These models, which are similar in their structure to pig growth models, contain various combinations of empirical and conceptual/ mechanistic equations (Black, 1995). They are well adapted to predict, on a daily basis, body composition changes of the sow from nutrient supplies. However, that level of approach is insufficient to describe the maternal–fetal exchange of nutrients, which is required for a better quantification of short- and long-term effects of nutrient supplies on the development of the fetuses.

## Partition of Nutrients During Pregnancy

The process of reproduction from conception to weaning can be considered as directed to buffer the developing progeny from nutritional distress (Oldham, 1991). It involves both homeostatic[1] and homeorhetic[2] control of nutrient partitioning (Bauman and Currie, 1980). The maintenance of body integrity is the main homeostatically controlled process, whereas the development of fetuses and udder, and at least to some extent, the constitution of body reserves in maternal tissue are regulated through homeorhetic controls. Changes in spontaneous feed intake, generally as the consequence of nutrient utilization by body tissues, play an important role in that homeorhesis. However, pregnant sows are generally restrictedly fed, suggesting that homeorhesis can act only partially in practical conditions. In particular, the constitution of body reserves, which can be considered to be a homeorhetic mechanism for ensuring sufficient nutrient supplies at the end of gestation and during the following lactation, is only partially achieved compared to a situation where animals are fed *ad libitum*.

A simplified representation of nutrient partitioning in pregnant sows is given in Fig. 6.1. Nutrient supplies are described as metabolizable energy and ileal digestible amino acids (or protein). The sow is represented as five

---

[1] Homeostasis – maintenance of physiological equilibrium, i.e. constant condition in the internal environment.
[2] Homeorhesis – orchestrated changes for the priorities of a physiological state, i.e. coordination of metabolism in various tissues to support a physiological status.

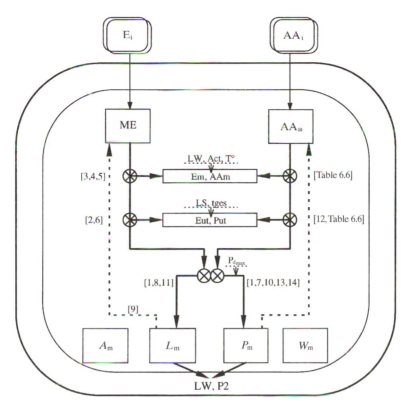

**Fig. 6.1.** Model of metabolizable energy and amino acid utilization in pregnant sows. Numbers indicated between brackets refer to equation numbers in Table 6.6. $E_i$: energy intake; ME: metabolizable energy; Em: energy for maintenance; Eut: energy retained in uterus; $AA_i$: amino acids intake; $AA_{ai}$: amino acids available at ileal level; AAm, AA for maintenance; Put: protein retained in uterus contents; $A_m$, $L_m$, $P_m$ and $W_m$: ash, lipids, protein and water in maternal body; $Pd_{max}$: maximal protein retention; LW: live weight; P2: backfat thickness; Act: activity; T°: ambient temperature; LS: litter size; tges: gestation stage.

compartments: uterus and conceptus, body lipids, body proteins, body water and body ash. For more practical applications, a simplified description of body components in three compartments (uterine contents, live weight and backfat depth) is also often used. The highest priority for nutrient utilization is given to maintenance and to the synthesis of products of conception and the development of the udder. The remaining nutrients are used for body protein and fat deposition. When nutrient supplies are not sufficient, body proteins or lipids are mobilized to support conceptus growth and maintenance requirements. Thus, the partition of nutrients during pregnancy can be considered as driven by two different processes:

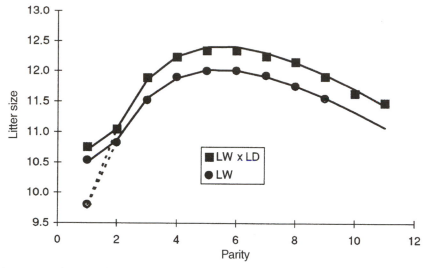

**Fig. 6.2.** Evolution of litter size (total born) with parity number in Large White (●, *n* = 8000) and crossbred Large White × French Landrace (■, *n* = 70,000) sows. For parity 1 the dotted line represents the expected litter size when gilts are mated at puberty.

**1.** The development of the conceptus which mainly depends on the number of fetuses, each fetus acquiring its own growth regulation early in pregnancy.

**2.** The availability of absorbed nutrients.

In the following sections of this chapter we shall focus first on the development of the litter throughout pregnancy. The utilization of nutrients for maintenance and maternal gain will then be quantified, in order to predict, on a daily basis, the body composition changes of the sow over pregnancy, as proposed in Fig. 6.1.

## Development of the Fetuses

### Ovulation and litter size

Results from commercial farms obtained through the Technical Sow Herd Management System (Dagorn *et al.*, 1992) on about 90,000 sows in France (Le Cozler *et al.*, 1997) indicate that average litter size increases with sow parity to reach a peak at the fifth parity for the total number of pigs born (Fig. 6.2), and at the fourth parity for the number born alive. The evolution of litter size with parity number is a characteristic of the genotype as illustrated in Fig. 6.2 by the difference between Large White and crossbred Large White × Landrace sows. The increase of litter size over the first three parities is not as

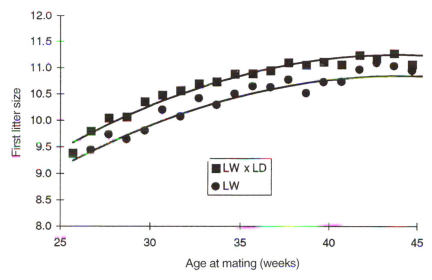

**Fig. 6.3.** Evolution of first litter size (total born) with age at mating in Large White (●, *n* = 8000) and crossbred Large White × French Landrace (■, *n* = 70,000) sows.

progressive as generally found in previous studies (Black *et al.*, 1986). Litter size is higher at the first parity and lower at the second parity than expected for a smooth evolution with parity number. This has already been noticed in many studies (Lucbert and Lavorel, 1984), but the biological bases have not yet been elucidated. The lower litter size at the second parity could be the consequence of processes occurring in the first parity. But a specific regulation for the size of the first litter, which appears to be highly affected by the age at first farrowing (Fig. 6.3), can also be suggested. Indeed, in Large White × Landrace crossbred gilts, the total number of piglets born per litter increases from 9.5 for gilts mated at 26 weeks of age to reach a maximum of 11.1 when first mating occurs later than 40 weeks of age. At puberty, which occurs around 27 ± 2.6 weeks for Large White × Landrace sows (Martinat-Botté *et al.*, 1995), the total number of piglets born per litter is about 9.8 and the number born alive amounts to 9.3. The difference between the actual size of the first litter and the litter size expected at puberty is then explained by the delayed first mating (Fig. 6.3). An adaptation of the model used by Black *et al.* (1986), for predicting total litter size (Lt) and number of piglets born alive (La) according to parity (N) is proposed. The following empirical relationships were adjusted on mean litter size per parity number, obtained from Large White (LW) (*n*=8000) and crossbred Large White × Landrace (LW × LD) (*n*=70,000) sows:

$$Lt = Lt_{max} - (Lt_{max} - Lt_{pub}) \, e^{-0.62(N-1)^{1.2}} - 0.005 \, Lt_{max} \, \beta^{1.60} \qquad (6.1)$$

where $Lt_{max}$ is the maximum litter size expected at the fifth parity (12.5 and 12.1 piglets for LW × LD and LW sows, respectively), $Lt_{pub}$ is the expected litter size at puberty, and $\beta = N - 5$ and is zero when $N < 6$. Litter size at first parity ($Lt_1$) and litter size at puberty are predicted from age at mating ($A_m$, wk) or expected age at puberty for the given genotype using the expression:

$$Lt_1 = (0.13 + 0.035\ A_m - 0.0004\ A_m^2)\ Lt_{max} \qquad (6.2)$$

Similar equations are obtained for the number of piglets born alive (La):

$$La = La_{max} - (La_{max} - La_{pub})\ e^{-0.81(N-1)^{1.5}} - 0.02\ La_{max}\ \beta^{2.16} \qquad (6.3)$$
$$La_1 = (0.17 + 0.035\ A_m - 0.0004\ A_m^2)\ La_{max} \qquad (6.4)$$

where $La_{max}$ is the maximum litter size (born alive) expected at the fourth parity (11.5 and 11.1 piglets born alive for LW × LD and LW sows, respectively), $La_{pub}$ is the expected litter size at puberty (9.3), and $\beta = N - 4$ and is zero when $N < 5$.

Litter size at farrowing is the result of both ovulation rate (OR) at mating and embryo survival during pregnancy. It increases in a curvilinear manner with OR, up to a maximum value which is higher in multiparous than in primiparous sows. Bazer *et al.* (1969) described this limit of litter size as the uterine capacity of the sows. This maximal limit is also affected by the genotype of the sow, but the mechanisms involved in that regulation are not well known (Père *et al.*, 1997). From the results of Legault (1978) and personal unpublished results it was possible to calculate a relationship between average litter size and ovulation rate (OR) in primiparous ($LS_p$) and multiparous ($LS_m$) Large White sows (Fig. 6.4):

$$LS_p = 11.5\ (1 - e^{-0.018\ OR^{1.73}}) \qquad (6.5)$$
$$LS_m = 15.3\ (1 - e^{-0.008\ OR^{1.93}}) \qquad (6.6)$$

Ovulation rate is affected by numerous factors including parity number and genotype. In gilts, ovulation rate is also clearly related to body weight at puberty and to the number of oestrous cycles since puberty. Flushing of gilts has a positive effect on number of ova shed. From literature studies, den Hartog and van Kempen (1980) found that OR increased by 1.9 ova in gilts flushed for a period of about two weeks prior to ovulation. However, according to the relationship between litter size and OR, the effect on litter size is much lower. In multiparous sows, ovulation occurs after weaning following a delay which is dependent on body condition of the sows at weaning, and feeding level during lactation. King (1987) showed in a review that in first litter sows, both the absolute levels and the rates of reduction of protein and lipid reserves are implicated in the length of the weaning to oestrus interval (WOI). He gave the following equations for primiparous sows:

$$WOI\ (days) = 81.5 - 3.58\ (kg\ body\ protein\ at\ weaning),\ R^2 = 0.63 \qquad (6.7)$$
$$WOI\ (days) = 38.6 - 0.63\ (kg\ body\ fat\ at\ weaning),\ R^2 = 0.46 \qquad (6.8)$$

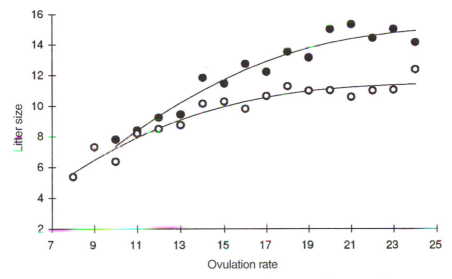

**Fig. 6.4.** Effect of ovulation rate on litter size (total born) in multiparous (●, $n = 246$) and primiparous (○, $n = 674$) Large White sows. (Adapted from Legault, 1974 and personal data.)

$$\text{WOI (days)} = 9.6 + 3.44 \text{ (kg body protein loss in lactation)}, R^2 = 0.63$$
$$(6.9)$$

$$\text{WOI (days)} = 9.4 + 0.59 \text{ (kg body fat loss in lactation)}, R^2 = 0.43$$
$$(6.10)$$

With a similar approach, but on sows from parity 1 to 4 (25 literature studies), Charette *et al.* (1995) showed that WOI and percentage of sows in oestrus after weaning were non-linear functions of live weight and/or protein mass at weaning. Their data were used to fit an exponential relationship between WOI and body weight at weaning:

$$\text{WOI} = 6.2 + 7.9 \, e^{-0.026 \text{ (body weight} - 130)}, R^2 = 0.55 \tag{6.11}$$

According to these results and to the evolution of litter size with parity number, it can be suggested that the overall efficiency of the reproductive function increases with body weight and protein mass up to the attainment of maturity.

In recent studies, the length of the interval from weaning to conception (WCI) was also found to affect the subsequent litter size (Dewey *et al.*, 1994; Vesseur *et al.*, 1994; Le Cozler *et al.*, 1997). When conception occurs within 1 to 4 days after weaning, or 22 to 25 days after weaning, litter size is 1.7 piglets higher than when conception occurs 8 to 11 days after weaning (Le Cozler *et al.*, 1997). This is consistent with the reduced litter size and the extended WCI observed when lactation length decreases. According to the

**Table 6.1.** Effect of stage of pregnancy (t, days) and number of fetuses (n) on fresh weight (g) and protein (g) and energy contents (kJ) of uterine tissues. (From Noblet et al., 1985; Noblet, 1990.)[a]

| Compartment | No. | Equation |
|---|---|---|
| Empty uterus | 1 | $Ln(weight) = 7.141 - 3.667\ e^{-0.0326\ t} + 0.0829\ n$ |
|  | 2 | $Ln(energy) = 9.853 - 3.098\ e^{-0.00691\ t} + 0.0654\ n$ |
|  | 3 | $Ln(protein) = 7.653 - 4.207\ e^{-0.00448\ t} + 0.0724\ n$ |
| Fetus | 4 | $Ln(weight) = 9.095 - 17.69\ e^{-0.0305\ t} + 0.0878\ n$ |
|  | 5 | $Ln(energy) = 11.64 - 13.94\ e^{-0.0182\ t} + 0.0822\ n$ |
|  | 6 | $Ln(protein) = 7.951 - 14.05\ e^{-0.0188\ t} + 0.0867\ n$ |
| Conceptus[b] | 7 | $Ln(weight) = 8.621 - 21.02\ e^{-0.0530\ t} + 0.1114\ n$ |
|  | 8 | $Ln(energy) = 11.72 - 8.62\ e^{-0.0138\ t} + 0.0932\ n$ |
|  | 9 | $Ln(protein) = 8.090 - 8.71\ e^{-0.0149\ t} + 0.0872\ n$ |
| Total uterus | 10 | $Ln(weight) = 8.837 - 16.07\ e^{-0.0497\ t} + 0.1061\ n$ |
|  | 11 | $Ln(energy) = 13.84 - 8.041\ e^{-0.0070\ t} + 0.0831\ n$ |
|  | 12 | $Ln(protein) = 10.01 - 7.573\ e^{-0.0075\ t} + 0.0817\ n$ |
| Mammary gland | 13 | $Ln(weight) = 6.47 + 0.080\ e^{0.04576\ (t-45)}$ |
|  | 14 | $Ln(energy) = 2.92 + 6.898\ e^{0.00185\ (t-45)}$ |

[a] These equations were obtained on Large White sows. They can be used on other genotypes by considering the actual live weight of the litter at farrowing, for the given genotype.
[b] Fetus + placenta + fluids.

results of Le Cozler et al. (1997) an increase of 0.4 to 0.6 piglets per litter can be expected for a 10-day increase in lactation length.

Thus, it appears that LS depends on numerous biological parameters such as ovulation rate, lactation length and WCI. Empirical relationships (equations 6.1 to 6.4) can be used to predict its average evolution with parity number, according to maximum LS and expected LS at puberty. However, these different criteria explain only 10–15% of the observed variability (Vesseur et al., 1994; Le Cozler et al., 1997), suggesting that LS remains a very difficult parameter to predict precisely.

## Uterine growth and development of fetuses

Fetuses develop very slowly during the first third of pregnancy, and about 60% of fetal growth or energy deposition in the uterus occurs during the last 30 days of pregnancy. Many relationships have been proposed to describe the development of uterus contents according to gestation stage and litter size (Pomeroy, 1960; Moustgaard, 1962; Noblet et al., 1985; Walker and Young, 1992). Noblet (1990) presented a detailed description of the progressive increase in weight, total DM, energy and protein in fetuses, fetal fluids, placenta, uterus and udder (Table 6.1, Fig 6.5). Partition of weight and

**Table 6.2.** Partition of uterine gain between compartments of uterus over pregnancy (114 days) for 12 fetuses. (From de Wilde, 1980; Noblet *et al.*, 1985; Beyer, 1986).

| Compartment | Weight (kg) | (%) | DM (g) | (%) | Protein (g) | (%) | Energy (MJ) | (%) |
|---|---|---|---|---|---|---|---|---|
| Fetus | 14.8 | 61 | 2760 | 74 | 1550 | 69 | 53.1 | 73 |
| Placenta | 4.3 | 18 | 390 | 10 | 277 | 12 | 7.9 | 11 |
| Fluids | 2.1 | 9 | 173 | 5 | 108 | 5 | 3.0 | 4 |
| Empty uterus | 3.0 | 12 | 420 | 11 | 326 | 14 | 8.3 | 11 |
| Total | 24.2 | 100 | 3742 | 100 | 2261 | 100 | 72.3 | 100 |

**Table 6.3.** Estimates of daily amino acid requirement for maintenance, and amino acid composition of maternal protein and fetal protein.

| | Maintenance requirement[a] | | Composition of | | | |
|---|---|---|---|---|---|---|
| | | | Body protein[b] | | Fetuses[c] | |
| | g kg$^{-1}$ BW$^{0.75}$ | % Lysine | g 16 g N$^{-1}$ | % Lysine | g 16 g N$^{-1}$ | % Lysine |
| Lysine | 36 | 100 | 7.0 | 100 | 5.9 | 100 |
| Methionine | 9 | 25 | 1.8 | 26 | 1.4 | 24 |
| Total sulphur amino acids | 49 | 139 | 3.4 | 49 | 2.7 | 46 |
| Threonine | 53 | 147 | 4.0 | 57 | 3.5 | 59 |
| Tryptophan | 11 | 31 | 1.0 | 14 | nd | |
| Isoleucine | 16 | 44 | 3.5 | 50 | 3.0 | 51 |
| Leucine | 23 | 64 | 7.0 | 100 | 6.2 | 105 |
| Valine | 20 | 55 | 4.7 | 67 | 4.6 | 78 |
| Phenylalanine | 18 | 50 | 4.4 | 63 | 3.4 | 58 |
| Total aromatic amino acids | 37 | 103 | 6.7 | 96 | 5.8 | 98 |
| Histidine | 0 | 0 | 2.3 | 33 | 2.3 | 39 |
| Arginine | 0 | 0 | 6.7 | 96 | 6.8 | 115 |

[a] From Fuller *et al.* (1989).
[b] From Sève (1994).
[c] From Duée (1984); Everts (1994).

nutrient content of the uterus between different compartments at the end of pregnancy is given in Table 6.2. Based on the data of De Wilde (1980), Noblet *et al.* (1985) and Beyer (1986), energy and protein deposited in the uterus over pregnancy (fetuses, placenta, fluids and gain of empty uterus) averages

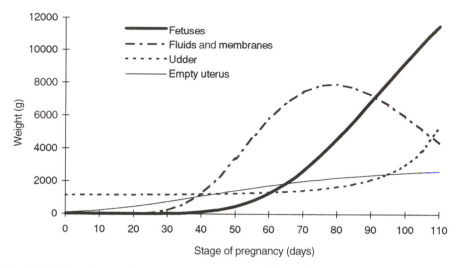

**Fig. 6.5.** Effect of stage of pregnancy (days) on weight (g) of fetuses, fluids and membranes, empty uterus and udder. (Adapted from Noblet *et al.*, 1985; Noblet, 1990.)

4.9 MJ and 150 g kg$^{-1}$ of fetus at farrowing. Amino acid content of fetal protein is given in Table 6.3.

Such relationships can be used to derive daily energy and protein requirements from the development of uterine contents according to litter size or litter weight at farrowing (Fig. 6.1). But it is more difficult to predict the effects of nutrient supplies on fetal growth. From a review of literature, Henry and Etienne (1978) estimated the increase in piglet weight (Wfe) at birth to be to 3.5 and 8.9 g for each supplemental megajoule of DE per day, in primiparous and multiparous sows, respectively. Pluske *et al.* (1995) found that the response is curvilinear in primiparous sows, whereas birth weight continues to increase linearly at very high levels of energy supplies in multiparous sows (i.e. > 45 MJ DE day$^{-1}$). Severe protein restriction in early or late gestation is also detrimental to fetal growth, but this is generally not observed when moderate restrictions are applied (Pond *et al.*, 1992; Pluske *et al.*, 1995). Walker and Young (1992) developed an equation to represent total fetal weight (Wfe) with litter size (LS) and nutrition (Energy: Ei, MJ DE and Nitrogen: Ni, g) as limiting factors of maximal total fetal weight (Wfe, g).

$$\text{Wfe} = \text{LS } e^{8.496 \; - \; 9.494e^{-0.017t}}/(1 + (4.83/\text{Ei}) - (4.55/\text{Ni})) \tag{6.12}$$

With this relationship, the effect on fetal weight of increasing Ei is not linear. It is higher for low energy intakes, but on average it is consistent with the results of Henry and Etienne (1978) (5 to 8 g per piglet for each megajoule of supplemental energy above 30 and 25 MJ day$^{-1}$, respectively).

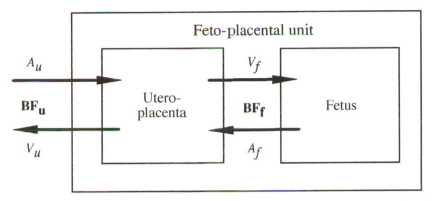

**Fig. 6.6.** Schematic description of nutrient exchanges in the gravid uterus. (Adapted from Père, 1995.)

### Nutrient uptake by the uterus and the fetuses

However, although very interesting for practical purposes, this global approach is insufficient to explain the biological consequences of maternal nutrition on growth and development of fetuses. It is necessary to quantify the nutrient uptake by the fetuses or the gravid uterus through the determination of substrate concentrations in artery (A) and vein (V) and the measurement of blood flow (BF). The schematic description of maternal–fetal exchanges of nutrients is given in Fig. 6.6. According to the Fick principle, net uterine uptake is calculated as $BF_u \times (A_u - V_u)$ and net fetal uptake as $BF_f \times (V_f - A_f)$, with $BF_u$ being the uterine blood flow, and $BF_f$ the fetal blood flow. However, due to technical difficulties in the pig species, related to abortion after surgery, only few estimations of maternal–fetal exchanges are available (Père, 1995). They are based on limited number of data obtained under general anaesthesia (Comline *et al.*, 1979; Reynolds *et al.*, 1985; Thulin *et al.*, 1989), or in conscious pigs for substrate balances across the feto-placental unit (Duée *et al.*, 1987) or across the fetus (Randall, 1982; Père, 1995).

From these different studies, glucose appears to be the major energy substrate for the fetus. The arteriovenous difference $(A_u - V_u)$ in the uterus amounts to about 0.2 to 0.4 mmol $l^{-1}$ (5 to 8% extraction), and appears to be relatively constant, at least over the last third of pregnancy (Reynolds *et al.*, 1985; Duée *et al.*, 1987; Père, 1995). In the conscious fetus, Père (1995) measured a fetal extraction $(V_f - A_f)$ of glucose of 0.31 mmol $l^{-1}$ (13% extraction). Glucose concentration in fetal blood was approximately 2.5 times lower than in the dam. For total amino acids, Duée *et al.* (1987) measured an arteriovenous difference across the gravid uterus of 0.18 mmol $l^{-1}$, representing an extraction of about 4.7%. For most of the amino acids, the concentration in fetal blood was equal to or higher than in the maternal

**Table 6.4.** Estimated daily uptake of major nutrients by the uterus of a gestating sow at about 110 days of pregnancy.

| | Arterio-venous difference mmol l$^{-1}$ | Extraction coefficient % | Uptake (BF$_u$ per fetus = 0.4 l min$^{-1}$)$^d$ | | |
| | | | per fetus mmol day$^{-1}$ | for 12 fetuses mmol day$^{-1}$ | for 12 fetuses kJ day$^{-1}$ |
|---|---|---|---|---|---|
| O$_2$$^a$ | 1.01 | 20 | 580 | 6980 | — |
| Glucose$^b$ | 0.26 | 8 | 97 | 1168 | 3260 |
| Amino acids$^c$ | 0.15 | 5 | 57 | 690 | 1625 |
| Total | — | — | — | — | 4885 |

$^a$ From Duée et al. (1987).
$^b$ From Reynolds et al. (1985); Duée et al. (1987); Père (1995).
$^c$ From Duée et al. (1987).
$^d$ From Père et al. (1996).

blood. This suggests, at least for some amino acids, an active transportation by the placental tissue (Duée, 1984). The extraction of other substrates (non-esterified fatty acids, ketone bodies) is generally not significant in most studies. Total blood flow in the uterine horn increases linearly with gestation stage (Père et al., 1996). It increases also with the number of fetuses in the uterine horn, but the average blood flow per fetus decreases in large litters. Thus, the modification of the uterine blood flow according to gestation stage or litter size appears to be a major adaptation of the sow to the rising requirement of its progeny. Few results are available on umbilical BF, and they were obtained in anaesthetized fetuses only. Reynolds et al. (1985) measured an umbilical BF similar to that of uterine BF when expressed per fetus. From these different results, an attempt was made to quantify the major exchanges of nutrients between the sow and the uteroplacental unit (Table 6.4). At 110 days of pregnancy, it can be calculated that daily glucose and amino acid uptakes by the uterus amount to 1168 and 690 mmol, respectively, for a litter of 12 fetuses. On an energetic basis, this corresponds to about 4885 kJ per day. That value can be compared to the daily energy accretion in uterine tissues which amounts to 1500 kJ at 110 days of gestation according to Noblet (1990; see also Table 6.1). The efficiency of energy for the development of uterine tissues is then about 31%. As proposed by Noblet et al. (1990), the efficiency of use of energy for energy deposition in the reproductive tract can also be calculated from oxygen uptake as the following ratio: energy accretion / (energy accretion + oxidative metabolism). An energy efficiency of 32% can be calculated on that basis with a heat production of 470 kJ mol$^{-1}$ O$_2$ (Noblet et al., 1990). Efficiency of use of energy for energy deposition in the conceptus ($k_c$) can also be determined from more classical energy balance trials. A first method which consists in partitioning metabolizable energy (ME) intake between maintenance, maternal energy deposition and uterine energy gain provides estimates of $k_c$ with rather large SD since most ME is used for sow maintenance and maternal gain; the mean value obtained with this approach was 50% (Close *et al.*, 1985; Noblet and Etienne, 1987). In such calculations, maintenance energy requirements of the uterus are included in sow energy requirements, and $k_c$ corresponds to a 'net' efficiency. In a second approach, the gross efficiency of uterine energy gain is calculated from the partition of ME intake between maintenance of maternal tissues, maternal energy deposition and uterine energy deposition. The $k_c$ value estimated here is 30% (Noblet and Etienne, 1987), which is comparable to the values estimated from nutrient utilization by the uterus.

## Nutrient Requirements for Maintenance

Obligatory losses of amino acids from the body have to be replaced in order to maintain body conditions. They can be evaluated according to the metabolic weight ($BW^{0.75}$) of the sow (Table 6.3). Requirements for metabolizable energy for maintenance ($ME_m$) are usually obtained from energy balance trials on sows fed different energy levels, and regression equations between energy retention and energy intake (Close *et al.*, 1985; Noblet and Etienne, 1987). Under thermoneutral conditions and with moderate physical activity, $ME_m$ varies between 400 and 440 kJ $kg^{-1}$ $BW^{0.75}$ (Beyer, 1986; Noblet and Etienne, 1987; Noblet *et al.*, 1989; Everts, 1994); the average value being 420 kJ $kg^{-1}$ $BW^{0.75}$. The comparison of pregnant and non-pregnant gilts indicates that pregnancy has no effect on $ME_m$ (Close *et al.*, 1985). In contrast with the preliminary study of Verstegen *et al.* (1971), recent experiments demonstrate that $ME_m$ can be considered as constant over pregnancy (Noblet and Etienne, 1987; Everts, 1994). Our data suggest that, when expressed per kilogram $BW^{0.75}$, $ME_m$ is also very similar in primiparous and multiparous sows. Data of Beyer (1986) showed an increased $ME_m$ with parity number, while Everts (1994) obtained slightly reduced $ME_m$ in older sows. These discrepancies may be explained by probable differences in physical activity over successive parities which were not measured by these two authors.

From a literature survey, it was calculated that the energy cost of standing in sows ranged between 0.25 and 0.30 kJ $kg^{-1}$ $BW^{0.75}$ $min^{-1}$, which is equivalent to doubling the instantaneous heat production during standing when compared to lying down. These values are four to five times higher than in ruminant species (Noblet *et al.*, 1993b). Under practical conditions, levels of physical activity can vary greatly between housing systems (indoor vs outside keeping) and between sows (stereotypic behaviour; Cronin *et al.*, 1986). In addition, some data indicate a higher level of activity in older or poor condition sows or under adverse climatic conditions (Cariollet and Dantzer, 1984). In all cases, physical activity can represent a source of variability of energy requirements between herds or between sows since, according to our data, a 100 min difference per day in the duration of standing is equivalent to the energy supplied by about 110 g of feed.

Literature data suggest that the lower critical temperature (LCT) of individually housed sows ranges between 20 and 23°C (Noblet *et al.*, 1989). This relatively high value of LCT is mainly the consequence of the low energy levels fed during the pregnancy period. Under some unfavourable climatic conditions, LCT is even higher; it can also be higher in very thin sows (Hovell *et al.*, 1977). On the other hand, straw bedding, higher feeding levels and group housing will decrease LCT (Verstegen and Curtis, 1988). For instance, LCT was 6°C lower in group-housed than in individually housed sows (Geuyen *et al.*, 1984). Compilation of literature data shows that the daily increment of heat production with temperature reduction varies with many factors: values range from 8 to 10 kJ $°C^{-1} kg^{-1} BW^{-0.75}$ (in group-housed sows or with sows close to LCT or with high feeding levels) to 15 to 18 kJ $°C^{-1} kg^{-1} BW^{-0.75}$ (in individually housed sows or at low ambient temperatures). In the latter case, this is equivalent to about 4% of $ME_m$ per degree centigrade below the LCT, or 70 g of feed per day and per degree centigrade for 200 kg BW sows. The amount of feed required for compensating the increased heat production can be calculated according to an efficiency of utilization of ME for energy gain of 85% which takes into account the partial contribution of heat increment of feed to the thermoregulatory demand (Noblet *et al.*, 1989).

## Weight Gain and Body Composition Changes of the Sow

### Maternal gain

In connection with usual mobilization of body reserves over lactation and progressive attainment of mature live weight, maternal weight has to increase during pregnancy, in addition to the weight gain of uterus and conceptus. The maternal weight gain to be achieved during pregnancy is then dependent on the composition and the amount of the previous lactation weight loss, and the strategy for increased live weight and body fatness over successive parities. From a review of literature, Henry and Etienne (1978) proposed empirical equations to predict net change in live weight in pregnancy (net gain, kg) from daily DE intake (MJ):

$$\text{Net gain} = -13.99 + 1.86 \text{ DE (primiparous sows)}, R^2 = 0.77 \quad (6.13)$$
$$\text{Net gain} = -23.03 + 1.69 \text{ DE (multiparous sows)}, R^2 = 0.78 \quad (6.14)$$

Whittemore and Morgan (1990) proposed similar equations for predicting the change in live weight (kg) and backfat depth (P2, mm) according to total food intake in pregnancy (FI, kg)

$$\text{Maternal weight gain} = -27.2 + 0.215 \text{ FI} \quad (6.15)$$
$$\text{Change in P2 (mm)} = -9.3 + 0.036 \text{ FI} \quad (6.16)$$

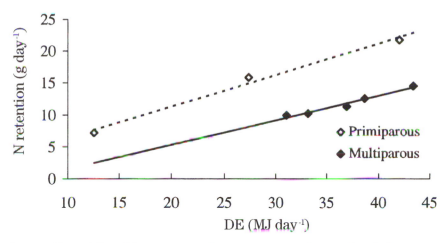

**Fig. 6.7.** Effect of digestible energy supply on average protein retention during pregnancy, in primiparous and multiparous sows. (Adapted from Etienne, 1991; Dourmad *et al.*, 1996.)

According to these equations 28 kg of food are required for 1-mm increment of P2 backfat depth, and 4.7 kg food for each 1-kg increment of maternal liveweight gain.

Maternal changes can also be predicted in a more deterministic way according to nutrient partitioning presented in Fig. 6.1. The nutrients available above maintenance and growth of the reproductive tissues are utilized for maternal gain, and partitioned among protein and lipid synthesis. When protein and amino acid supplies are not limiting, protein retention (PR) is mainly dependent on energy supply and maximum potential protein deposition ($Pd_{max}$). In growing pigs this was generally described by a linear/plateau relationship between protein retention and metabolizable energy intake. Williams *et al.* (1985) suggested a similar relationship in pregnant sows. But the results of the experiments, in which the response of pregnant gilts (Kemm, 1974; Willis and Maxwell, 1984; Etienne, 1991; King and Brown, 1993) or multiparous sows close to their mature body weight (parity 4 on average, Dourmad *et al.*, 1996) to increasing energy intake up to 42 MJ DE day$^{-1}$ was measured, failed to provide any evidence of a plateau (Fig. 6.7). This suggests that the energy level fed to pregnant sows in practical conditions is usually below the level required for maximum N retention. For a given energy supply, higher protein retention is generally measured in gilts than in multiparous sows. This is partly explained by the lower energy requirement for maintenance in gilts in relation with their lower body weight.

Total daily protein retention in pregnant sows is also affected by pregnancy stage. It increases after 50 to 60 days of gestation in primiparous (Elsley *et al.*, 1966; Willis and Maxwell, 1984; Noblet and Etienne, 1987; King and

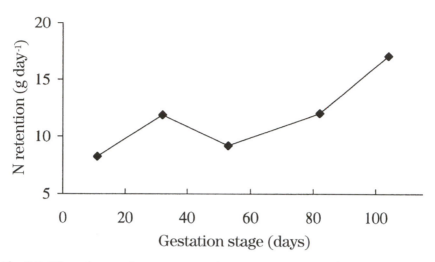

**Fig. 6.8.** Effect of stage of pregnancy on nitrogen retention in multiparous sows receiving on average 36.6 MJ DE day$^{-1}$. (Adapted from Dourmad *et al.*, 1996.)

Brown, 1993; Everts and Dekker, 1994) and in multiparous sows (Rombauts, 1962; Salmon-Legagneur, 1965; Everts and Dekker, 1994). Measurement of N retention at five stages of pregnancy by Dourmad *et al.* (1996) showed a significant and transitory increase in N retention around 32 days after mating (Fig. 6.8). It was suggested that this evolution of N retention with gestation stage would be driven through hormonal regulation. A transient rise in oestrogen secretion occurs from 20 to 40 days of pregnancy, with a peak of secretion around 30 days (Fèvre *et al.*, 1968; Edgerton and Erb, 1971; Robertson and King, 1974). Thereafter, plasma oestrogen concentration remains very low until day 70 and increases in a curvilinear manner from day 70 until parturition (Robertson and King, 1974). Thus, the prediction of protein retention in pregnant sows according to energy supply appears to be much more complex than generally assumed. Protein retention is linearly affected by energy supply, and for a given energy supply by gestation stage and litter size.

The data of Etienne (1991), King and Brown (1993) and Dourmad *et al.* (1996), based on a total of 46 experimental groups and 472 N balance measurements, were used to fit a relationship for the prediction of N retention (NR) according to gestation stage (*t*, days from mating), litter size (LS), maintenance ME requirement at mating (ME$_m$, MJ) and ME supply. NR was partitioned among NR in maternal tissues (NR$_m$) and NR in conceptus (NR$_c$), the last being predicted by derivation of equation 9 in Table 6.1. The following relationship was found:

$$NR = NR_c + 45.9\,(t/100) - 105.3\,(t/100)^2 + 64.4\,(t/100)^3 - 0.4$$
$$+\ cp\,(ME - ME_m),\ (n = 46,\ R^2 = 0.88,\ rsd = 1.9) \qquad (6.17)$$

Where $cp = 0.571$ in the first pregnancy, and $0.366$ later.

The data used for determining equation 6.17 issued from studies where NR was measured by the balance technique (collection of urine and faeces), which is known for overestimating the retention by about 15% (Just *et al.*, 1982). Thus, the real protein retention is about 15% lower than calculated by equation 6.17.

When dietary protein supply is below the requirement for maximal retention, NR increases linearly with protein intake until it reaches the maximum retention which depends on gestation stage, energy supply, litter size and parity number. The response of NR to N intake during this N-limiting phase ($NR_{lim}$) was evaluated by Williams *et al.* (1985) from different literature studies in gilts and sows:

$$NR_{lim} = 0.37 \text{ NI} - 2.47, R^2 = 0.83 \tag{6.18}$$

However the 0.37 coefficient for the slope has a limited biological significance, since it is affected by both the protein quality and the efficiency of amino acid utilization. For instance, with high quality protein (soybean meal and skim milk powder) King and Brown (1993) measured slopes of 0.40 to 0.52. In fact, protein retention is generally not limited by total protein supply, but by the supply of the most limiting amino acids. In the study of King and Brown (1993), threonine appeared to be the most limiting amino acid. Assuming a threonine content of retained protein of 4% (Table 6.3) and a true digestibility of threonine of 88% (calculated from ingredients, ITCF, 1995), the following relationship can be derived from the results of King and Brown (1993), for the utilization of threonine in the second half of pregnancy:

$$\text{Threonine retention (g day}^{-1}) = 0.58 \text{ digestible threonine} - 0.41 \tag{6.19}$$

This suggests an efficiency of use of digestible threonine for pregnancy of about 55 to 60%. For digestible lysine, Dourmad and Etienne (unpublished results) measured an efficiency of 65 to 70%, in agreement with the 65% value that can be derived from the study of King and Brown (1993).

From these different equations, it is possible to determine the amount of energy deposited as protein in maternal tissues. The efficiency of utilization of ME has then to be evaluated to determine the corresponding ME requirement. Literature estimates of ME efficiency for maternal gain vary between 70 and 85% (Close *et al.*, 1985; Noblet and Etienne, 1987; Everts and Dekker, 1994). These high values are consistent with the high proportion of energy retained as fat in material tissues ($k_f = 80\%$; Noblet *et al.*, 1990). With the advancement of pregnancy, a larger proportion of energy is retained as protein (Dourmad *et al.*, 1996). Because the corresponding efficiency ($k_p = 60\%$; Noblet *et al.*, 1990) is smaller than $k_f$, a lower efficiency for maternal gain in late pregnancy may be expected. For that reason, a mean efficiency of 75% can be suggested. According to $k_p$ and $k_f$, and the amount of energy

**Table 6.5.** Prediction of protein, fat or energy content ($Y$ as kg or MJ) of sows from live weight (LW, kg) and backfat thickness (BT, mm at P2 site) (Model: $Y = a + b$ LW $+ c$ BT).

| Constituent | a | b | c | Rsd | Authors |
|---|---|---|---|---|---|
| Protein | −2.3 | 0.186 | −0.22 | 1.8 | Whittemore and Yang, 1989 |
| | 1.7 | 0.175 | −0.38 | 1.4 | Everts, 1994 |
| | 2.3 | 0.178 | −0.33 | 1.9 | Dourmad et al., 1997 |
| Fat | −21.5 | 0.258 | 1.24 | 2.7 | King et al., 1986 |
| | −20.4 | 0.205 | 1.48 | 5.2 | Whittemore and Yang, 1989 |
| | −10.4 | 0.110 | 2.00 | 3.1 | Everts, 1994 |
| | −26.4 | 0.221 | 1.33 | 6.1 | Dourmad et al., 1997 |
| Energy | −870 | 12.6 | 54.2 | 186 | Whittemore and Yang, 1989 |
| | −408 | 8.9 | 68.1 | 126 | Everts, 1994 |
| | −1074 | 13.7 | 45.9 | 198 | Dourmad et al., 1997 |
| Water | 12.4 | 0.486 | −0.77 | 5.0 | Whittemore and Yang, 1989 |
| | 6.9 | 0.615 | −1.69 | 4.0 | Everts, 1994 |
| | 23.6 | 0.551 | −0.92 | 5.2 | Dourmad et al., 1997 |
| Ash | −3.4 | 0.066 | −0.124 | 2.1 | Whittemore and Yang, 1989 |
| | 0.6 | 0.037 | −0.082 | 0.5 | Dourmad et al., 1997 |

retained as protein, it is then possible to calculate the amount of fat in maternal tissues.

## Prediction of body composition changes of sows during pregnancy

The process of nutrient partitioning in the sow during pregnancy, as described above, allows the prediction of protein and fat retention in maternal tissues and in uterine contents. However, for practical application, it is necessary to predict more simple criteria, accessible at the farm level, such as body weight and backfat depth. These criteria are used by the farmer to determine targets of body condition according to parity and physiological stage, in order to optimize the reproductive performance of each sow. Relationships between protein and fat or energy contents in the body and live weight, and backfat thickness have been proposed in the literature (Table 6.5). These equations can be used to predict body weight and backfat thickness form either lipids and protein, or energy and protein. However, such predictions have to be used with caution, mainly because the value of backfat thickness can vary with the site of measurement and the equipment. For instance, lower values of backfat thickness are generally measured *in vivo* with ultrasonic devices ($USP_2$, mm) than measured directly by callipers or introscope on the carcass ($USP_2 = 0.84 P_2 + 2.6$; King et al., 1986).

**Table 6.6.** Main equations used for nutrient partitioning in pregnant sows.

| | | |
|---|---|---|
| Energy utilization in pregnant sows | $ME = ME_m + ER_c/k_c + ER_{mf}/k_f + ER_{mp}/k_p$ | [1] |
| | $ME_m$: ME for maintenance | |
| | $ER_c$: energy retained in conceptus (Table 6.1, equation 8), | [2] |
| | $k_c$: efficiency of ME for uterine growth | |
| | $ER_{mf}$: energy retained as fat in maternal tissues | |
| | $k_f$: efficiency of ME for lipids deposition | |
| | $ER_{mp}$: energy retained as protein in maternal tissues | |
| | $k_p$: efficiency of ME for protein deposition | |
| ME for maintenance | *In thermoneutral conditions* | |
| | $ME_m = 420$ kJ, kg$^{-1}$ BW$^{-0.75}$ for moderate activity (240 min day$^{-1}$ standing) | [3] |
| | PA (physical activity) = 0.30 kJ, kg$^{-1}$ BW$^{-0.75}$ min$^{-1}$ standing | [4] |
| | *Below low critical temperature (LCT)* in individually housed sows: | |
| | LCT = 20–23°C and HP increases by 15 to 18 kJ kg$^{-1}$ BW$^{-0.75}$ °C$^{-1}$ | [5] |
| | in group-housed sows: | |
| | LCT = 15–16°C and HP increases by 8 to 10 kJ kg$^{-1}$ BW$^{-0.75}$ °C$^{-1}$ | [5'] |
| Efficiency of use of ME | $k_c = 0.50$ (marginal efficiency), $k'_c = 0.30$ (total efficiency) | [6] |
| | $k_p = 0.60$ (efficiency for maternal gain as protein) | [7] |
| | $k_f = 0.80$ (efficiency for maternal gain as lipids) | [8] |
| | $k_r = 0.80$ (average efficiency of mobilization of energy body reserves) | [9] |
| Energy retained in maternal body | $ER_{mp} = 23.8 \times 6.25$ (NR − NR$_c$) | [10] |
| | $ER_{mf} = 39.7$ lip | [11] |
| Nitrogen retention | NR: total daily N retention | |
| | NR$_c$: N retention in conceptus (Table 6.1, equation 9) | [12] |
| | *When protein and AA are not limiting* | |
| | NR = NR$_c$ − 0.4 + 45.9 ($t$/100) − 105.3 ($t$/100)$^2$ + 64.4 ($t$/100)$^3$ + $a$ (ME − ME$_{mm}$) | [13] |
| | where $a = 0.571$ in the first pregnancy and is 0.366 later, ME$_{mm}$ = ME$_m$ at mating | |
| | *When amino acid supply is limiting* | |
| | (NR × 0.065) = −0.36 + 0.65 digestible lysine | [14] |
| | (NR × 0.040) = −0.41 + 0.58 digestible threonine | |

The different equations which can be used for nutrient partitioning in gestating sows are summarized in Table 6.6 according to the model presented in Fig. 6.1. They were used to predict nutrient utilization and body composition changes in different practical situations (Table 6.7) for primiparous and multiparous animals.

# Conclusion

Empirical relationships, adapted for each genotype, allow the prediction of litter size according to parity number or ovulation rate, but only a small part of the variability in prolificacy is explained by these criteria. From litter size and gestation stage, it is possible to quantify the development of the fetuses and the gravid uterus, and to derive specific energy and protein requirements for gestation. However, the effects of the different nutrients on fetal growth are not well known. More basic studies on nutrient exchanges at the uterine or fetal level, and the mechanisms involved in their regulation, should assist a

**Table 6.7.** Nutrient utilization in primiparous and multiparous sows fed 30 or 35 MJ ME day$^{-1}$ with non-limiting supplies of amino acids: prediction of body composition changes.

|  | Primiparous | Multiparous | |
| --- | --- | --- | --- |
| Energy supply, MJ ME day$^{-1}$ | 30 | 30 | 35 |
| Initial condition | | | |
| 　Body weight (kg) | 140 | 230 | |
| 　Backfat depth (mm) | 18 | 17 | |
| Expected litter size | 11.0 | 12.5 | |
| ME for maintenance (MJ day$^{-1}$) | 20.7 | 28.5 | 29.6 |
| Energy retained in conceptus (MJ day$^{-1}$) | 0.5 | 0.6 | 0.6 |
| Energy retained in maternal body (MJ day$^{-1}$) | 6.2 | 0.5 | 3.2 |
| Protein retention (g day$^{-1}$) | | | |
| 　Conceptus | 15.3 | 17.4 | 17.4 |
| 　Maternal body | 55.6 | 26.5 | 36.2 |
| Total weight gain (kg) | 58.9 | 33.0 | 48.3 |
| Net weight gain (kg) | 40.9 | 11.0 | 27.1 |
| P2 change (mm) | 3.1 | −2.4 | 2.4 |
| Crude lysine requirement (%)[a] | | | |
| 　at 30 days of pregnancy | 0.35 | 0.31 | 0.31 |
| 　at 110 days of pregnancy | 0.69 | 0.45 | 0.44 |
| ME required for 5°C below thermoneutrality | 3.6 | 4.8 | 4.8 |

[a] For 12.5 MJ ME kg$^{-1}$.

better understanding and quantification of these effects. This should be possible in the near future thanks to the development of new experimental procedures combining the measurement of blood flow, and arterial and venous concentrations of nutrients, at uterine and umbilical level.

The quantitative approach of nutrient utilization in pregnant sows, described in this chapter, allows the prediction of the required nutrients (energy and amino acids) for achieving targets of body condition at farrowing, according to parity number and expected litter size. The recommendations derived from this approach are much more flexible than former recommendations obtained from more empirical or field studies. It is possible to consider the effects of housing conditions and ambient temperature, and to some extent the behaviour of the animals. In gestating sows, these parameters explain, with body weight, most of the variations in energy requirements. The partition of energy among the different energy sources (glucose, fatty acids, acetic acid, propionic acid, etc.) should improve the prediction of the effect of nutrient supplies. But more research is needed for a precise quantification of the utilization of these different nutrients at digestive and metabolic levels.

# References

ARC (1981) *The Nutrient Requirement of Pigs*. Commonwealth Agricultural Bureau, Slough, UK, p. 50

Bauman, D.E. and Currie, W.B. (1980) Partitioning of nutrients during pregnancy and lactation: a review of mechanisms involving homeostasis and homeorhesis. *Journal Dairy Science* 63, 1514–1529.

Bazer, F.W., Clawson, A.J., Robinson, O.W. and Ulberg, L.C. (1969) Uterine capacity in gilts. *Journal et Reproduction and Fertility* 18, 121–132.

Beyer, M. (1986) Untersuchungen zum Energie- und Stoffumsatz von graviden und laktierenden Sauen sowie Saugferkeln – ein Beitrag zur Präzisierung des Energie- und Proteinbedarfes. *Promotionsarbeit aus dem Forschungszentrum für Tierproduktion* – Rostock.

Black, J.L. (1995) The evolution of animal growth models. In: Moughan, P.J., Verstegen, M.V.A. and Visser-Reyneveld, M.I. (eds), *Modelling Growth in the Pig*. EEAP, Wageningen, pp. 3–9.

Black, J.L., Campbell, R.G., Williams, I.H., James, K.J. and Davies, G.T. (1986) Simulation of energy and amino acid utilisation in the pig. *Research and Development in Agriculture* 3, 121–145.

Cariollet, R. and Dantzer, R. (1984) Motor activity of tethered sows during pregnancy. *Annales de Recherches Vétérinaires* 15, 257–261.

Charette, R., Bigras-Poulin, M. and Martineau, G.P. (1995) Une méta-analyse de l'anoestrus nutritionnel chez la truie. *Journées de la Recherche Porcine en France* 27, 31–36.

Close, W.H. and Cole, D.J.A. (1986) Some aspects of the nutritional requirements of sows: the relevance in the development of a feeding strategy. *Livestock Production Science* 15, 39–52.

Close, W.H., Noblet, J. and Heavens, R.P. (1985) Studies on the energy metabolism of the pregnant sow. The partition and utilization of metabolizable energy in pregnant and non-pregnant animals. *British Journal of Nutrition* 53, 267–279.

Comline, R.S., Fowden, A.L. and Silver, M. (1979) Carbohydrate metabolism in the foetal pig during late gestation. *Quarterly Journal of Experimental Physiology* 64, 277–289.

Cronin, G.M., Van Tartwijk, J.M.F.M., Van der Hel, W. and Verstegen, M.W.A. (1986) The influence of degree of adaptation to thether-housing by sows in relation to behaviour and energy metabolism. *Animal Production* 42, 257–268.

Dagorn, J., Badouard, B. and Legault, C. (1992) Performance control of pig farms in France. In: *Satellite Symposium on Pig Management Information Systems.* Spanish Ministry of Agriculture (ed.), EAAP, Madrid, pp. 7–20.

den Hartog, L.A. and van Kempen, G.J.M. (1980) Relation between nutrition and fertility in pigs. *Netherlands Journal of Agricultural Science* 28, 211–227.

Dewey, C.E., Martin, S.W., Friendship, R.M. and Wilson, M.R. (1994) The effect on litter size of previous lactation length and previous weaning-to-conception interval in Ontario Swine. *Preventive Veterinary Medicine* 18, 213–223.

De Wilde, R.O. (1980) Protein and energy retention in pregnant and non-pregnant gilts. *Livestock Production Science* 7, 505–510.

Dourmad, J.Y. (1987) Composition du gain de poids de la truie gestante: prévision en fonction des apports énergétiques et protéiques. *Journées de la Recherche Porcine en France* 19, 203–214.

Dourmad, J.Y., Etienne, M., Prunier, A. and Noblet, J. (1994) The effect of energy and protein intake of sows on their longevity: a review. *Livestock Production Science* 40, 87–97

Dourmad, J.Y., Etienne, M. and Noblet, J. (1996) Reconstitution of body reserves in multiparous sows during pregnancy. Effect of energy intake during pregnancy and mobilization during the previous lactation. *Journal of Annual Science* 74, 2211–2219.

Dourmad, J.Y., Etienne, M. and Noblet, J. (1997) Prédicton de la composition chimique de la truie reproductrice à partir du poids vif et de l'épaisseur de lard dorsal. *Journées de la Recherche Porcine en France* 29, 255–262.

Duée, P.H. (1984) Influence de la nutrition azotée sur le métabolisme et la reproduction chez la truie. Phd thesis, Université Pierre et Marie Curie Paris VI, Paris, France.

Duée, P.H., Simoes-Nunes, C., Pégorier, J.P., Gilbert, M. and Girard, J. (1987) Uterine metabolism of the conscious gilt during late pregnancy. *Pediatric Research* 22, 587–590.

Edgerton, L.A. and Erb, R.E. (1971) Metabolites of progesterone and oestrogen in domestic sow urine. I. Effect of Pregnancy. *Journal of Animal Science* 32, 515–524.

Elsley, F.W.H., Anderson, D.M., MacDonald, I., MacPherson, R.M. and Smart, R. (1966) A comparison of live weight changes, nitrogen retention and carcass composition of pregnant and non-pregnant gilts. *Animal Production* 8, 391–410.

Etienne, M. (1991) Apports énergétiques de gestation et accrétion de protéines chez la truie nullipare. *Journées de la Recherche Porcine en France* 23, 69–74.

Everts, H. (1994) Nitrogen and energy metabolism of sows during several reproductive cycles in relation to nitrogen intake. PhD thesis, Wageningen Agricultural University, The Netherlands.

Everts, H. and Dekker, R.A. (1994) Effect of nitrogen supply on the excretion of nitrogen and on energy metabolism of pregnant sows. *Animal Production* 59, 293–302.

Fèvre, J., Léglise, P.C. and Rombauts, P. (1968) Du rôle de l'hypophyse et des ovaires dans la biosynthèse des oestrogènes au cours de la gestation chez la truie. *Annales de Biologie Animale Biochimie Biophysique* 8, 225–233.

Fuller, M.F., McWilliam, R., Wang, T.C. and Giles, R. (1989) The optimum dietary amino acid pattern for growing pigs. 2. Requirements for maintenance and for tissue protein accretion. *British Journal of Nutrition* 62, 255–267.

Geuyen, T.P.A., Verhagen, J.M.F. and Verstegen, M.W.A. (1984) Effect of housing and temperature on metabolic rate of pregnant sows. *Animal Production* 38, 477–485.

Henry, Y. and Etienne, M. (1978) Alimentation énergétique du porc. *Journées de la Recherche Porcine en France* 10, 119–167.

Hovell, F.D. de B., Gordon, J.G. and Mac Pherson, R.M. (1977) Thin sows. 2. Observations on the energy and nitrogen exchanges of thin and normal sows in environmental temperatures at 20 and 5°C. *Journal of Agricultural Science* 89, 523–533.

ITCF (1995) In: Jondreville, C., Van den Broecke, J., Gatel, F. and Van Cauwenberghe S. (eds), Digestibilité iléale des acides aminés des matières premières, tables de digestibilité iléale. Paris, France, pp. 1–53.

Just, A., Fernandez, J.A. and Jorgensen, H. (1982) Nitrogen balance studies and nitrogen retention. *Les Colloques de l'INRA* 12, 111–122.

Kemm, E.H. (1974) A study of the protein and energy requirements of the pregnant gilt (*Sus scrofa domesticus*). PhD thesis, University of Stellenbosch, Stellenbosch, South Africa.

King, R.H. (1987) Nutritional anoestrus in young sows. *Pig News and Information* 8, 15–22.

King, R.H. and Brown, W.G. (1993) Interrelationships between dietary protein level, energy intake and nitrogen retention in pregnant gilts. *Journal of Animal Science* 71, 2450–2456.

King, R.H., Speirs, E. and Eckerman, P. (1986) A note on estimation of chemical body composition of sows. *Animal Production* 43, 167–170.

Le Cozler, Y., Dagorn, J., Dourmad, J.Y., Johansen, S. and Aumaitre, A. (1997) Effect of previous weaning to conception interval and lactation length on subsequent litter size in sows. *Livestock Production Science* 51, 1–11.

Le Dividich, J., Mormède, P., Catheline, M. and Caritez, J.C. (1991) Body composition and cold resistance of the neonatal pig from European (Large White) and Chinese (Meishan) breeds. *Biology of the Neonate* 59, 268–277.

Legault, C. (1978) Génétique et reproduction chez le porc. *Journées de la Recherche Porcine en France* 10, 43–62.

Lucbert, J. and Lavorel, O. (1984) Baisse de la prolificité de la truie en seconde portée analyse des données de deux élevages expérimentaux. *Journées de la Recherche Porcine en France* 16, 115–124.

Martinat-Botté, F., Forgerit, Y., Bussière, J., Maurel, M.C., Plat, M., Macar, C., Poirier, P., Nolibois, P. and Terqui, M. (1995) Utilisation d'un progestagène (régumate) au moment du tarissement de la primipare. *Journées de la Recherche Porcine en France* 27, 51–56.

Moustgaard, J. (1962) Foetal nutrition in the pig. In: Morgan, J.T. and Lewis, D. (eds),

*Nutrition of Pigs and Poultry*. Butterworth, London, pp. 189–206.

Noblet, J. (1990) Bases d'estimation du besoin énergétique de la truie au cours du cycle de reproduction. PhD thesis, Université Pierre et Marie Curie Paris VI, Paris, France.

Noblet, J. and Etienne, M. (1987) Metabolic utilization of energy and maintenance requirements in pregnant sows. *Livestock Production Science* 16, 243–257.

Noblet, J., Close, W.H., Heavens, R.P. and Brown, D. (1985) Studies on the energy metabolism of the pregnant sow. 1. Uterus and mammary tissue development. *British Journal of Nutrition* 53, 251–265.

Noblet, J., Dourmad, J.Y., Le Dividich, J. and Dubois, S. (1989) Effect of ambient temperature and addition of straw or alfalfa in the diet on energy metabolism in pregnant sows. *Livestock Production Science* 21, 309–324.

Noblet, J., Dourmad, J.Y. and Etienne, M. (1990) Energy utilization in pregnant and lactating sows: modeling of energy requirements. *Journal of Animal Science* 68, 562–572.

Noblet, J., Shi, X.S. and Dubois, S. (1993) Energy cost of standing activity in sows. *Livestock Production Science* 34, 127–136.

Oldham, J.D. (1991) AFRC technical committee on responses to nutrients. Theory of response to nutrients by farm animal. Pregnancy and lactation. *Nutrition Abstracts and Reviews* 61, 683–722.

Père, M.C. (1995) Maternal and fetal blood levels of glucose, lactate, fructose, and insulin in the conscious pig. *Journal of Animal Science* 73, 2994–2999.

Père, M.C., Dourmad, J.Y. and Etienne, M. (1996) Variation du débit sanguin utérin au cours de la gestation chez la truie. *Journees de la Recherche Porcine en France* 28, 371–378.

Père, M.C., Dourmad, J.Y. and Etienne, M. (1997) Effect of number of pig embryos in uterus on their survival and development, and on maternal metabolism. *Journal of Animal Science* 75, 1337–1342.

Pluske, J.R., Williams, I.H. and Aherne, F.X. (1995) Nutrition of the neonatal pig. in: Varley, M.A. (ed.), *The Neonatal Pig. Development and Survival*. CAB International, Wallingford, UK.

Pomar, C., Dewey, L.H. and Minvielle, F. (1991) Computer simulation model of swine production systems: II Modeling body composition and weight of female pigs, fetal development, milk production, and growth of suckling pigs. *Journal of Animal Science* 69, 1489–1502.

Pomeroy, R.W. (1960) Infertility and neonatal mortality in the sow. 3. Neonatal mortality and foetal development. *Journal of Agricultural Science* 54, 31–56.

Pond, W.G., Maurer, R.R., Mersmann, H.J. and Cummins, S. (1992) Response of fetal and newborn piglets to maternal protein restriction during early or late pregnancy. *Growth, Development and Aging* 56, 115–127.

Randall, G.C.B. (1982) Changes in fetal and maternal blood at the end of pregnancy and during parturition in the pig. *Research in Veterinary Science* 32, 278–282.

Reynolds, L.P., Ford, S.P. and Ferrel, C.L. (1985) Blood flow and steroid and nutrient uptake of the gravid uterus and fetus of sows. *Journal of Animal Science* 61, 968–974.

Robertson, H.A. and King, G.J. (1974) Plasma concentrations of progesterone, oestrone, oestradiol-17b and oestrone sulphate in the pig at implantation, during pregnancy and at parturition. *Journal of Reproduction and Fertility* 40, 133–141.

Rombauts, P. (1962) Evolution de l'anabolisme gravidique chez la truie en fonction de l'âge de l'animal. *Annales de Zootechnie* 11, 39–51.

Salmon-Legagneur, E. (1965) Quelques aspects des relation nutritionnelles entre la gestation et la lactation chez la truie. *Annales de Zootechnie* 14, 1–137.

Sève, B. (1994) Alimentation du porc en croissance: intégration des concepts de protéine idéale, de disponibilité digestive des acides aminés et d'énergie nette. *INRA Production Animal* 7, 275–291.

Thulin, A.J., Allee, G.L., Harmon, D.L. and Davis, D.L. (1989) Utero-placental transfer of octanoic, palmitic and linoleic acids during late gestation in gilts. *Journal of Animal Science* 67, 738–745.

Verstegen, M.W.A., Van Es, A.J.H. and Nijkamp, H.J. (1971) Some aspects of energy metabolism of the sow during pregnancy. *Animal Production* 13, 677–681.

Verstegen, M.W.A. and Curtis, S.E. (1988) Energetics of sows and gilts in gestation crates in the cold. *Journal of Animal Science* 66, 2865–2875.

Vesseur, P.C., Kemp, B. and den Hartog, L.A. (1994) The effect of weaning to oestrus interval on litter size, live born piglets and farrowing rate in sows. *Journal of Animal Physiology and Animal Nutrition* 71, 30–38.

Walker, B. and Young, B.A. (1992) Modeling the development of uterine components and sows body composition in response to nutrient intake during pregnancy. *Livestock Production Science* 30, 251–264.

Whittemore, C.T. and Morgan, C.A. (1990) Model components for the determination of energy and protein requirements for the breeding sows: a review. *Livestock Production Science* 26, 1–37.

Whittemore, C. and Yang, H. (1989) Physical and chemical composition of the body of breeding sows with different body subcutaneous fat depth at paturition, different nutrition during lactation and differing litter size. *Animal Production* 48, 208–215.

Williams, I.H., Close, W.H. and Cole, D.J.A. (1985) Strategies for sow nutrition: predicting the response of pregnant animals to protein and energy intake. In: Cole, D.J.A. (ed.), *Recent Advances in Animal Nutrition*, Nottingham, UK, pp. 133–147.

Willis, G.M. and Maxwell, C.V. (1984) Influence of protein intake, energy intake and stage of gestation on growth, reproductive performance, nitrogen balance and carcass composition in gestating gilts. *Journal of Animal Science* 58, 647–656.

# Lactation and Neonatal Growth

**7**

## R.H. King,[1] J. Le Dividich[2] and F.R. Dunshea[1]

[1]*Victorian Institute of Animal Science, Sneydes Road, Werribee 3030, Australia;* [2]*INRA Station de Recherches Porcine, 35590 Saint-Gilles, France*

## Introduction

Neonatal growth in pigs is primarily dependent upon the milk production of sows. The relationships between piglet growth and nutrient output by lactating sows (King *et al.*, 1989; Noblet and Etienne, 1989) indicate that sows suckling 10 piglets which attain an acceptable preweaning growth rate of 220 g day$^{-1}$ would produce in excess of 10 kg of milk each day of lactation. Recent estimates of milk production determined by the isotope dilution method would confirm the relatively high levels of milk production in lactating sows (Schoenherr *et al.*, 1989; Tokach *et al.*, 1992). Substantial quantities of nutrients are required by sow mammary tissue for milk synthesis: for example, daily feed intakes in excess of 7 kg will be required to supply sufficient nutrients to match the needs for the nutrients in 10 kg of milk (King, 1994).

Nutrient intake by the sow during lactation is unlikely to be sufficient to meet the needs of milk production. Any nutritional shortfall will result in catabolism of body protein and body fat to supply precursors for milk constituents. However, as milk supply outstrips nutrient intake, more and more nutrients are derived from body reserves and both reproductive and lactation performances are compromised. Information is required on milk yield of sows, the extent to which dietary nutrients are utilized directly for milk synthesis, or for the deposition and subsequent catabolism of body reserves to support milk synthesis and the degree to which the sow will protect her body reserves during lactation. Models of sow metabolism have been proposed which incorporate this knowledge of metabolism in lactating sows. These models have the potential to predict the lactation performance of sows in many different situations. Certainly, growth simulation models have been used successfully to predict the growth performance of pigs in commercial herds (Mullan *et al.*, 1993).

Models have been used to predict pig performance and often these simulation models are mixtures of empirical relationships and biological mechanisms. The lactating sow model described by Pettigrew *et al.* (1992) has empirical components but is more of a metabolic model based on the rates of metabolic movement in and out of various metabolic pools. The sow models proposed by Whittemore and Morgan (1990) and Black *et al.* (1986) are much less mechanistic and are primarily based upon nutrient partitioning and empirical relationships derived from experimental data. The greater the extent to which the model relies upon inputs at the level of nutrient partitioning and metabolism rather than empirical relationships, the more likely it will operate satisfactorily over a wide range of conditions. In pigs, there is much less information about the biological mechanisms involved in the lactation response to nutrition than to the growth response. Consequently, sow lactation models are often much more empirical in their structure than growth models.

This chapter will include empirical relationships between sow milk yield and components of piglet nursing demand, sow body reserves and sow nutrient intake, as well as quantitative aspects of metabolism, mammary uptake and efficiency of utilization of nutrients and precursors for milk synthesis in the sow. The final section will examine the potential for the growth by the neonatal piglet and relate the utilization of nutrients in sow milk to piglet growth performance.

## Empirical Relationships in Sow Lactation

Milk yield is dependent upon:

**1.** the lactational ability of the sow, which is influenced by genotype, amount of body reserves and nutrient intake during lactation; and
**2.** the nursing demand of the litter which is influenced by litter size, piglet weight and suckling frequency (Auldist and King, 1995).

Milk yield of the sow has typically been shown to increase through the first three to four weeks postpartum and then slowly decrease thereafter (Elsley, 1971). However, more recent data indicate that lactation peaks earlier in lactation (Schoenherr *et al.*, 1989; Toner *et al.*, 1996). Whittemore and Morgan (1990) described the upper boundary of the lactation curve of sows by the following equation:

$$MY = a \times e^{-0.025t} \times e^{-e[0.5 - 0.1 (t)]} \tag{7.1}$$

where MY is the limit to potential milk yield, *t* is the day of lactation and *a* is a variable ranging from 18 to 30 depending upon sow genotype. The potential milk yield is unlikely to be reached in most commercial situations because of suboptional nursing demand. The nursing demand of the litter is dependent upon litter size, piglet body weight and suckling frequency and the impact of these factors on the milk yield of sows is discussed further in the following section.

## Nursing demand of the piglets

The linear response of milk yield to litter size is mainly due to increased number of functional glands. Elsley (1971) compiled four earlier studies which investigated the response of milk yield to litter size, ranging from 4 to 12 pigs and found a strong linear relationship between litter size (LS) and milk yield (MY, kg day$^{-1}$):

$$MY = 0.581 \text{ LS} + 1.81 \tag{7.2}$$

The relationships presented by King *et al.* (1989) and Auldist *et al.* (1994) indicated that both the slope and intercept of the relationship had increased since the work of Elsley (1971). Improved milk yields of modern genotypes, together with more accurate measurement of milk yield by dilution techniques, are likely to be responsible for the greater milk yields reported in these recent studies. The influence of litter size upon milk yield of the sow was described by the equations:

$$MY = 0.796 \text{ LS} + 2.00 \text{ (King } et\ al.,\ 1989) \tag{7.3}$$
$$MY = 0.689 \text{ LS} + 5.98 \text{ (Auldist } et\ al.,\ 1994) \tag{7.4}$$

It is clear that the number of piglets is a major factor which determines milk yield but the size or weight of the piglets also plays a role. Hartman *et al.* (1962) were first to report a positive association between birth weight and milk intake of the piglet. They suggested that heavier piglets may be more efficient at draining the teats than lighter piglets and therefore may stimulate a greater subsequent milk flow. While van der Steen and de Groot (1992) established that piglet body weight influenced milk yield the relative response is often small. For example, increasing body weight from 1.9 to 4.4 kg or decreasing body weight from 4.5 to 1.9 kg in the fostering experiment of King *et al.* (1996) caused a short-term increase of 26% and decrease of 22%, respectively, in milk yield. Thus, although there is a relationship between piglet body weight and milk consumption, the relatively small range in individual piglet body weights observed in early lactation is unlikely to significantly influence sow milk yield in commercial piggeries. Increasing the suckling frequency will also stimulate higher milk yield (Auldist and King, 1995) but the major variable factor of nursing demand which influences milk yield is the number of piglets in the litter. Relationships such as those described by equations (7.3) and (7.4) may be used to predict sow milk yield if piglet growth rate data are not available to estimate milk yield using the appropriate milk : gain ratios that are described later.

## Lactational ability of the sow

Milk yield can vary substantially depending upon the dietary supply of protein and energy and the endogenous supply of precursors or the body reserves of sows. The effect of body reserves on milk production is well illustrated by the data of Mullan and Williams (1989), who manipulated gilts during pregnancy so that body reserves at parturition were either high, medium or low. At restricted energy intakes during lactation, sows with higher body reserves produced more milk; for each kilogram increase in sow body weight (at parturition) average daily litter growth increased by 5.4 g (Mullan and Williams, 1989). King and Eason (1998) also observed that average litter growth rate responded to increased sow maternal body weight at parturition by 7.0 g kg$^{-1}$ when sows were restricted in intake during lactation. A lower response of 4.8 g kg$^{-1}$ was reported by Yang *et al.* (1989) when sows were fed *ad libitum* during lactation. When Mullan and Williams (1989) offered diets *ad libitum* to sows during lactation, the sows with lower body reserves consumed more feed which enabled them to produce similar amounts of milk to those sows with the higher body reserves at parturition. Thus the influence on milk production of body reserves of sows at parturition may be limited to sows which receive low feed intakes during lactation.

Although milk production is well buffered by body reserves, it still responds to dietary energy intake during lactation. Williams (1995) presented the results of a number of studies which emphasized the variability of empirical relationships and indicated that there is a variable but positive response of milk output (measured by piglet growth rate) to dietary energy intake of sows during lactation. Nevertheless, for maternal energy intakes between 20 and 70 MJ ME day$^{-1}$ during lactation average litter growth rate responds by about 9.0 g MJ$^{-1}$. These earlier data summarized by Williams (1995) are supported by the recent results of Koketsu *et al.* (1994) and King *et al.* (1995) who observed that average litter growth rate responds to maternal ME intake by 4.2 g MJ$^{-1}$ and 10.2 g MJ$^{-1}$ respectively.

The response reported by Williams (1995) is equivalent to an increase of 37 g milk for each megajoule increase in daily maternal ME intake (Whittemore and Morgan, 1990). Milk output is relatively unaffected by the manipulation of energy or protein content of the sow lactation diet (Tritton *et al.*, 1996) except at relatively low protein concentrations; milk yield decreased by 15.6% when dietary protein was reduced from 13.3% to 6.3% CP (King *et al.*, 1993b). Milk output is responsive to dietary manipulation, either directly during lactation by altering maternal energy intake, or indirectly through the influence of previous nutritional history on the body reserves of sows commencing lactation.

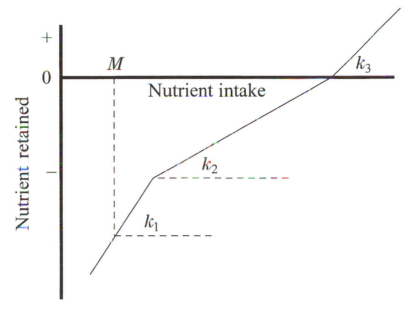

**Fig. 7.1.** Effect of nutrient intake on retention of nutrients in lactating sows: $M$ is the requirement for maintenance; $k_1$ is the efficiency when dietary nutrient intake is insufficient to meet the requirements and body tissue is mobilized to provide the deficit; $k_2$ is the efficiency of nutrient utilization for milk production; and $k_3$ is the efficiency of nutrient utilization when dietary nutrient is in excess of that for milk production and the excess is used for tissue deposition.

## Partition of Nutrients

The factorial approach has often been used to predict the dietary require-ments of sows during lactation (ARC, 1981; NRC, 1988). The factorial approach requires estimates of the partition of nutrients between the require-ment for maintenance, for tissue deposition within the body, and for products such as milk, faeces and urine which are subsequently lost from the body. Studies in the partition of nutrients require the determination of the efficiency of nutrient utilization. During lactation, not only the nutrient utilization for milk production is required, but also the determination of the extent and origin of catabolism of body reserves, as often sows mobilize significant amounts of body reserves to support milk production.

A simplified schematic representation of a proposed relationship between nutrient retention and nutrient intake is illustrated in Fig. 7.1. Milk nutrient output is often linearly related to changes in nutrient intake, partic-ularly at lower levels of nutrient intake. Thus the efficiencies of dietary nutrient and body tissue nutrient utilization for milk production and tissue deposition described in Fig. 7.1 not only include the direct nutrient efficiencies but are rather complex combinations of the nutrient efficiencies and the nutrient output response to nutrient intake.

## Quantitative estimates of nutrient efficiency

Nutrient efficiencies of milk production are difficult to measure accurately due to errors in determination of milk production and the maintenance requirement by sows. With these concerns in mind, relevant estimates of energy and protein utilization for milk production are more fully described below.

### Energy utilization

MILK SYNTHESIS FROM DIETARY SOURCES. Literature estimates of the energetic efficiency of milk synthesis from dietary sources are variable, but much of the variation has been due to the method of calculation and difficulties in estimating milk production. Verstegen *et al.* (1985) reported that the energetic efficiency of milk production varied from 62% to 72% depending upon the method used for calculation. Using the preferred method, which included both energy output in milk and dietary energy input as independent variables, Verstegen *et al.* (1985) found that the efficiency of conversion of ME into milk was 67–69%. The results of Burlacu *et al.* (1985) indicated an efficiency of 71.2% while Noblet and Etienne (1987a) estimated that the efficiency of utilization of ME for milk production was 71–72% throughout lactation. Based on this more recent work, a value of 70% appears to be a reasonable estimate for the efficiency of utilization of dietary ME for milk production when sows are offered standard lactation diets.

MILK SYNTHESIS FROM BODY TISSUE RESERVES. Dietary energy intake often does not meet the energy requirements for milk, and the remaining energy is derived from maternal body reserves which is provided primarily from metabolism of fat and to a much lesser extent, protein reserves (Noblet and Etienne, 1987a). There have been few studies conducted to study the efficiency of utilization of body reserves to support milk production of sows. It is generally recognized that energy in milk is derived more efficiently from tissues than from ME given in the feed. De Lange *et al.* (1980) estimated that milk energy was synthesized from body tissues with an efficiency of up to 84%. Noblet and Etienne (1987a) reported values between 86 and 89% for the efficiency of energy in body reserves for milk production. This compares favourably with other species (Moe *et al.*, 1971; Partridge *et al.*, 1983) and the high values, to a large extent, reflect the efficiency of transfer of energy in fat reserves to milk.

The efficiency of ME for deposition of maternal tissue in pregnant sows is about 80% (Close *et al.*, 1985; Noblet and Etienne, 1987b). The relatively high value for this energetic efficiency in pregnant sows is partly explained by the low feeding level used in the experiments, but also the energy deposited in the maternal body is predominantly fat. Combining the efficiency of ME for deposition of maternal tissue in sows with that of the conversion efficiency of energy from maternal tissue to milk energy results in a value close to 70% for overall efficiency of energy storage during pregnancy and its subsequent mobilization during lactation. This suggest that the deposition of energy reserves during pregnancy and the subsequent loss of reserves during the following lactation results in an efficiency that is only marginally less than the energy efficiency of direct utilization of dietary energy for milk. A similar overall energetic efficiency for milk synthesis for both pathways has been established for the dairy cow (Moe *et al.*, 1971).

### Protein utilization

MILK SYNTHESIS FROM DIETARY SOURCES. Compared to energy metabolism in lactating sows, there are fewer data available on the efficiency of utilization of digestible protein and lean tissue for milk protein secretion. Burlacu *et al.* (1985) reported that the utilization efficiency of digestible N in lactating sows into assimilated N, which was predominantly milk N, was 71%. The efficiency of conversion of digestible N intake into milk N output was in excess of 66% when King *et al.* (1995) offered protein-adequate diets to lactating sows. Similarly, Everts (1994) confirmed an efficiency of utilization of digestible N to milk N of 68%. Thus it is reasonable to assume that 70% of digestible N is converted to milk N provided that the diet has been correctly formulated and the dietary amino acid profile is similar to the balance of essential amino acids in sow milk.

MILK SYNTHESIS FROM BODY TISSUE RESERVES. There are few quantitative data on the efficiency of utilization of dietary protein for tissue deposition and the consequent mobilization of these lean tissue reserves to supply protein to support milk production. The amino acid balance in sow milk is very similar to that in the lean tissue of pigs (King *et al.*, 1993a). Thus if lean tissue reserves are mobilized to supply protein, the resultant amino acids will become available in a similar balance to that required for synthesis of milk protein. Consequently there should be little wastage of amino acids and the efficiency of conversion of tissue protein into milk protein is likely to be high and in excess of 85% (Beyer *et al.*, 1988).

Obviously the efficiency of conversion of digestible protein into lean tissue will depend upon the biological value and amino acid balance of dietary protein (Williams *et al.*, 1985). Under conditions where dietary protein was in excess, the efficiency of conversion of dietary protein into maternal and uterine tissue protein was between 54 and 56% in mid to late pregnancy (King and Brown, 1993). Earlier Burlacu *et al.* (1985) reported an efficiency of 75% for retained protein as a proportion of digestible protein. But many of the data suggest that the conversion of digestible protein in pregnant sows is often below 60% (Williams *et al.*, 1985). Consequently the deposition of

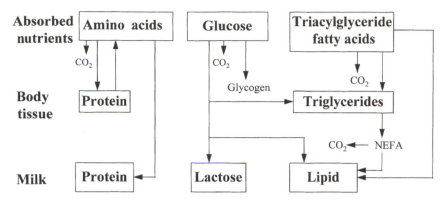

**Fig. 7.2.** Schematic representation of metabolism in lactating sows. (Adapted from Pettigrew *et al.*, 1992.)

protein reserves during pregnancy and the subsequent use of reserves during the following lactation is likely to be a less efficient system than the direct utilization of dietary protein into milk. The optimum feeding strategy for lactating sows should ensure that dietary protein and amino acid intake during lactation is sufficient to meet all the needs of milk protein output.

## Sow Metabolism

While there is extensive information on the digestion of feedstuffs in sows, surprisingly little is known about mammary uptake and metabolism of specific nutrients in the lactating sow. This is despite the fact that the needs of mammary tissues for milk precursors is so dominant in the lactating sow.

The metabolism of the lactating sow is shown in Fig. 7.2 and includes use of absorbed amino acids, glucose, fatty acids, acetate and propionate for the synthesis of milk components and accretion of body tissues. Quantitative information of tissue metabolism and the precursor product relationships outlined in Fig. 7.2 is limited by the application of techniques for estimation of nutrient uptake by the sow mammary gland.

The common method of estimating nutrient uptake relies upon simultaneously measuring blood flow rate through the mammary gland ($F$) and the metabolite concentration in arterial ($A$) and venous ($V$) blood in the mammary gland. The $A - V$ difference in metabolite concentration provides a quantitative estimation of metabolite uptake when used in conjunction with blood flow rate through the mammary system.

$$\text{Metabolite uptake} = F(A - V) \tag{7.5}$$

However, there have been difficulties in collecting the venous sample and in measuring blood flow rate in lactating sows. Blood flow rate can be measured by dye dilutions (Metcalf *et al.*, 1991) by implantation of transit-time blood flow probes (Giles, 1992) and by the Fick principle (Linzell, 1974). The Fick principle involves measuring the $A - V$ differences and milk output of a substance which is entirely transferred from blood into synthesized milk.

$$F \text{ (ml h}^{-1}) = \text{output (mg h}^{-1}) / A - V \text{ (mg ml}^{-1}) \tag{7.6}$$

The Fick principle has been successfully used to measure blood flow rate in dairy cows and has been adopted for use in the lactating sow by using lysine as the precursor indicator (Trottier, 1994).

Until the recent work by Trottier (1994), studies on the role of metabolites in milk synthesis were confined to the classic work of Linzell *et al.* (1969) and Spincer *et al.* (1969) who measured $A - V$ differences for glucose, fatty acids, amino acids and other blood precursors of milk nutrients in the pig. More recently, a technique for measuring blood flow based on dye dilution has been developed for the lactating sow (Gannon *et al.*, 1997)

## Amino acid metabolism

Experimentation to directly investigate amino acid utilization for milk synthesis has only recently begun to appear in the literature. An effort at direct quantification of amino acid uptake by the mammary gland in the sow has been recently reported by Trottier and Easter (1995) who measured plasma amino acid concentrations from blood entering and leaving the mammary gland. Calculation of the $A - V$ difference is a useful tool for estimating amino acid uptake by mammary tissue and may lead to a greater understanding of the amino acid kinetics of milk production.

Low $A - V$ differences of most non-essential amino acids (Linzell *et al.*, 1969; Spincer *et al.*, 1969), suggest that they can be synthesized within the sow mammary gland. However, there are substantial $A - V$ differences for all essential amino acids (Table 7.1). The $A - V$ differences for individual amino acids observed by Trottier (1994) are the lowest, but compare favourably with the previous data of Linzell *et al.* (1969) whereas the values of Spincer *et al.* (1969) appear unrealistically high (Table 7.1).

The $A - V$ difference can provide a quantitative estimation of amino acid uptake when used in conjunction with blood flow through the mammary gland. Lysine is the amino acid that has the highest transfer rate (Trottier, 1994) and was used as the precursor for the Fick method of estimating blood flow to enable uptake and output data to be compiled for the essential amino acids (Table 7.2). Use of the Fick principle assumes that lysine is not metabolized in the mammary gland and is entirely transferred from blood into milk. While this may not be strictly correct, the data in Table 7.2 indicate that the degree of balance between uptake and output was fair for most amino acids, suggesting that their use is almost exclusively for milk protein synthesis. It appears from the data of Linzell *et al.* (1969) that over twice as much

**Table 7.1.** Arteriovenous differences of amino acids across the sow mammary gland ($\mu$mol $l^{-1}$).

| Amino acid | Linzell *et al.* (1969) | Spincer *et al.* (1969) | Trottier (1994) |
|---|---|---|---|
| Lysine | 54.0 | 73.9 | 37.3 |
| Methionine | 12.7 | 23.5 | 10.2 |
| Threonine | 31.9 | 61.3 | 31.4 |
| Tryptophan | – | – | 9.9 |
| Leucine | 53.8 | 145.6 | 64.9 |
| Isoleucine | 43.5 | 90.0 | 32.8 |
| Arginine | 47.1 | 66.6 | 42.0 |
| Histidine | 26.4 | 39.3 | 11.5 |
| Valine | 66.6 | 133.2 | 42.6 |
| Phenylalanine | 26.6 | 46.0 | 22.0 |
| Glutamate | 126.4 | 194.4 | 85.0 |

**Table 7.2.** Estimates of the amino acid uptake from plasma and output in milk.

| | Linzell *et al.* (1969)[a] | | | Trottier (1994) | | |
|---|---|---|---|---|---|---|
| | Uptake (g day$^{-1}$) | Output (g day$^{-1}$) | Uptake/ output (%) | Uptake[b] (g day$^{-1}$) | Output[c] (g day$^{-1}$) | Uptake/ output (%) |
| Lysine | 1.56 | 2.14 | 73 | 23.3 | 22.3 | 104 |
| Methionine | — | — | — | 6.5 | 5.4 | 120 |
| Threonine | 0.96 | 1.27 | 75 | 16.0 | 12.4 | 129 |
| Tryptophan | — | — | — | 8.6 | — | — |
| Leucine | 2.52 | 2.52 | 100 | 36.5 | 25.5 | 143 |
| Isoleucine | 1.32 | 1.49 | 89 | 18.4 | 12.0 | 153 |
| Arginine | 1.90 | 1.42 | 134 | 31.3 | 14.5 | 215 |
| Histidine | 0.77 | 0.77 | 100 | 7.6 | 7.9 | 96 |
| Valine | 3.48 | 1.46 | 238 | 21.3 | 15.6 | 137 |
| Phenylalanine | 1.15 | 1.18 | 97 | 15.5 | 11.3 | 137 |

[a] Uptake and output are on an individual gland basis.
[b] Uptake of nutrients by the mammary gland was calculated assuming plasma flow was 4275 l day$^{-1}$ (Trottier, 1994).
[c] (N.L. Trottier, Illinois, 1966, personal communication).

valine was absorbed as was secreted in milk protein. Similarly, the output of arginine was substantially less than that of uptake. If the excess uptake is genuine then valine, and to a lesser extent arginine, may be oxidized in the mammary tissue. These amino acids may be oxidized to supply nitrogen and carbon for the non-essential amino acids. However, the possibility of specific mammary requirements for cellular remodelling and structural protein synthesis should also be considered. In the goat, there is evidence that valine and arginine are oxidized to an appreciable extent (Mepham and Linzell, 1966).

This early work by Linzell *et al.* (1969) on mammary uptake of valine is supported by the results of recent empirical studies on the valine requirements of high producing sows. Tokach *et al.* (1993) and Richert *et al.* (1995, 1996) reported that the dietary valine requirements during lactation were substantially greater than the level suggested by the balances of amino acids in sows' milk, particularly during high nursing demand. However, examination of the recent data collected by Trottier (1994) suggests that the ratio of uptake : output for valine was no different from the majority of other essential amino acids. These results question the validity of a dietary requirement for valine in excess of that suggested by the amino acid profile in sows' milk. Further fundamental studies are required to determine whether valine is used by the mammary gland for metabolic or maintenance needs as well as for milk protein synthesis.

## Carbohydrate and lipid metabolism

There have been relatively few studies where mammary gland carbohydrate and lipid metabolism have been measured in the pig, and none of these studies have been conducted with contemporary pigs which exhibit a high milk yield. Glucose is the major precursor for milk lactose which in turn is the major osmotic determinant of milk volume.

The arterio-venous differences for the major carbohydrates and lipids in the sow are given in Table 7.3. The principal sources of non-amino-acid carbon uptake are glucose, triglyceride and lactate. Surprisingly, there appears to be very little uptake of non-esterified fatty acids (NEFA) although without the use of isotope dilution techniques this may be masked by simultaneous uptake of NEFA and release from hydrolysed triglycerides (Pethick and Dunshea, 1993). To date there has been only a single measure of mammary blood flow in a single sedated sow providing an estimate of 41 ml $min^{-1}$ per 100 g tissue while milk production was 134 ml $day^{-1}$ per 100 g tissue (Linzell *et al.*, 1969). The resultant ratio between blood flow and milk yield was 440 : 1 which compares favourably with mean values of 430 : 1 for the lactating cow and 480 : 1 for the lactating goat (Linzell, 1974). The value of 440 : 1, combined with the average haematocrit (31.7%) observed by Linzell *et al.* (1969) has been used to provide a ratio of plasma flow : milk yield of 300 : 1 to estimate carbohydrate and lipid uptake by the mammary gland (Table 7.3). Thus the average glucose, triglyceride and lactate $A-V$ differences were 28.7, 8.6 and 6.1 mg $dl^{-1}$ or the equivalent uptakes of 8.6, 2.6 and

**Table 7.3.** Arteriovenous differences (mg per 100 ml) and average uptake of plasma carbohydrates and lipids in the lactating sow.

|  | Linzell et al. (1969)[a] | Spincer et al. (1969) | Spincer and Rook (1971) | Average | Average uptake[b] |
|---|---|---|---|---|---|
| Glucose | 19.8 | 37.5 | 28.7 | 28.7 | 8.6 |
| Triglyceride | 7.6 | 8.1 | 10.2 | 8.6 | 2.6 |
| Lactate | 10.8 | 1.4 | ND | 6.1[e] | 1.8[e] |
| VFA | 1.0 | 0.6[c] | ND | 0.8 | 0.24 |
| NEFA | 0.2 | ND[d] | ND | 0.2 | 0.06 |
| β-OHB | 0.2 | 0.2 | ND | 0.2 | 0.06 |

[a] Adjusted from whole blood to plasma using the mean haematocrit of 31.7%.
[b] Uptake (g per 100 ml of milk) assuming that the ratio of blood flow : milk yield is 440 : 1 as determined in a single sow by Linzell et al. (1969) and a haematocrit of 31.7%.
[c] Value is for acetate only.
[d] Not determined.
[e] The mean of the data from Linzell et al. (1969) and Spincer et al. (1969) is presented despite the large differences in the individual values.

$1.8$ g dl$^{-1}$ milk. These average uptake values compare to milk lactose and lipid concentrations of 3.9 and 5.2 g dl$^{-1}$, respectively, for the sows of Linzell *et al.* (1969) or values closer to 5.4 and 7.1 for milk lactose and lipid concentrations in the milk of current genotypes (Table 7.5). Quite clearly there is quantitatively more glucose taken up than lactose secreted whereas the converse is true for lipid. The suggestion is therefore that some of the glucose is used for *de novo* fatty acid synthesis.

The metabolic fate of glucose and triglyceride fatty acids taken up by the mammary gland has been investigated using isotope dilution techniques (Linzell *et al.*, 1969; Spincer and Rook, 1971). Based on precursor and end-product specific radioactivity in a single sow infused with [$^{14}$C]glucose at least 40% of milk triglyceride glycerol, 54% of CO$_2$ and 59% of milk lactose were derived from plasma glucose (Linzell *et al.*, 1969). Spincer and Rook (1971) provided similar data when they estimated that at least 38% and 70% of milk triglyceride glycerol and lactose, respectively, were derived from plasma glucose. There was some incorporation of glucose into triglyceride fatty acids but this was very low during the course of the infusions. However, $^{14}$C from glucose did continue to accumulate in milk triglyceride fatty acids for many hours after cessation of the infusion suggesting a slowly turning over mammary tissue pool of fatty acids which are synthesized *de novo* (Linzell *et al.*, 1969; Spincer and Rook, 1971). On the other hand, glucose incorporation into the glycerol moiety of milk triglyceride is extremely rapid as is the transfer of plasma-borne triglycerides (Spincer and Rook, 1971). Indeed, using the transfer quotient technique, Spincer and Rook (1971) estimated

**Table 7.4.** Balance sheet for carbohydrate and lipid plasma precursors and milk end products (g carbon day$^{-1}$)$^a$.

| Precursor | Plasma$^b$ | End product | Milk$^c$ | Milk$^d$ |
|---|---|---|---|---|
| Glucose | 224 | Lactose | 102 | 140 |
| Triglyceride | 129 | Triglyceride | 257 | 351 |
| Lactate | 43 | | | |
| Total | 396 | Total | 359 | 491 |

$^a$ Based on a milk yield of 6.5 l day$^{-1}$ estimated from a litter size of 8.3 (Elsley, 1971).
$^b$ Based on data from Table 7.3 and a ratio of plasma flow : milk yield of 300 : 1.
$^c$ Milk composition of 39 and 52 g l$^{-1}$ of milk lactose and lipid, respectively (Linzell *et al.*, 1969).
$^d$ Milk composition of 54 and 71 g l$^{-1}$ (Table 7.5).

that 62% and 51% of milk triglyceride palmitic and palmitoleic acids were derived from plasma triglyceride palmitic acid while 71% and 42% of milk triglyceride stearic and oleic acids were derived from plasma triglyceride stearic acid.

The data in Table 7.4 are an attempt to provide a balance sheet for the fate of non-amino-acid carbon in the mammary gland. While it cannot be discounted that milk amino acid carbon can arise from plasma glucose and that milk lactose and triglyceride carbon can arise from plasma amino acids, these contributions are likely to be small and, without further evidence, have been ignored. Since the only mammary uptake data were obtained in unimproved genotypes in the late 1960s and little detail has been provided on milk yield, it has been necessary to estimate milk yield from the average litter size cited by Linzell *et al.* (1969) of 8.3 and the relationship between litter size and milk yield in contemporaries of these pigs (Elsley, 1971), which provides an estimate of 6.5 l day$^{-1}$. Based on the estimated milk composition of Linzell *et al.* (1969) it is evident that the plasma glucose, triglyceride and lactate carbon can essentially provide the carbon secreted as milk lactose and triglyceride. However, it should be noted that no attempt has been made to account for oxidation, which in the case of glucose can be quite considerable and was estimated by Linzell *et al.* (1969) to be 34%. When an attempt is made to reconcile these data against a more expected milk lactose and triglyceride composition and estimated output, the plasma carbon sources fall well short of the mark (Table 7.4). However, to a large extent, this is a moot point since today's sows produce much more milk per sucking pig (Toner *et al.*, 1996).

While it can be concluded that glucose is the predominant source of carbon for lactose synthesis and is incorporated into milk triglycerides, there is a dearth of information on the quantitative importance of plasma-derived glucose, triglycerides and NEFA as sources of milk carbon. This is particularly so given that today's sows are producing more milk, are consuming higher fat diets and probably rely more heavily on body fat reserves for nutrients than the sows of the late 1960s.

# Neonatal Growth

During the period immediately after birth the piglet experiences dramatic changes in its growth pattern. For example, during the first postnatal week, the pig's growth rate is five- to sevenfold higher than that achieved during the last prenatal week. These changes are concomitant with quantitative and qualitative changes in diet. However, other factors such as birth weight and thermal conditions are also important since, to some extent, they determine the ability of the newborn to survive and the availability of dietary energy for growth. This section describes the preweaning growth pattern of pigs. Piglets are assumed to be weaned between 2 and 4 weeks of age as is practised in major pig producing countries.

## Pattern of growth during the suckling period

During the 2 to 4 weeks' suckling period, piglets from conventional geno-types grow at a mean rate of 180–200 g day$^{-1}$ (Fig. 7.3). By comparison, those from modern genotypes grow faster (250–270 g day$^{-1}$) while both patterns of growth rate are very similar. In fact two periods are clearly distinguishable.

The early period of 5 to 7 days corresponds to the establishment of lactation in the sow and to the formation of the nursing order in the litter and the associated events. During this period, growth rate is very variable (Thompson and Fraser, 1988) and increases gradually to a plateau ranging from 190–210 g day$^{-1}$ to 250–270 g day$^{-1}$, depending on the environmental conditions, parity, litter size, genotype, successful establishment of lactation and other factors. Growth rate of sucking pigs during the remaining lactation period is essentially constant. However, large variations in growth can occur depending on consumption of milk, and to a lesser extent, creep food. Supplemental heating is routinely supplied in the pigs' area so that thermal conditions have little direct effect on the growth of the sucking pigs.

The results of artificial rearing studies suggest that the biological potential for neonatal pig growth is much greater than that achieved by the sucking piglet. Postnatal live weight gains of over 500 g day$^{-1}$ are possible during the first weeks of life, through the supply of sufficient nutrients to artificially reared piglets (Hodge, 1974). Commercial pig production has not yet been able to exploit fully the potential of early growth that can be offered by the young piglet.

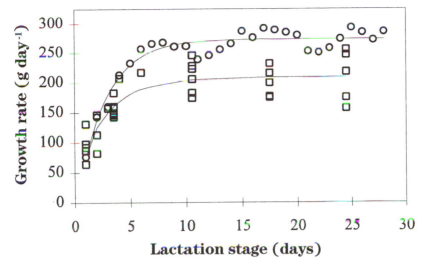

**Fig. 7.3.** Pattern of growth rate of piglets during the suckling period. Data on conventional genotypes (□) are from Salmon-Legagneur and Aumaitre (1962), Elliot and Lodge (1977), van Kempen *et al.* (1985), Okai *et al.* (1977), Bishop *et al.* (1985), Noblet and Etienne (1986, 1989), Thompson and Fraser (1986, 1988), King and Dunkin (1986), King *et al.* (1993b), Beyer and Jentsch (1994). Data on modern genotype (○) (265 piglets from 23 litters weighed daily) were provided by Teurnier (unpublished data). The least-squares fitted equations were as follows (± SE):

$$Y(\square) = 209\,(\pm\,7.3) - 170\,(\pm\,25)\,e^{-0.35\,(\pm\,0.08)x}$$

and

$$Y(\bigcirc) = 273\,(\pm\,3.6) - 289\,(\pm\,28)\,e^{-0.38\,(\pm\,0.05)x}$$

with $Y$ = growth (g day$^{-1}$) and $x$ = day postnatal.

## Factors affecting growth during the suckling period

### Birth weight

Birth weight depends on several factors including genotype, parity, position in the uterine horn, litter size and nutrition of the gestating sow. However, as selection for prolificacy has resulted in greater litter sizes the average birth weight will likely decrease whereas the variation in birth weight and the proportion of weak piglets are increased with the consequent effects on mortality and growth.

Birth weight is important in terms of both survival and postnatal growth. For example, only 44% of piglets weighing less than 1.0 kg survive to weaning (Caugant and Gueblez, 1993). In addition, weaning weight at 3–4 weeks is generally positively correlated to birth weight (Fig. 7.4), with $R^2$ being in the order of 0.2 to 0.3 (Caugant and Gueblez, 1993; Rousseau *et al.*, 1994). Based

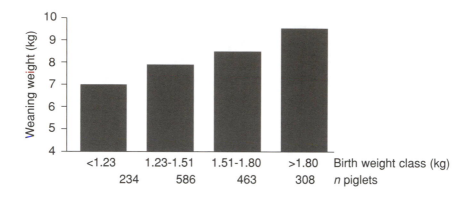

**Fig. 7.4.** Relationship between birth weight and body weight at weaning at 27 days. (According to Rousseau *et al.*, 1994.)

upon the data in Fig. 7.4, the weight of individual pigs at 27 days of age increased by 0.29 kg for every 100 g increase in birth weight. This response of piglet growth rate to birth weight is in part due to the relationship between piglet body weight and milk consumption described earlier in the chapter, but is also an effect of the likely greater nursing demand of the larger, more viable, neonatal piglet. This effect of birth weight on subsequent growth becomes progressively less with increase in age, but it can persist up to slaughter. Caugant and Gueblez (1993) reported that piglets of low birth weight ($< 1.18$ kg) gained 4.3% less on 4.1% more feed between birth and slaughter than did piglets that weighed more than 1.87 kg at birth.

### Composition of colostrum and milk
Once suckling is established in the newborn, nutrition and therefore growth depends mostly on the availability of colostrum and milk. This is convincingly attested by the high relationship ($R_2 = 0.87$ to 0.90) between preweaning growth of pigs and sow milk nutrient output (Noblet and Etienne, 1989). Nutrients available for growth depend on both the composition and the amount of colostrum and milk produced and on their efficiency of conversion into gain. The composition of colostrum (Table 7.5) is characterized by rapid changes to that of milk during the 24–48 h transition period after parturition (Aumaitre and Sève, 1978; Hartmann and Holmes, 1989). Compared with milk, colostrum is higher in dry matter and crude protein and lower in fat and lactose. Fat is the main source of energy accounting for 30–40% and 55–60% of the total energy of colostrum and milk, respectively. Both colostrum and milk are essentially devoid of medium chain triglycerides and fats are mostly composed of long chain triglycerides (C > 14). Of the major colostrum and

**Table 7.5.** Average composition (g kg$^{-1}$) of sow colostrum and milk.

|  | Colostrum | Milk |
|---|---|---|
| Dry matter | 215 | 187 |
| Crude protein (N×6.38) | 105 | 54 |
| Fat | 54 | 71 |
| Lactose | 38 | 54 |
| Gross energy (MJ kg$^{-1}$) | 5.4 | 5.0 |
| Essential amino acid (g 16 g N$^{-1}$) |  |  |
|    Lysine | 7.32 | 7.33 |
|    Methionine + cystine | 3.24 | 3.30 |
|    Tryptophan | 1.86 | 1.24 |
|    Threonine | 5.46 | 4.52 |
|    Leucine | 9.81 | 8.43 |
|    Isoleucine | 3.87 | 3.98 |
|    Valine | 5.95 | 5.25 |
|    Histidine | 3.17 | 3.23 |
|    Phenylalanine | 4.56 | 4.09 |
|    Tyrosine | 5.63 | 4.26 |

According to Salmon-Legagneur (1961), Klobasa *et al.* (1987), Elliott *et al.* (1971), Dourmad *et al.* (1991), King *et al.* (1993a), Etienne *et al.* (1993), Herpin and Le Dividich (1995).

milk constituents, fat is the most readily influenced by diet. For example, feeding more fat to the sow during late gestation or during lactation is associated with a higher content of fat in colostrum (Pettigrew, 1981) and milk (Shurson *et al.*, 1986). But, increased weight losses and increased body fat mobilization in the sow also result in an enhanced milk fat content (van Kempen *et al.*, 1985; Noblet and Etienne, 1986). In contrast, milk protein and lactose are marginally dependent on the composition of the sow diet (Lewis and Speer, 1973; King *et al.*, 1993b), while amino acid composition (g 16 g N$^{-1}$) of colostrum and milk is remarkably constant. Finally, colostrum and to a lesser extent milk contain a variety of polypeptide growth factors (insulin, EGF, IGFs). Their main target organ is the developing intestine, but they also stimulate protein synthesis in skeletal muscle if they are absorbed in the active form (Xu and Wang, 1995).

Colostrum is remarkably well utilized by the newborn piglet with a ratio of ME/GE between 95 and 97% (Le Dividich *et al.*, 1997) and a ratio of N retained/N absorbed of 91% (Le Dividich *et al.*, 1994). Similarly, the digestibility of milk nutrients is very high, being in the order of 97–99% for major constituents (Lucas and Lodge, 1961).

The efficiency of use of milk decreases linearly with age because maintenance requirement accounts for an increasing fraction of total energy intake (Table 7.6). For example, milk : gain ratio averages 3.75 in piglets

**Table 7.6.** Growth rate and efficiency of utilization of sow milk in piglets between birth and 28 days of age. (From Beyer and Jentsch, 1994.)

|  | Week of lactation | | | |
|---|---|---|---|---|
|  | 1 | 2 | 3 | 4 |
| Body weight gain (BWG) (g day$^{-1}$) | 143 | 174 | 177 | 157 |
| Milk conversion rate |  |  |  |  |
| g milk : g BWG | 3.48 | 3.99 | 4.32 | 4.96 |
| kJ milk energy : g BWG | 18.0 | 19.4 | 20.7 | 23.4 |
| g milk protein : g BWG | 0.23 | 0.20 | 0.21 | 0.25 |

weaned at 3 weeks of age and 4.08 in those weaned at 4 weeks (Salmon-Legagneur and Aumaitre, 1962; Beyer and Jentsch, 1994). Milk protein : gain ratio (g g$^{-1}$) ranges from 0.17 to 0.22 (Campbell and Dunkin, 1983; Noblet and Etienne, 1986; Beyer and Jentsch, 1994).

Despite its nutritional importance, there is little known about efficiency of milk nutrients for growth. The efficiency of utilization of colostral ME for energy retention in the pig is 90% (Le Dividich *et al.*, 1994). Noblet and Etienne (1987c) used comparative slaughter to determine the efficiency of use of sow's milk by piglets between birth and three weeks of age. They found that 52–57% of milk ME and 85% of milk protein were retained in the empty body. If endogenous loss and digestibility of protein are taken into account, the biological value of protein in sow's milk must approach 1.0. This is not surprising as the amino acid balance in sow's milk is very similar to that in the lean tissue of pigs (King *et al.*, 1993a).

### Is sow milk the best food for the suckling piglet?

This question arises from the observation that piglets reared by the sow gain less than those artificially reared (Leece, 1969; Braude *et al.*, 1970). From data of Hodge (1974) and Harrell *et al.* (1993), the potential for neonatal pig growth is between 400 and 500 g day$^{-1}$ from birth to 3–4 weeks of age providing evidence that sow-reared piglets have unrealized growth potential (Boyd *et al.*, 1995; Pluske *et al.*, 1995). Both availability and composition of milk are the major reasons for the slower growth rate of sow-reared piglets. Firstly, the hypothesis that the milk produced by the sow may be limiting is substantiated by the observation that piglets have a higher capacity for milk ingestion than usually recorded. Thus, in bottle-fed piglets, colostrum consumption during the first 24 h averages 490 g kg$^{-1}$ birth weight (Le Dividich *et al.*, 1997) which is much higher than the 325–345 g kg$^{-1}$ birth weight reported in natural suckling (Le Dividich and Noblet, 1981; Milon *et al.*, 1983). Similarly, piglets fed artificial milk from 4 to 28 days of age consumed 24% more than the control sow-reared piglets which resulted in a 30% increase in growth rate (Ferreira *et al.*, 1988). The results of the artificial rearing studies by Harrell *et al.* (1993) suggest that the biological potential for neonatal pig growth is at least 450 g day$^{-1}$ between birth and 21 days of age. Milk energy output by high producing sows failed to satisfy the energy needs of piglets beyond about day 7 of lactation (Boyd *et al.*, 1995). The disparity progressively increased so that nutrient output limited overall preweaning growth by at least 40% (Boyd *et al.*, 1995). To meet the energy requirements of a 10 pig litter at 21 days of age, the sow would have to produce in excess of 18 kg milk day$^{-1}$. Although the modern sow has a high capacity to produce this level of milk, its full expression is likely to be limited by the nursing demand of the piglets.

Secondly, the growth potential of the suckling pig may be limited by composition of milk and notably by the amount of milk protein. This hypothesis is substantiated by data of Noblet and Etienne (1987c) indicating that the protein to energy ratio in sow milk is lower than that which provides maximum growth in the young pig (Lecce and Coalson, 1976; Campbell and Dunkin, 1983; Williams, 1995). The dietary lysine requirement for piglets during their first three weeks of life is about 0.9 g lysine MJ$^{-1}$ GE (Auldist *et al.*, 1997) whereas sow milk contains only 0.79 g lysine MJ$^{-1}$ GE (Table 7.5). Sow's milk is likely to have evolved to enhance the survival of the baby pig. The newborn piglet is small in body size and has a large surface area to body weight ratio relative to other farm animals. The piglet has less than 2% body fat (Mellor and Cockburn, 1986) and must rely upon the small energy stores associated with glucose in the liver and protein in skeletal muscle in cold environmental temperatures and if dietary energy intake is restricted. Sow's milk is well designed for survival of piglets because it is high in fat (Table 7.5) and it is delivered at frequent intervals by the sow. Because of the relatively low protein to energy ratio, sow's milk encourages the piglet to deposit body fat which can serve both as an energy store and an insulation layer. Although fat deposition may help the survival of piglets outdoors, it is doubtful whether pigs need to deposit fat so early in life if raised in modern commercial facilities. Manipulation of sow's milk to match more closely the dietary protein and amino acid requirements of the piglet would increase protein deposition by 4.3 g day$^{-1}$ for each 0.1 g MJ$^{-1}$ increase in lysine : GE ratio (Auldist *et al.*, 1997) and improve the protein : fat ratio in the body of young pigs. However, sow milk protein concentration is only marginally dependent on the composition of the sow diet (King and Dunkin, 1986; King *et al.*, 1993b) and it is unlikely that the protein to energy ratio in sow milk can be manipulated by conventional means.

### Creep food

Creep food has two potential roles for the unweaned piglets: to maintain a high GR as the availability of sow milk declines and to prepare the piglets to consume and to digest complex carbohydrate and protein of the weaning diet. As reviewed by Pluske *et al.* (1995), preweaning creep food intake accounts for about 1.4 to 5.4% of daily ME intake in piglets weaned at 21 days, and for about 1.2 to 7.3% of daily ME intake in those weaned at 28 days. In fact, contribution of creep feed to preweaning growth rate is quite variable, depending on the age at weaning, being of marginal benefit for pigs weaned at 21 days of age (Hampson and Kidder, 1986) but accounting for 28% of variation in weight gain in the week preceding weaning at 28 days of age (Fraser *et al.*, 1994). However, growth rate in the immediate period following weaning is often poorly related to preweaning creep food intake (Barnet *et al.*, 1989; Fraser *et al.*, 1994).

# Conclusion

Considerable evidence indicates that the availability and composition of milk are the major factors limiting neonatal growth. Milk production in sows requires substantial amounts of nutrients which may be supplied from dietary sources or from the mobilization of body reserves. Information on the efficiencies with which dietary energy and protein are used for milk production, either directly, or indirectly via body tissues, is available and has been used in simulation models to predict the dietary requirements of sows during lactation. However, any further advances in the development of simulation models which operate satisfactorily over a wide range of conditions, will require additional information on metabolism and mammary uptake of specific nutrients.

Presently, milk yield may be predicted from empirical relationships or from the relationships between piglet growth and estimates of the gross efficiency of milk utilization by the neonatal pig. Milk yield, despite apparent improvements in the past 20 years, still fails to satisfy the potential for piglet growth. Nursing demand of the litter appears to be the major influence on milk yield. Nevertheless, both efficiencies of the utilization of milk nutrients for growth and factors controlling the nursing demand have to be determined before milk output from sows will reach levels necessary to achieve the potential growth of neonatal pigs.

# References

Agricultural Research Council (1981) *The Nutrient Requirements of Pigs.* Commonwealth Agricultural Bureaux, Slough, UK.

Auldist, D.E. and King, R.H. (1995) Piglets role in determining milk production in the sow. In: Hennessy D.P. and Cranwell P.D. (eds) *Manipulating Pig Production V.* Australasian Pig Science Association, Werribee, Australia, pp. 114–118.

Auldist, D.E., Morrish, L., Thompson, M. and King, R.H. (1994) Response of sows to varying litter size. *Proceedings of The Nutrition Society of Australia* 18, 175.

Auldist, D.E., Stevenson, F.L., Kerr, M.G., Eason, P. and King, R.H. (1997) Lysine requirements of pigs from 2 to 7 kg live weight. *Animal Science* 65, 501–507.

Aumaitre, A. and Sève, B. (1978) Nutritional importance of colostrum in the piglet. *Annales de Recherches Vétérinaires* 9, 181–192.

Barnet, K.L., Kornegay, E.T., Risley, C.R., Linderman, M.D. and Schurig, C.R. (1989) Characterization of creep feed consumption and its subsequent effects on immune response, scouring index and performance of weanling pigs. *Journal of Animal Science* 67, 2698–2708.

Beyer, M. and Jentsch, W. (1994) Relationship between milk production of the sow and growth performance in piglets In: Souffrant, W.B. and Hagemeister, H. (eds), *Proceedings of the VIth International Symposium on Digestive Physiology in Pigs*, pp. 226–229.

Beyer, M., Hoffman, L., Schiemann, R., Jentsch, W., Burlacu, G., Iliescu, M., Machajew, E.C., Babinszky, L., Gundel, H., Lassota, L., Walach-Janiak, M. and Zeman, L. (1988) Biological basics for the factorial derivation of the energy and protein requirements for pregnant and lactating sows and suckling piglets. *Fifth International Symposium on Protein Metabolism and Nutrition*. EAAP Publication No. 35. Wissenschaftliche Zeitschrift der Wilhelm-Pieck-Universität Rostock, 37, 92–93.

Bishop, T.C., Stahly, T.S. and Cromwell, G.L. (1985) Effects of dietary fat and triamcinolone additions during late gestation on the body energy reserves of neonatal pigs. *Journal of Animal Science* 61, 1476–1484.

Black, J.L., Campbell, R.G., Williams, I.H., James, K.J. and Davies, G.T. (1986) Simulation of energy and amino acid utilisation in the pig. *Research and Development in Agriculture* 3, 121–145.

Boyd, D.R., Kensinger, R.S., Harrell, R. and Bauman, D.E. (1995) Nutrient uptake and endocrine regulation of milk synthesis by mammary tissue of lactating sows. *Journal of Animal Science* 73(Suppl. 2), 36–56.

Braude, R., Mitchell, K.G., Newport, M.J. and Porter, J.W.G. (1970) Artificial rearing of pigs. I. Effect of frequency and level of feeding on performance and digestion of milk proteins. *British Journal of Nutrition* 24, 501–516.

Burlacu, G., Iliescu, M. and Caramida, P. (1985) Efficiency of food utilisation by pregnant and lactating sows. *Proceedings 10th International Symposium on Energy Metabolism*, 15–21 Sept. Airlie, Virginia, Abstr, p. 81.

Campbell, R.G. and Dunkin, A.C. (1983) The effects of energy intake and dietary protein on nitrogen retention, growth performance, body composition and some aspects of energy metabolism of baby pigs. *British Journal of Nutrition* 49, 221–230.

Caugant, A. and Guéblez, R. (1993) Influence of piglet weight at birth on subsequent production traits. *Journées de la Recherche Porcine en France* 25, 123–128.

Close, W.H., Noblet J. and Heavens, R.P. (1985) Studies on energy metabolism of the pregnant sow. 2. The partition and utilisation of metabolizable energy intake in pregnant and non-pregnant animals. *British Journal of Nutrition* 53, 267–279.

de Lange P.G.B., van Kempen, G.J.M., Klaver, J. and Verstegen, M.W.A. (1980) Effect of condition of sows on energy balances during 7 days before and 7 days after parturition. *Journal of Animal Science* 50, 886–891.

Dourmad, J.Y., Etienne, M. and Noblet, J. (1991) A contribution to the study of amino acid requirement for lactation in sows. *Journées de la Recherche Porcine en France* 23, 61–68.

Elliott, J.L. and Lodge, G.A. (1977) Body composition and glycogen reserves in the neonatal pig during the first 96 hours postpartum. *Canadian Journal of Animal Science* 57, 141-150.

Elliott, R.F., Vander Noot, G.W., Gilbreath, R.L. and Fisher, N. (1971) Effect of dietary protein level on composition changes in sow colostrum and milk. *Journal of Animal Science* 32, 1128-1137.

Elsley, F.W.H. (1971) Nutrition and lactation in the sow. In: Falconer I.R. (ed.), *Lactation*. Butterworths, London, pp. 393-411.

Etienne, M., Dourmad, J.Y. and Noblet, J. (1993) Production laitière de la truie et croissance du porcelet In: Association Française de Médecine Vétérinaire (ed.), *Bien-être de la truie et Productivité*, pp. 19-30.

Everts, H. (1994) Nitrogen and energy metabolism of sows during several reproductive cycles in relation to nitrogen intake. PhD thesis, The University of Wageningen, Wageningen, The Netherlands.

Ferreira, A.S., Costa, P.M.A., Gomes, J.C. and Rostagno, H.S. (1988) Natural suckling vs artificial rearing of pigs using sow's milk, cow's milk and soyabean extract. *Revista de Sociedada Bresileira de Zootechnia* 17, 227-237.

Fraser, D., Feddes, J.J.R. and Pajor, E.A. (1994) The relationship between creep feeding behavior of piglets and adaptation to weaning: effect of diet quality. *Canadian Journal of Animal Science* 74, 1-6.

Gannon, N.J., Parr, R.A., Kerton, D.J. and Dunshea, F.R. (1997) A technique for measuring arterio-venous difference, including blood flow, in the mammary gland of the sow. *Proceedings of the Nutrition Society* 56, 167A.

Giles, L.R. (1992) Energy expenditure of growing pigs exposed to high ambient temperature. PhD thesis, University of Sydney, Sydney, Australia.

Hampson, D.J. and Kidder, D.E. (1986) Influence of creep feeding and weaning on brush border enzyme activities in the piglet small intestine. *Research in Veterinary Science* 40, 24-31.

Harrell, R.J., Thomas, M.J. and Boyd, R.D. (1993) Limitations of sow milk yield on baby pig growth. *Proceedings, Cornell Nutrition Conference for Feed Manufacturers*, pp. 156-164.

Hartman, D.A., Ludwick, T.M. and Wilson, R.F. (1962) Certain aspects of lactation performance in sows. *Journal of Animal Science* 21, 883-886.

Hartmann, P.E. and Holmes, M.A. (1989) Sow lactation. In: Barnett, J.L. and Henesssy, D.P. (eds) *Manipulating Pig Production II*. Australasian Pig Science Association, Werribee, Australia, pp. 72-97.

Herpin, P. and Le Dividich, J. (1995) Thermoregulation and the environment. In: Varley, M.A. (ed.), *The Neonatal Pig. Development and Survival*. CAB International, Wallingford, UK, pp. 57-95.

Hodge, R.M.W. (1974) Efficiency of food conversion and body composition of the preruminant lamb and the young pig. *British Journal of Nutrition* 32, 113-126.

King, R.H. (1994) Feeding gilts and sows of the new genotypes to optimise reproductive performance. *North Carolina Pork Producers 38th Annual Conference*, pp. 14-26.

King, R.H. and Brown, W.G. (1993) Interrelationships between dietary protein level, energy intake, and nitrogen retention in pregnant gilts. *Journal of Animal Science* 71, 2450-2456.

King, R.H. and Dunkin, A.C. (1986) The effects of nutrition on the reproductive performance of first-litter sows. 4. The relative effects of energy and protein intake during lactation on the performance of sows and their piglets. *Animal Production* 43, 319–325.

King, R.H., and Eason, P.J. (1998) The effect of body weight of sows on the response to dietary lysine during lactation. *Journal of Animal Science* 76 (Suppl. 1), 162 (Abst.).

King, R.H., Toner, M.S. and Dove, H. (1989) Pattern of milk production in sows. In: Barnett, J.L. and Hennessy D.P. (eds), *Manipulating Pig Production 11*. Australasian Pig Science Association: Werribee, Australia, p. 98.

King, R.H., Rayner, C.J. and Kerr, M. (1993a) A note on the composition of sow's milk. *Animal Production* 57, 500–502

King, R.H., Toner, M.S., Dove, H., Atwood, C.S. and Brown, W.G. (1993b) The response of first-litter sows to dietary protein during lactation. *Journal of Animal Science* 71, 2457–2463.

King, R.H., Toussaint, J., Eason, P.J. and Morrish, L. (1995) The effect of food intake during lactation on nitrogen metabolism of first-litter sows. In: Hennessy D.P. and Cranwell P.D. (eds), *Manipulating Pig Production V*. Australasian Pig Science Association, Werribee, Australia, p. 133.

King, R.H., Dunshea, F.R., Mullan, B.P. and Dove, H. (1996) The influence of piglet body weight on milk production of sows. *Livestock Production Science* 47, 169–174.

Klobasa, F., Werhahn, E. and Butler, J.E. (1987) Composition of sow milk during lactation. *Journal of Animal Science* 64, 1458–1466.

Koketsu, Y., Dial, G.D., Marsh, W.E., Pettigrew, J.E. and King, V.L. (1994) Feed intake during lactation and subsequent reproductive performance of sows. *Journal of Animal Science* 72 (Suppl. 1), 333 (Abstr).

Lecce, J.G. (1969) Rearing colostrum-free pigs in an automatic feeding device. *Journal of Animal Science* 28, 27–33.

Lecce, J.G. and Coalson, J.A. (1976) Diets for rearing colostrum-free piglets with an automatic feeding device. *Journal of Animal Science* 42, 622–629.

Le Dividich, J. and Noblet, J. (1981) Colostrum intake and thermoregulation in the neonatal pig in relation to environmental temperature. *Biology of the Neonate* 40, 167–174.

Le Dividich, J., Herpin, P. and Rosario-Ludovino, R.M. (1994) Utilization of colostral energy by the newborn pig. *Journal of Animal Science* 72, 2082–2089.

Le Dividich, J., Herpin, P., Paul, E. and Strullu, F. (1997) Effect of fat content of colostrum on voluntary colostrum intake and fat utilization in newborn pigs. *Journal of Animal Science* 75, 705–713.

Lewis, A.J. and Speer, V.C. (1973) Lysine requirements of the lactating sow. *Journal of Animal Science* 37, 104–110.

Linzell, J.L. (1974) Mammary blood flow and methods of identifying and measuring precursors of milk. In: Larson B.L. and Smith V.R. (eds), *Lactation, a Comprehensive Treatise*, vol. 1. *The Mammary Gland, Development and Maintenance*. Academic Press, New York, pp. 143–225.

Linzell, J.L., Mepham, T.B., Annison, E.F. and West C.E. (1969) Mammary metabolism in lactating sows: arteriovenous differences of milk precursors and the mammary metabolism of [$^{14}$C]glucose and [$^{14}$C]acetate. *British Journal of Nutrition* 23, 319–332.

Lucas, I.A.M. and Lodge, G.A. (1961) *Nutrition of the Young Pig*. Technical Bulletin No. 22. Commonwealth Agricultural Bureaux, Farnham Royal, UK, 106 pp.

Mellor, D.J. and Cockburn, F. (1986) A comparison of energy metabolism in the new born infant, piglet and lamb. *Quarterly Journal of Experimental Physiology* 71, 361–379.

Mepham, T.B. and Linzell, J.L. (1966) A quantitative assessment of the contribution of individual plasma amino acids to the synthesis of milk proteins by the goat mammary gland. *Biochemical Journal* 101, 76–83.

Metcalf, J.A., Sutton J.D., Cockburn J.E., Napper, D.J. and Beever, D.E. (1991) The influence of insulin and amino acid supply on amino acid uptake by the lactating bovine mammary gland. *Journal of Dairy Science* 74, 3412–3420.

Milon, A., Aumaitre, A., Le Dividich, J., Franz, J. and Metzger, J.J. (1983) Influence of prematurity on colostrum composition and subsequent immunity of piglets. *Annales de Recherches Vétérinaires* 143, 533–540.

Moe, P.W., Tyrell, H.F. and Flatt, W.P. (1971) Energetics of body tissue mobilization. *Journal of Dairy Science* 54, 548–553.

Mullan, B.P. and Williams, I.H. (1989) The effect of body reserves at farrowing on the reproductive performance of first-litter sows. *Animal Production* 48, 449–457.

Mullan, B.P., Davies, G.T. and Charles, M. (1993) Formulating commercial grower diets using Auspig. In: Batterham E.S. (ed.), *Manipulating Pig Production IV*. Australasian Pig Science Association, Attwood, Australia, p. 220.

National Research Council (1988) *Nutrient Requirements of Swine*, 9th edn. National Academy Press, Washington DC.

Noblet, J. and Etienne, M. (1986) Effect of energy level in lactating sows on yield and composition of milk and nutrient balance of piglets. *Journal of Animal Science* 63, 1888–1896.

Noblet, J. and Etienne, M. (1987a) Metabolic utilization of energy and maintenance requirements in lactating sows. *Journal of Animal Science* 64, 774–781.

Noblet, J. and Etienne, M. (1987b) Metabolic utilisation of energy and maintenance requirements in pregnant sows. *Livestock Production Science* 16, 243–257.

Noblet, J. and Etienne, M. (1987c) Body composition, metabolic rate and utilization of milk nutrients in suckling piglets. *Reproduction, Nutrition and Development* 27, 829–839.

Noblet, J. and Etienne, M. (1989) Estimation of sow milk nutrient output. *Journal of Animal Science* 67, 3352–3359.

Okai, D.B., Aherne, F.X. and Hardin, R.T. (1977) Effects of sow nutrition in late gestation on the body composition and survival of the neonatal pig. *Canadian Journal of Animal Science* 57, 439–448.

Partridge, G.C., Fuller, M.F. and Pullar, J.D. (1983) Energy and nitrogen metabolism of lactating rabbits. *British Journal of Nutrition* 49, 507–516.

Pethick, D.W. and Dunshea, F.R. (1993) Fat metabolism and turnover. In: Forbes, J.M. and France, J. (eds), *Quantatitive Aspects of Ruminant Digestion and Metabolism*. CAB International, Wallingford, UK, pp. 291–314.

Pettigrew, J.E. (1981) Supplemental dietary fat for peripartial sows: a review. *Journal of Animal Science* 53, 107–117.

Pettigrew, J.E., Gill, M., France, J. and Close, W.H. (1992) A mathematical integration of energy and amino acid metabolism of lactating sows. *Journal of Animal Science* 70, 3742–3761.

Pluske, J.R., Williams, I.H. and Aherne, F.X. (1995) Nutrition of the neonatal pig. In: Varley, M.A. (ed.), *The Neonatal Pig. Development and Survival*. CAB International, Wallingford, UK, pp. 187–235.

Richert, R.D., Goodband R.D., Tokach M.D. and Nelssen J.L. (1995) Valine and lysine independently improve sow productivity during lactation. *Journal of Animal Science* 73 (Suppl. 1), 85 (Abstr.).

Richert, B.T., Tokach, M.D., Goodband, R.D., Nelssen, J.L., Pettigrew, J.E., Walker, R.D. and Johnston, L.J. (1996) Valine requirements of the high-producing lactating sow. *Journal of Animal Science* 74, 1307–1313.

Rousseau, P., Chatelier, C., Dutertre, C. and Lévêque, J.C. (1994) Heating systems for piglets nests: comparison between IR lamp and electrically heated floor. Piglets performance and behaviour and energy cost. *Journées de la Recherche Porcine en France* 25, 47–54.

Salmon-Legagneur, E. (1961) La composition du lait de truie. Relations entre les variations des teneurs du lactose et des autres constituants. *Annales de Biologie Animales Biochimie Biophysique* 1, 295–303.

Salmon-Legagneur, E. and Aumaitre, A. (1962) Influence de la quantité de lait et sa composition sur la croissance du porcelet sous la mère. *Annales de Zootechnie* 11, 181–196.

Schoenherr, W.D., Stahly, T.S. and Cromwell, G.L. (1989) The effects of dietary fat or fibre addition on yield and composition of milk from sows housed in a warm or hot environment. *Journal of Animal Science* 67, 482–495.

Shurson, G.C., Hogberg, M.G., Defever, N., Radecki, S.V. and Miller, E.R. (1986) Effect of adding fat to the sow lactation diet on lactation and rebreeding performance. *Journal of Animal Science* 62, 672–680.

Spincer, J. and Rook, J.A.F. (1971) The metabolism of [U-$^{14}$C]glucose, [1-$^{14}$C]palmitic acid and [1-$^{14}$C]stearic acid by the lactating mammary gland of the sow. *Journal of Dairy Research* 38, 315–322

Spincer, J., Rook, J.A.F. and Towers, K.G. (1969) The uptake of plasma constituents by the mammary gland of the sow. *Biochemical Journal* 111, 727–732.

Thompson, B.K. and Fraser, D. (1986) Variation in piglet weights: Development of within-litter variation over 5-week lactation and effect of farrowing crate design. *Canadian Journal of Animal Science* 66, 361–372.

Thompson, B.K. and Fraser, D. (1988) Variation in piglet weight: weight gains in the first days after birth and their relationship with later performance. *Canadian Journal of Animal Science* 68, 581–590.

Tokach, M.D., Pettigrew, J.E., Crooker, B.A., Dial, G.D. and Sower, A.F. (1992) Quantitative influence of lysine and energy intake on yield of milk components in the primiparous sows. *Journal of Animal Science* 70, 1864–1872.

Tokach, M.D., Goodband, R.D., Nelssen, J.L. and Kats, L.J. (1993) Valine – a deficient amino acid in high lysine diets for the lactating sow. *Journal of Animal Science* 71 (Suppl. 1), 68 (Abstr).

Toner, M.S., King, R.H., Dunshea, F.R., Dove, H. and Attwood, C.S. (1996). The effect of exogenous somatotropin on lactation performance of first-litter sows. *Journal of Animal Science* 74, 167–172.

Tritton, S.M., King, R.H., Campbell, R.G., Edwards, A.C. and Hughes, P.H. (1996). The effect of dietary protein and energy levels of diets offered during lactation on the lactational and subsequent reproductive performance of first-litter sows. *Animal Science* 62, 573–579.

Trottier, N.L. (1994) Protein metabolism in the lactating sow. PhD thesis, University of Illinois, Urbana-Champaign, USA.

Trottier, N.L. and Easter, R.A. (1995) Daily amino acid uptake by the mammary gland in the lactating sow: A new approach for estimating amino acid requirements. *Journal of Animal Science* 73 (Suppl. 1), 85 (Abstr).

van der Steen, H.A.M. and de Groot, P.N. (1992) Direct and maternal breed effects on growth and milk intake of piglets: Meishan versus Dutch breeds. *Livestock Production Science* 30, 361–373.

van Kempen, G.J.M., Geerse, C., Verstegen, M.W.A. and Mesu, J. (1985) Effect of feeding level on milk production of sows during four weeks of lactation. *Netherlands Journal of Agricultural Science* 33, 23–34.

Verstegen, M.W.A., Mesu, J., van Kempen, G.J.M. and Geerse, C. (1985) Energy balances of lactating sows in relation to feeding level and stage of lactation. *Journal of Animal Science* 60, 731–740.

Whittemore, C.T. and Morgan, C.A. (1990) Model components for the determination of energy and protein requirements for breeding sows: a review. *Livestock Production Science* 26, 1–37.

Williams, I.H. (1995) Sow milk as a major nutrient source before weaning. In: Hennessy, D.P. and Cranwell, P.D. (eds), *Manipulating Pig Production V.* Australasian Pig Science Association, Werribee, Australia, pp. 107–113.

Williams, I.H., Close, W.H. and Cole, D.J.A. (1985) In: Haresign, W. and Cole, D.J.A. (eds), *Recent Advances in Animal Nutrition - 1985*. Butterworths, London, pp. 133–147.

Xu, R.J. and Wang, T. (1995) Absorption of insulin-like growth factor 1 in neonatal pigs is independent of gut closure. In: Hennessy, D.P. and Cranwell, P.D. (eds), *Manipulating Pig Production V.* Australasian Pig Science Association, Werribee, Australia, p. 82.

Yang, H., Eastham, P.R., Phillips, P. and Whittemore, C.T. (1989). Reproductive performance, body-weight and body condition of breeding sows with differing body fatness at parturition, differing nutrition during lactation, and differing litter size. *Animal Production* 48, 181–201.

# Growth and Body Composition

**8**

## G.C. Emmans and I. Kyriazakis

*Animal Biology Division, Scottish Agricultural College, West Mains Road, Edinburgh EH9 3JG, UK*

## Introduction

The idea that an animal needs to progress to maturity, and then to reproduce, in order to create the next generation, goes back to Aristotle. He saw an animal as having the *potential* to grow. In more modern evolutionary theory it is also understood that a genotype must be able to progress to maturity in order to reproduce; if it did not then its genes would not be transmitted to the next generation. Arguments such as this lead to the view that the idea of an animal having a *potential* to grow is a useful one. Whittemore and Fawcett (1974) and Whittemore (1976) emphasized the importance of protein growth and saw that the potential rate of protein retention was an important part of describing a pig. The rate was needed in order that the pig's performance could be predicted on controlled feeding. On controlled feeding the question of the potential lipid growth does not arise. But where feeding is *ad libitum* it is necessary to consider the *potential* lipid, as well as the *potential* protein, growth of the animal (Emmans, 1988, 1995, 1997) in order to predict intake.

As it can be assumed that there are only four major chemical components in the empty body of a pig – protein, ash, water and lipid – the description of growth in chemical terms is almost bound to be simpler than one in anatomical terms where many tissues and organs of changing composition would need to be considered. Even if the anatomical body is simplified into butcher's terms, such as total carcass weight and carcass lean, bone and fat, it is still a more complex description than one in chemical terms. As the composition of these components changes as the animal matures it is not possible to connect requirements for energy and amino acids directly with growth in such terms. Because the chemical approach is simpler it has a greater chance of success. A continuing problem is that of predicting the anatomy from the weights of the chemical components.

In this chapter the general description of potential growth in animals, as developed by Emmans (1988), is made the basis for describing the potential chemical growth of pigs. There are other approaches as described in Chapters 2 and 13 in this book, and in that of Black (1995). It is for the reader to choose from these approaches.

The forms for *potential* growth will be described. There will of course be data sets on the *actual* growth of pigs that are not well described by these forms. This is to be expected whenever *actual* growth is not the *potential*. Such divergencies should act as a spur to finding the reason for them, rather than as a reason for rejecting the forms of function that have such utility for describing potential growth.

## The Form of the Potential Growth Curve

A growth curve, such as that shown in Fig. 8.1, is usually drawn so that it has an asymptote, or the final size, $S_m$, where the subscript 'm' denotes maturity; the curve is S-shaped or sigmoid in form. The cumulative growth curve can be seen as being the result of the accumulation of the instantaneous growth rates, $dS/dt$, which occur at each size, $S$. The relationship between $dS/dt$ and $S$ can therefore be seen as being the causal one which produces the growth curve. It gives an answer to the question: how fast can this particular pig grow, i.e. change its size, at the particular size, $S$, that it now has? The broad form of the relationship that must apply if the growth curve is to be sigmoid in shape is shown in Fig. 8.2. Growth rate must increase from a low absolute value when the pig is very small towards a maximum value as the pig gets bigger and must then decline towards zero as the pig progresses towards its final mature size, $S_m$.

The exact shape of the curve which expresses $dS/dt$ in terms of $S$ is of importance. It needs to be noted that the change in size, measured on the scale $S$, is related to the current size, measured on the same scale. It is an example of the rule that a rate of change of state needs to be expressed as a function of the current state. This is clearly preferable to relating a rate of change in one kind of size, protein weight for example, to another measure of current size such as live weight, although this is often done.

To determine what the shape of the relationship will be between $dS/dt$ and $S$ under non-limiting conditions, where the *actual* growth is the *potential*, there are two possibilities. Either we can assert that the relationship has a certain form, which may be derived from some theoretical set of assumptions (see, for example France *et al.*, 1996), or we can collect data on the growth of real pigs and try to see what the relationship is. It may seem that the

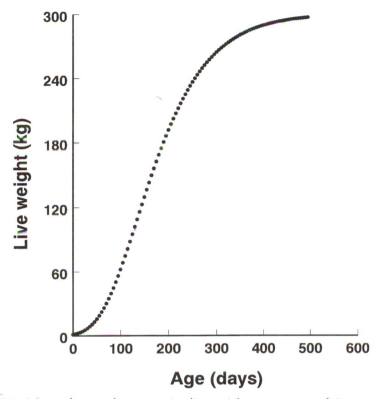

**Fig. 8.1.** A form of a growth curve, using live weight as a measure of size.

second way must be better, but this is, in fact, not so. In science the purpose of collecting data in the real world is to test our ideas about the world. Initially the tests are of our qualitative views as expressed in the forms of the functions chosen. Later tests concern the ability to make these relationships quantitative. We therefore need to have ideas before we get data to test them. Apart from not being scientific, the approach of relying on data has the difficulty that the real data that we collect on *actual* growth may well not be reflecting the *potential* of the pigs to grow. This may be because of deficiencies in either their feeding or in the environments supplied. It is their *potential* to grow that is of importance in making predictions into the future and which needs to be described. Where data exist on the growth of a particular kind of pig in a particular environment the data themselves are sufficient. Fitting some functional form through such data will not give us any more information about the growth of that particular kind of pig in that particular environment and can have no generality. Using such data as a description of the pig for the future may well have serious consequences.

The first question to be asked is: what will the form of the relationship be

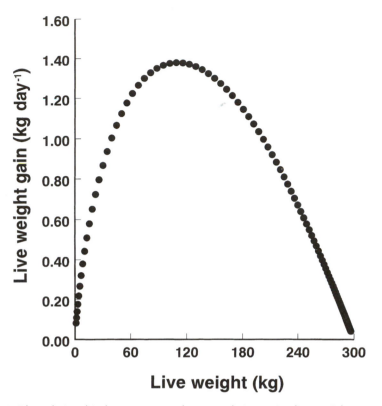

**Fig. 8.2.** The relationship between growth rate and size, using live weight as a measure of size. The numbers apply to a good 1998 boar (maximum growth rate: 1400 g day$^{-1}$).

between $dS/dt$ and $S$ in non-limiting conditions? We need to be able to have a mathematical function which describes this relationship. The assumption is made that the same form of the function will apply to all pigs, but that different kinds of pigs will have different values for the parameters of the function. If different kinds of pig differ in the form of their growth function then the problem of describing potential growth becomes unmanageable. Each pig becomes a 'universe of one', not only in the quantitative value of its growth, but also in the mathematical form needed to describe it.

There have been many ideas about what the form of the function should be. These views have been based sometimes on observations and sometimes on theories of growth. One way to approach the problem is to consider the relative, or specific, growth rate, $R$. The value of $R$ is given by $R = (dS/dt)/S$, i.e. it is the rate at which the pig grows *relative* to its current size. It is measured in units such as kg kg$^{-1}$ day$^{-1}$. A reasonable assumption is that $R$ will decrease monotonically as $S$ increases, towards a value of zero as $S$

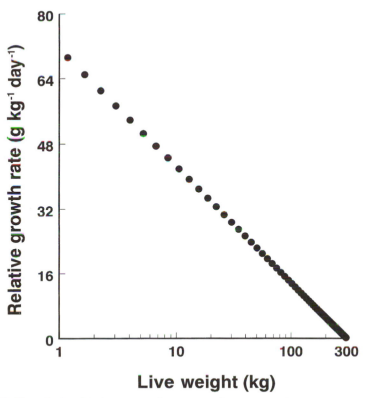

**Fig. 8.3.** The relationship between relative growth rate, *R*, and live weight on a logarithmic scale. Live weight is used as a measure of size.

approaches $S_m$. A monotonic decrease means that the value of *R* will never increase as *S* increases. Many forms could be proposed for this relationship. A simple one, which has some nice properties and which has been widely used, is that which results from the assumption that *R* will decline linearly as the logarithm of *S* increases. The equation which describes this relationship is:

$$R = a - B.\log_e S \qquad (8.1)$$

The form of the function is shown in Fig. 8.3. If the pig progresses towards a final size, $S_m$, then it follows that $R = 0$ when $S = S_m$. By substituting this pair of values in the equation, we find that:

$$a = 0 + B.\log_e S_m \qquad (8.2)$$

Substituting for the value of '*a*' in the original equation leads to:

$$R = B(\log_e S_m - \log_e S) \qquad (8.3)$$

which can be written in the alternative form of:

$$R = B.\log_e(S_m/S) \tag{8.4}$$

From the definition of $R$ as $(dS/dt)/S$ this can be rearranged to give the absolute growth rate, $dS/dt$, as:

$$dS/dt = B.S.\log_e(S_m/S) \tag{8.5}$$

The potential absolute growth rate thus depends only on the current size, $S$, which is a state variable, and on the values of two parameters, $S_m$, the final or mature size, and a rate parameter, $B$. The values of the two parameters can be seen as describing the kind of pig for a particular measure of size. In chemical growth the measures of size are the weights of the chemical components currently and at maturity and there will be as many measures of size as there are independent components.

It is useful to consider the idea of a degree of maturity which is defined as $u = S/S_m$ and its current size as a proportion of mature size. Writing the growth rate equation in this form gives:

$$du/dt = B.u.\log_e(1/u) \tag{8.6}$$

where $du/dt$ is the absolute rate of maturing. It is the $[u.\log_e(1/u)]$ term which has the shape shown in Fig. 8.2. The value of $B$ sets the height of the curve as shown in Fig. 8.4.

The position taken here is that the major chemical components of the pig's body grow according to the form of function described above where conditions are non-limiting. The growth equation, which has the form shown in Fig. 8.1, is the Gompertz (1825) growth equation and is:

$$u = \exp[-\exp(G_0 - B.t] \tag{8.7}$$

where $G_0$ is a transformed initial degree of maturity at $t = 0$. It is calculated as $G_0 = \log_e(-\log_e u_0)$ where $u_0$ is $S_0/S_m$ and size is $S_0$ when $t = 0$. For describing the potential protein growth of the pig the value of the function relative to others is discussed by Emmans and Kyriazakis (1997). Growth functions are often commended for having the quality of 'flexibility'. By adding more parameters functions can always be made more flexible but we cannot see that this is an advantage. For *potential* growth it seems to us that simple rules and functions are likely to apply. If complex flexible functions are needed to describe data sets it is likely that they are not describing *potential* but *actual* growth which is failing to be the potential.

## Chemical Composition

In chemical terms the size of the pig is measured as the weights of the chemical components that are of interest. The classical chemical description of the body of animals is a useful one. The live weight is the sum of the fill (of the gut and the bladder) and the empty body weight, $EBW$. The empty body weight is the sum of the weights of the lipid, $L$, and the lipid-free empty body, $LFEB$. The lipid-free empty body is the sum of the weights of water, $WA$, and

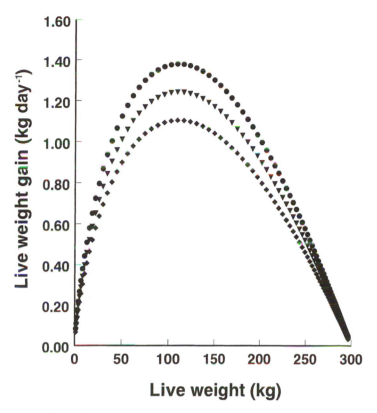

**Fig. 8.4.** The relationship between growth rate and size, using live weight as a measure of size, for three pig genotypes. All have a mature live weight of 300 kg; the *B* values are 0.0100 (♦), 0.01125 (▼) and 0.0125 (●) per day.

the lipid-free dry matter, *D*. The composition of the lipid-free dry matter appears to differ little between kinds of animal and degrees of maturity. It has about 0.78 protein, 0.20 ash with the remainder as carbohydrate (Emmans, 1988). The carbohydrate content will be much affected by the time between last feeding and slaughter.

Looked at in this way the chemical body of the pig is made up of *D* (which is about 1.23 times the protein weight, *P*, and can be assumed to be of constant composition in potential growth), *WA* and *L*. It is generally the case that the proportion of the empty body as water decreases, and that of lipid increases, as the pig grows in non-limiting conditions. The way in which the protein content of the body changes depends on the extent of the increase in fatness. Whittemore *et al.* (1988) assumed, wrongly, that protein weight was a simple power function of live weight in order to generate the rate of protein retention from the rate of total growth. As seen in Chapter 2 of this volume, to approach protein growth in this way can be misleading.

As the protein content of the lipid-free dry matter is constant the growth of this component can be predicted as a multiple of the growth of protein. If the weights of water and lipid could also be predicted from protein growth in some simple way in non-limiting conditions, then the potential growth of the entire chemical body could also be predicted in a simple way and few numbers would be needed to describe a pig sufficiently in order to predict its potential chemical growth.

## Growth of the Chemical Components

Under conditions that are non-limiting the actual growth will be the potential growth. It is possible that the potential growth curves of the major chemical components (protein, from which the lipid-free dry matter can be calculated, water and lipid) are of the same shape as each other, i.e. they follow the S-shaped curve described above. In this case we need to know the values of the parameters for mature size and the rate of maturing for each of the components, in order to predict the potential growth of a given kind of pig. With three components, and two parameters for each, the values of six parameters then need to be estimated in total to give a sufficient description of the potential chemical growth of a given kind of pig.

However, the actual position appears to be simpler than this for two reasons: (i) the values of the rate parameters, $B$, for each of the three components can be assumed to be the same for all of the three components (Emmans, 1988); and (ii) the relationship between water and protein, when properly expressed (Emmans and Kyriazakis, 1995) can be assumed to be the same for all pigs. This means that only three parameters need to be estimated to give a sufficient description of a given kind of pig. The three parameters are: (i) its mature size, expressed as protein, $P_m$; (ii) its mature fatness, expressed as the ratio of mature lipid weight, $L_m$, to mature protein weight which is $L_m/P_m = LPR_m$; and (iii) the general rate parameter, $B$. Pigs will differ substantially in their values of the three parameters and the values would be expected to change under selection. These values are combined with three constants which are the water : protein ratio at maturity, the allometric exponent relating water to protein weights and the ash : protein ratio (Emmans and Kyriazakis, 1997).

Ferguson and Gous (1993a,b) have shown how the values of the parameters which describe the potential chemical growth of the pig can be estimated. The method follows from Fig. 8.3 which describes the expected relationship between the relative growth rate and size. On a logarithmic scale of size the relationship is expected to be linear. This means that estimates of the relative growth rate at only two sizes are needed to estimate the values of the two parameters. The further away the sizes are the better will be the estimates. Of course more data may be useful and can be used to test if the relationship is linear.

The assumption is that, at maturity, all pigs have the same composition of their lipid-free body irrespective of their mature size or fatness. The consequences of this assumption have been described by Emmans and Kyriazakis (1997).

## Allometric Relationships

A widely used method for relating the weights of chemical components to each other is to make the weight of one a simple power function of the weight of the other (Needham, 1934). The form of the equation is:

$$L = a.P^b \tag{8.8}$$

using protein and lipid weights as an example. If this relationship holds then $L$ can be predicted from $P$ providing the values of the two parameters '$a$' and '$b$' are known. We expect $P$ to have a final value, $P_m$. It follows from the power relationship that $L$ must also have a final value to be called $L_m$. Putting these vales into the equation gives:

$$L_m = a.P_m^b \tag{8.9}$$

so that:

$$a = L_m/P_m^b \tag{8.10}$$

Substituting back into the original equation gives:

$$L = L_m(P/P_m)^b \tag{8.11}$$

It was pointed out earlier that a degree of maturity is the weight of a component as a proportion of its mature, or final, value. The degrees of maturity of protein, $u_p$, and of lipid, $u_l$, are defined as $P/P_m$ and $L/L_m$ respectively. The equation for $L$ above can then be written as:

$$u_l = u_p^b \tag{8.12}$$

This equation states that the degree of maturity in lipid is a simple power function of the degree of maturity in protein with the single parameter '$b$'. In equation 8.11 we had:

$$L = L_m(P/P_m)^b \tag{8.13}$$

Differentiating gives the rate at which lipid changes relative to protein:

$$dL/dP = (L_m/P_m^b)P^{b-1} \tag{8.14}$$

Since, by definition, $dL/dt = (dL/dP).(dP/dt)$, the rate of lipid growth can be found from the value of the rate of protein growth, $dP/dt$, and $dL/dP$. Similar arguments apply for finding the rate of growth of water.

## Component Growth Curves

If the potential growth curves of two components are of the Gompertz form with the same value of the rate parameter, $B$, then it can be shown that the weights of the two components will be related to each other by a simple power function. The relationship, earlier *assumed* in equation 8.9, that the degree of maturity in one component is a simple power function of the degree of maturity of the other, is *proved* here. For two components, say protein and lipid, we have the Gompertz form of growth equation:

$$u_1 = \exp\{-\exp[-B(t-t_1^*)]\} \tag{8.14}$$
$$u_p = \exp\{-\exp[-B(t-t_p^*)]\} \tag{8.15}$$

using the assumption that the values of the rate parameters are the same for each component. Taking logarithms:

$$\log_e u_1 = -\exp[-B(t-t_1^*)] \tag{8.16}$$
$$\log_e u_p = -\exp[-B(t-t_p^*)] \tag{8.17}$$

Dividing equation 8.16 by 8.17, rearranging and simplifying the exponents, leads to:

$$\log_e u_1 = b.\log_e u_p \tag{8.18}$$

where $b = \exp[B(t_1^* - t_p^*)]$. This equation is equivalent to:

$$u_1 = u_p^b \tag{8.19}$$

as can be seen by taking the logarithms at each side of equation 8.19.

It thus does not need to be assumed *a priori* that one component is related to another by a simple power function. The relationship can be deduced from the assumed form of the growth curve, and the other necessary assumption that both components have the same value of the rate parameter. This deduction is a strong argument in favour of using the Gompertz form of the growth curve for potential growth. It does not appear to be the case that any other form of growth function leads to the consequence of allometry. The widespread observation of allometric relationships between the weights of components can thus be seen as an argument in favour of using the Gompertz form of growth equation.

## An Example of Pig Growth

Doornenbal (1971, 1972a, b) carried out a serial slaughter experiment on Lacombe castrated pigs and produced the data shown in Table 8.1. These data are probably the largest set from a serial slaughter experiment on pigs. The number of pigs at each slaughter point was still not large and there will therefore be some noise due to true variation between pigs as well errors of measurement. But the data can be used to illustrate the points made above about the potential growth of the chemical components. The assumed

**Table 8.1.** The weights (kg), and transformed degrees of maturity $(-\log_e(-\log_e(u)))$, of the chemical components of castrated Lacombe pigs. (From Doornenbal, 1971, 1972a,b.)

| Age (days) | Protein | | Lipid | | Remainder | |
|---|---|---|---|---|---|---|
| | Weight | $G_p$ | Weight | $G_l$ | Weight | $G_z$ |
| 44.5 | 1.52 | −1.081 | 1.28 | −1.558 | 6.52 | −0.985 |
| 78.8 | 3.02 | −0.816 | 3.48 | −1.322 | 12.76 | −0.696 |
| 93.8 | 4.62 | −0.608 | 6.15 | −1.157 | 18.57 | −0.489 |
| 116.8 | 5.83 | −0.473 | 9.90 | −0.995 | 23.01 | −0.348 |
| 121.8 | 6.84 | −0.368 | 13.50 | −0.873 | 25.96 | −0.259 |
| 136.0 | 8.44 | −0.211 | 18.20 | −0.740 | 31.64 | −0.093 |
| 141.8 | 8.78 | −0.178 | 20.46 | −0.682 | 33.11 | −0.051 |
| 150.5 | 9.54 | −0.106 | 24.46 | −0.588 | 35.37 | 0.014 |
| 166.2 | 11.03 | 0.034 | 28.42 | −0.501 | 39.84 | 0.142 |
| 177.8 | 11.53 | 0.081 | 34.50 | −0.376 | 42.13 | 0.209 |
| 199.5 | 14.64 | 0.380 | 45.40 | −0.167 | 51.65 | 0.498 |
| 211.0 | 15.30 | 0.447 | 52.13 | −0.043 | 52.82 | 0.536 |

mature weights of protein, $P_m$, lipid, $L_m$, and the remainder, $Z_m$, were assumed to be 29.0, 147.9 and 94.83 kg respectively. The remainder is the sum of the ash, water and carbohydrate. These values were arrived at by analysis of the data to produce equations with the same value of the rate parameter for each component. The actual weights in Table 8.1 can then be expressed as degrees of maturity as shown in the table. The potential growth equation:

$$u_t = \exp\{-\exp[-B(t-t^*)]\} \tag{8.20}$$

can be transformed by taking the logarithms of both sides, changing the signs, taking the logarithms again and changing the signs again, to:

$$-\log_e[-\log_e(u_t)] = G_t = B\,(t-t^*) \tag{8.21}$$

When $t = 0$, then:

$$\log_e[-\log_e(u_o)] = B.t^* = G_o \tag{8.22}$$

The transformed degree of maturity, $G_t$, is expected to be a linear function of time, $t$. The Gompertz variables (i.e. this transformation of the weights) for the three chemical components, given in Table 8.1, are plotted against age, $t$ days, in Fig. 8.5. It can be seen that the data are in good agreement with the expectation in that they lie on straight lines of similar slope. The mean value estimated for $B$ is 0.00928 day$^{-1}$.

Between animals, both between species and between animals within a species, it would be expected (Taylor, 1980a, b) that $B$ would fall as $P_m$ increased. Using Taylor's proposed value for the exponent relating rate parameters to mature size of 0.27 the scaled rate parameter, $B$, is estimated as $B^* = B.P_m^{0.27}$. For the pigs of Doornenbal $B^*$ is estimated as $0.00928 \times 29.0^{0.27}$

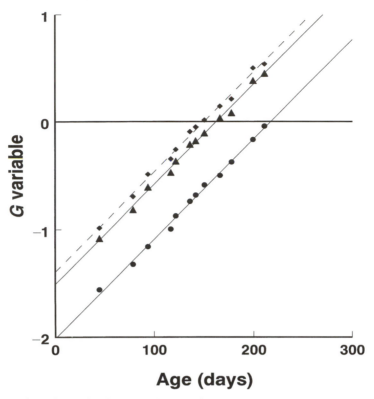

**Fig. 8.5.** The relationship between the transformed weights of the chemical components (protein, ▲, lipid, ● and remainder, ♦) and age for castrated Lacombe pigs of Doornenbal (1971, 1972a,b). See text and Table 8.2 for details.

= 0.0230. This is close to the value of 0.0233 estimated by Emmans (1997) using prenatal data for sheep and cattle indicating that these pigs were in line with interspecies expectations. While it cannot be demonstrated that the pigs of Doornenbal grew at their potential, and thus that their data are suitable for testing the form of the growth function used here, these data do indicate that the form is a reasonable one. The values of the growth parameters, $P_m$, $LPR_m$ and $B^*$ (or $B$) for pigs, and the possible effects of selection, are discussed below.

## Parameter Estimation and the Consequences of Genetic Selection

On the above arguments the sufficient set of parameters which describes the *potential* chemical growth of a pig are: mature protein weight, $P_m$, the mature fatness, $LPR_m$, and the rate parameter, $B$. To solve the equations to predict the growth rates of the components, presented above, the initial conditions, at birth for example, also need to be known. For Doornenbal's pigs the values of the parameters were 29.0 kg, 5.10 kg kg$^{-1}$ and 0.00928 day$^{-1}$, respectively. Ferguson and Gous (1993b) estimated the values for the boars that they were using to be very different; they were 38.7 kg, 2.60 kg kg$^{-1}$ and 0.0107 day$^{-1}$. Under genetic selection the mean values of the parameters in the population will change as selection proceeds. Ferguson *et al.* (1997) have discussed the levels of genetic variation that might be present in the values of the parameters. The way in which the values will change with time under selection will depend on this variation, the selection intensity and the details of the trait, or index, selected for. In the past decades there has been selection against fatness (usually at a live weight) and for growth rate and, sometimes, efficiency in commercial breeding programmes.

Selection for weight at an age will normally increase both $P_m$ and $B$ as animals with higher values of either of these parameters will grow faster. It will also be likely to increase the unscaled rate parameter $B$, but the extent to which this occurs will depend on the age at selection. Selection for weight at a late age will act mainly on mature size as this is the main determinant of weight at late ages. The earlier the age at selection the greater will be the emphasis on the rate parameter.

Selection against fatness at a weight will decrease $LPR_m$, because animals with higher values of this parameter will be fatter at all weights. It will also tend to increase $P_m$ as at a weight animals with larger mature weights will be less mature and hence leaner, other things being equal. Selection for efficiency, at least on *ad libitum* feeding, will increase growth rate because faster growing animals will be using less food for maintenance. It will also tend to decrease fatness because animals with leaner gain will be more efficient on a gain : feed basis. A consequence is that there will be an increase in both $P_m$ and $B$ and a decrease in $LPR_m$. Taking all of these effects of selection into account it would therefore be expected that, over time, as commercial selection proceeds, whatever the detailed index used, population mean values for $P_m$ and $B$ will increase, and that of $LPR_m$ will decrease. The rates of change will depend on the intensity of selection.

The only way to be certain of the values of the growth parameters for a given population is to measure them in a sample of the population using the kind of protocol described by Ferguson and Gous (1993a). This would seem to be a task for the breeders of pigs who would then communicate the values to their customers for use in performance prediction and modelling (Ferguson *et al.*, 1994). In the meantime possible starting values are given in Table 8.2.

**Table 8.2.** Possible values for the growth parameters for different kinds of pig. The parameters are: $P_m$, the mature protein weight; $LPR_m$, the ratio of lipid to protein at maturity; and B, the Gompertz rate of maturing parameter.[a]

| Kind of pig | Sex | $P_m$ (kg) | $LPR_m$ | B (day$^{-1}$) |
|---|---|---|---|---|
| Best of 1998 | Boar | 50.0 | 2.0 | 0.0140 |
| | Castrate | 45.0 | 3.6 | 0.0112 |
| | Gilt | 38.0 | 3.0 | 0.0140 |
| Moderate of 1998 | Boar | 45.0 | 2.8 | 0.0125 |
| | Castrate | 40.5 | 4.6 | 0.0100 |
| | Gilt | 34.0 | 4.2 | 0.0125 |
| Poor of 1998 | Boar | 40.0 | 3.6 | 0.0100 |
| | Castrate | 36.0 | 5.5 | 0.0088 |
| | Gilt | 30.0 | 4.5 | 0.0100 |

[a] In the future the values of $P_m$ and B will be likely to increase while that of $LPR_m$ will be likely to fall (see text for explanation).

# Growth in Limiting Conditions

*Actual* growth will be the *potential* only if the environment allows it. Among the necessary conditions are: (i) feed must be available *ad libitum*; (ii) the ratios of all nutrients to energy need to exceed the requirement; (iii) the bulk of the feed must not constrain intake; (iv) the environment must not be too hot. The possible effects of these conditions on feed intake, and hence growth, are discussed in Chapter 10.

The prediction of actual growth, where the conditions do not allow the potential to be attained, is an interesting and challenging task. Whittemore and Fawcett (1974) and Whittemore (1976) set out how this could be done on controlled feeding under certain assumptions. It is clear there can be no general way of describing *actual* growth in terms of mathematical functions. It will depend on the *actual* conditions, how these change with time and how they interact with the particular kind of animal being used. It is not sensible to try to find functional forms which describe *actual* growth across a range of conditions.

Where feeding is controlled the problem is to predict performance. On *ad libitum* feeding a further problem is to predict feed intake. In both cases the first problem to be solved is that of predicting the growth rate of protein (Kyriazakis, 1996). On *ad libitum* feeding the growth of lipid will follow from the predicted energy intake, an energy system and a description of the climatic environment. It is not the task of this chapter to describe such methods of prediction in detail. There is some discussion in Chapter 2.

# Discussion

Once the important distinction has been made between *potential* and *actual* growth it is possible to try to see how *potential* growth may be described. The approach taken here has been to focus attention on to the growth of the major chemical components. As (i) there are only four of them – ash, protein, water and lipid – and (ii) the weights of two of them – ash and water – can be estimated from those of protein in potential growth, the approach is simple. It is very much easier than trying to deal with the anatomy of growth, i.e. of trying to describe the growth of tissues and organs. So far there have been no successful attempts to do this. A further advantage of using the chemical description of growth is that it connects the growth of the animal with its requirements for the energy and nutrients coming from its food. It is an approach that has been widely used following Whittemore and Fawcett (1974) and has been extended by Emmans and Fisher (1986) and Emmans (1989).

The idea of a potential growth rate of protein has come to be widely accepted but there has been much less consideration of a potential growth rate of lipid. The reason for emphasizing that both components have potential growth curves is to allow the energy requirement, and hence feed intake in non-limiting conditions, to be calculated (Emmans, 1997). The potential growth curve of lipid does differ from that of protein in one important way. While the potential growth rate of protein sets an upper limit to protein gain the rate of gain of lipid can actually exceed that which is set by its potential. This, at first sight bizarre, result is because the pig is prepared to gain lipid at a faster rate than is set by its potential *if this helps it to get closer to its potential rate of protein growth*. On feeds with nutrient : energy ratios below the requirement, the pig will attempt to overeat energy and get fatter than it would do on balanced feeds. These effects are discussed in Chapter 10. The distinction is between the rate of lipid that the pig seeks – *the desired lipid growth rate* – which is described as its potential above, and the rate at which it can gain lipid if this helps with some other objective. Again the prediction of actual lipid growth will involve a wider set of rules than can be discussed here.

The equations presented here to describe *potential* growth are believed to be of the correct form. They are derived from only two assumptions which are: (i) for any one chemical component the relative growth rate declines linearly with the logarithm of the weight of that component; and (ii) that the rate of decline, the rate of maturing parameter $B$, is the same for all four components. There will of course be data sets on the actual growth of pigs that are not well described by these functions. This is to be expected whenever *actual* growth is not the *potential*. Such divergencies should act as a spur to finding the reason for them, rather than as a reason for rejecting the forms of function that have such utility.

The values in Table 8.2 indicate the mean values of the parameters that may be expected in some populations of pig. An interesting question is the degree of variation, and covariation, in these values among the individuals that comprise the population. The issue was raised by Emmans and Fisher (1986) and some work relevant to pigs has been published (Ferguson *et al.*, 1997). Variation and covariation are both important in describing populations. One fruitful consequence of the relatively simple description of chemical growth proposed here is that this problem remains a tractable one. With much more complex descriptions of growth the problems of variation and covariation rapidly become unsolvable.

# References

Black, J.L. (1995) Modelling energy metabolism in the pig – critical evaluation of a simple reference model. In: Moughan, P.J., Verstegen, M.W.A. and Visser-Reyneveld, M.I. (eds), *Modelling Growth in the Pig*. Wageningen Pers, Wageningen, The Netherlands, pp. 87–102.

Doornenbal, H. (1971) Growth, development and chemical composition of the pig. I. Lean tissue and protein. *Growth* 35, 281–295.

Doornenbal, H. (1972a) Growth, development and chemical composition of the pig. II. Fatty tissue and chemical fat. *Growth* 36, 185–194.

Doornenbal, H. (1972b) Growth, development and chemical composition of the pig. III. Bone, ash and moisture. *Growth* 39, 427–434.

Emmans, G.C. (1988) Genetic components of potential and actual growth. In: Land, R.B., Bulfield, G. and Hill, W.G. (ed), *Animal Breeding Opportunities*. British Society of Animal Production Occasional Publication 12, pp. 153–181.

Emmans, G.C. (1989) The growth of turkeys. In: Nixey, C. and Grey, T.C. (eds), *Recent Advances in Turkey Science*. Poultry Science Symposium No. 21. Butherworths, London, pp. 135–166.

Emmans, G.C. (1995) Energy systems and the prediction of energy and feed intakes. In: Moughan, P.J., Verstegen, M.W.A. and Visser-Reyneveld, M.I. (eds), *Modelling Growth in the Pig*. Wageningen Pers, Wageningen, The Netherlands, pp. 115–122.

Emmans, G.C. (1997) A method to predict the food intake of domestic animals from birth to maturity as a function of time. *Journal of Theoretical Biology* 186, 189–199.

Emmans, G.C. and Fisher, C. (1986) Problems in nutritional theory. In: Fisher, C. and Boorman, K.N. (eds), *Nutrient Requirements of Poultry and Nutritional Research*. Butterworths, London, pp. 9–39.

Emmans, G.C. and Kyriazakis, I. (1995) A general method for predicting the weight of water in the empty bodies of pigs. *Animal Science* 61, 103–108.

Emmans, G.C. and Kyriazakis, I. (1997) Models of pig growth: problems and proposed solutions. *Livestock Production Science* 51, 119–129.

Ferguson, N.S. and Gous, R.M. (1993a) Evaluation of pig genotypes. I. Theoretical aspects of measuring genetic parameters. *Animal Science* 56, 233–243.

Ferguson, N.S. and Gous, R.M. (1993b) Evaluation of pig genotypes. II. Testing experimental procedure. *Animal Science* 56, 233–249.

Ferguson, N.S., Gous, R.M. and Emmans, G.C. (1994) Preferred components for the construction of a new simulation model of growth, feed intake and nutrient requirements of growing pigs. *South African Journal of Animal Science* 24, 10–17.

Ferguson, N.S., Gous, R.H. and Emmans, G.C. (1997) Modelling the growth and nutrition of pigs. *Animal Science* 64, 513–522.

France, J., Dijkstra, J., Thornley, J.H.M. and Dhanoa, M.S. (1996) A simple but flexible growth-function. *Growth, Development and Ageing* 60, 71–83.

Gompertz, B. (1825) On the nature of the function expressive of the law of human mortality and on a new method of determining the value of life contingencies. *Philosophical Transactions of the Royal Society* 513–585.

Kyriazakis, I. (1996) A solution to the problem of predicting the response of an animal to its diet. *Proceedings of the Nutrition Society* 55, 155–166.

Needham, J. (1934) Chemical heterogony and the ground plan of animal growth. *Biological Reviews* 9, 70–108.

Taylor, St. C.S. (1980a) Genetic size – scaling rules in animal growth. *Animal Production* 30, 161–165.

Taylor, St. C.S. (1980b) Genetically standardised growth equations. *Animal Production* 30, 167–175.

Whittemore, C.T. (1976) A study of growth responses to nutrient inputs by modelling. *Proceedings of the Nutrition Society* 35, 383–391.

Whittemore, C.T. and Fawcett, R.H. (1974) Model responses of the growing pig to the dietary intake of energy and protein. *Animal Production* 19, 221–231.

Whittemore, C.T., Tullis, J.B. and Emmans, G.C. (1988) Protein growth in pigs. *Animal Production* 45, 437–445.

# Quantitative Regulation by Endocrine Systems

## 9

## J.P. McNamara[1] and R.D. Boyd[2]

*[1]Department of Animal Sciences, 233 Clark Hall, Washington State University, Pullman, WA 99164–6320, USA; [2]Pig Improvement Company, PO Box 348, Franklin, KY 42136, USA*

## Introduction

The study of hormonal control of metabolism has yielded a wealth of knowledge on functions and mechanisms of hormones. This has been used in applications for improvement of animal growth, milk production, health and reproductive efficiency. Examples include: the role of gonadal steroids in muscle and adipose growth; the role of pituitary hormones in lactation and growth; measurement of circulating steroids to predict ovulation or pregnancy; and the use of prostaglandins and derivatives to control the timing of ovulation and parturition. Another important aspect includes instances in which fundamental knowledge has prevented the use of various agents in agricultural production, either for lack of efficacy, safety, or both. The community of animal production, and in particular of pig production, can point to the practical application of biological principles with a measure of pride, because of unprecedented discovery and application of knowledge to improve food production and safety.

What research and development objectives now remain? Research efforts on the physiological mechanisms of hormones which affect metabolism and reproduction will continue and will lead to further important practical applications. However, we also require improvement in *quantitative* knowledge: (i) of the effects of hormones, singly and in combination and (ii) on kinetic rates of metabolic and reproductive processes. Knowledge of both physiological mechanisms and kinetic effects is important. If we possessed a quantitative description of hormonal regulation across several levels of biological organization, we would have an even more efficient production of animal products.

In this context, *quantitative* is defined as possessing an understanding of a process in sufficient detail that it can be described in a discrete and explicit mathematical formulation. Thus, this definition may apply more closely to the effects of hormones on the kinetics and interactions of biochemical pathways than to the chemical mechanism by which a hormone acts. However, we make the strong point that knowledge of both mechanisms and kinetics is required eventually for full understanding. A quantitative description of hormonal action requires the identification and estimation of parameters which describe the effect of a hormone on a process with a certain accuracy, precision, and estimate of variance. The concentration of a hormone, receptor binding and dissociation constants, change in concentrations of intracellular signalling molecules, and the net biological effect must all be determined. Again, in the context of animal models, the net effect includes the effects at the cellular, organ and whole animal levels, even though the hormone may act only on one receptor in one cell type.

Also implicit in this definition is the inclusion of dynamic effects of time. There are at least three such pertinent effects of time: the adaptations of a system (over time) to the effect of a hormone; the cumulative effects of hormones (over time) on a process or integrated processes; and the differential effects of hormones at different 'times' of the life cycle or physiological states. Thus, single time, static equations relating input to output are useful for some practical applications but are extremely limited in scope and cannot be extrapolated to the dynamic situation. Such static descriptions are inadequate for a quantitative description of hormonal effects.

As a general example of what we may need to know, consider the effect of a hormone $[H]$ on the rate of flux of a reaction from substrate $[S]$ to intermediate $[I]$, catalysed by an enzyme $[E]$, which can exist in an inactive $[E]$ or activated $[EA]$ state. Additionally, the activity of this enzyme may partially or totally control the rate of flux $(F)$ of substrates through a metabolic pathway consisting of two or more enzyme-catalysed reactions resulting in a final product $[P]$. The activity of $E$ may also be under the control of another hormone $[O]$. The flux $F$ may be further attenuated or enhanced by other hormones (controllers, $[C]$) acting at other points (enzymes) in the pathway. Lipogenesis, gluconeogenesis and muscle or milk protein synthesis are examples which follow this form. It is not required that the exact molecular mechanisms are known, but we must know the number and activity of the enzymes in a pathway, and concentrations of starting substrate(s), intermediates and the final product(s).

Thus, if we measure $[H]$, $[E]$, $[EA]$, $[S]$, $[I]$, $[P]$, $[O]$, $[C]$, and $[F]$ we can construct a mathematical equation which relates $[H]$ to the concentration of active and inactive enzyme, the concentrations and rate of change of concentrations of $[S]$ and $[P]$, the rate of total pathway flux, and the net effects of other factors $[O]$ and $[C]$. This is an example of a relatively *simple* biochemical system. Yet, it is usually the ability to measure these variables which limits improvement, not the mathematics involved. When one needs to estimate and validate parameter values for complex relationships, especially those involved in variable rates of change and non-steady-state conditions, the mathematics may be more daunting (Cornish-Bowden and Cardenas, 1990; Crabtree, 1993), however, it is still an accurate biological description which is the first need.

Another aspect of quantitative may imply a certain amount of 'predictability'. When all sources of variation are precisely defined, predictability is a reality. Humans standing on the moon demonstrate this statement to be true. In biology we are far from that level of definition. However, there are situations in which empirical effects of a hormone are known and a commercial product is already being used to directly alter rates of production in animals without either a mechanistic description of the effect or a quantitative description of the processes affected. In some cases such full definition may not be required, however further hormonal manipulations of animals will require ever more precise descriptions of the full scope of effects. The scientist must clearly distinguish between empirical or kinetic relationships (relating inputs and outputs at the same level of organization) and mechanistic ones (using characteristics or lower levels of organization to explain behaviour of higher levels). In our ideal situation, critical elements of both mechanism and kinetics are defined.

Our approach in this chapter is to concentrate on the biology involved in hormonal regulation of growth, pregnancy and lactation. As the community of 'swine science', we include the basic scientist through the producer, with a common goal of using basic information to practical advantage. We will compile a partial list of important areas in pig production in which we should benefit from a more quantitative description of hormonal control. This chapter will deal primarily with biology and not mathematics; however it is explicitly urged that biologists and mathematicians work ever more closely to meet our common needs. There are already several excellent descriptions of mathematical approaches and quantitative systems in biological control, from a molecular level to an ecological level and the reader in need of specific information is directed to them (Finkelstein and Carson, 1979; Carson *et al.*, 1981; France and Thornley, 1983; Crabtree, 1993; Hopkins and Leipold, 1996).

## Endocrine Control of Metabolic Processes

Critical hormonally controlled processes in pigs include muscle protein synthesis and degradation, fatty acid and triglyceride synthesis and hydrolysis in adipose tissue, and milk lactose, protein and fat synthesis. Figure 9.1 is a commonly used diagram of nutrient flux among major compartments (Boyd

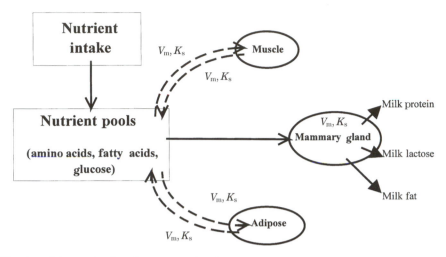

**Fig. 9.1.** Schematic identifying critical pathway control parameters which are changed by hormones during lactation. A schematic such as this has been used by several authors (Bauman and Vernon, 1993). This adds specific parameters needed for a quantitative analysis. See text for details and for a more sophisticated analysis of integrated control see Baldwin (1995).

and Bauman, 1989) and adds to it some basic kinetic parameters which describe rates of flux. Elimination of the mammary gland aspect represents control during growth. Also represented are kinetic parameters describing flux of nutrients among compartments, such as $K_s$ or sensitivity of a reaction or pathway to a substrate, and $V_{max}$, or maximal velocity of the pathway. The value of these parameters may vary with genotype, nutritional manipulation and other environmental influences and should not be viewed as having a constant value. The effect of hormones, whether of endogenous origin or exogenously administered, on the kinetics of flux of nutrients, acts by altering the values of these parameters. For example, the $K_s$ and $V_{max}$ of amino acids to mammary gland protein synthesis will be different among different genotypes, stages of lactation, nutritional states or exogenous hormone treatments.

Equation forms describing typical metabolic interactions and hormonal control are given in Table 9.1. Using the general flow among tissues, and basic equations, we can point out some specific quantitative aspects of control: (i) the body is not a closed system, and normally, nutrient pool size is eventually maintained, either by a draw on muscle and/or adipose pools, or an increase in nutrient intake or both (this adaptation may take place over a period of time from minutes to weeks); (ii) responses to new physiological states (increased or decreased muscle or adipose growth or mammary secretion) usually also include a change in intake, over a varying interval of time; (iii)

**Table 9.1.** Quantitative kinetic equations in metabolism and hormonal control.

| Equation form | Description, control by hormones |
| --- | --- |
| **1.** $v = V_{max} / (1 + (K_s/[S]))$ | Substrate sensitivity and maximal velocity curve (Michaelis–Menten kinetics). Applies to most synthetic reactions. A hormone may act by increasing or decreasing the value of $K_s$ or $V_{max}$ such as: |

$$v = ([H] \times V_{max}) / (1 + (K_s/[S])) \text{ or } v = (V_{max}) / [H]) / (1 + (K_s/[S]));$$
$$v = V_{max} / (1 + (K_{sH}/[S])); \text{ where } K_s = (K_s \times [H]) \text{ or } (K_s/[H]).$$

| | |
| --- | --- |
| | Where one $[H]$ may have opposite effects on the $V_{max}$ or $K_s$ of the reaction in different organs, this is a quantitative description of homeorhetic control. |
| **2.** $v = V_{max} / (1 + (K_s/[S]) + ([I]/K_i))$ | Same kinetic description as above, but with inhibition by factor $I$. Hormones may act on $K_s$ and $V_{max}$ or may also alter $K_I$, such that $K_{IH} = (K_I \times [H])$ or $(K_I / [H])$. |
| **3.** $v = V_{max} / (1 + (K_s/[S])^\theta$ | Maximal velocity and sensitivity curve, but sensitivity of the reaction to substrate is sigmoidal, described by a $\theta$. Hormones may alter $K_s$, $V_{max}$ or ($\theta$), such that in a very narrow substrate range a large effect on the reaction may result. The relationships in Fig. 9.2 are an example. |
| **4.** $H f ([S_i]/[S_o])$ | Representation that a hormone effect is a variable function of the concentration of substrate or metabolite ($[S_i]$ against a reference concentration $[S_o]$). When a substrate affects a hormone which then alters different reactions, this coordinates the hormonal effects among pathways or organs. As the value $f$ changes, this is a quantitative description of homeorhetic control. |
| **5.** $\% B = M/(1 + (K_d/[H]^\theta))$ | Hormonal binding curve: $\%B$ is per cent of hormone bound to receptor, $M$ is maximal binding, $K_d$ is sensitivity or association constant, $H$ is hormone concentration, and $\theta$ represents the sigmoidal, or exponential function. The relationship of receptor bound to biological effect can be linked from this equation to the ones above. |
| **6.** $[R] \times [H] / [RH] = K_d$ | This form indicates per cent of receptor bound to hormone as well, where $R$ is receptor, $RH$ is bound receptor and $K_d$ is the dissociation constant. With these six basic equations the loop of hormone with substrate can be completed, and similar or different reactions in different organs connected. Determination of parameter values must be done for different stages of growth and physiological state. |

coordination of control comes from differential reduction or increase in $K_s$ and $V_{max}$ in the various pathways of the different organs, eventually regaining a new pool in balance. For each tissue, these parameters may be under the control of one or more hormones. Additionally, these parameters will be altered by homeorhetic hormones which come into play in different physiological states. All these parameters can be estimated in *in vivo* or *in vitro* experiments and thus provide testable quantitative hypotheses. For a more sophisticated analysis of integrated control see Baldwin (1995) and Hopkins and Leipold (1996).

Protein hormones acting as quantitative controllers include somatotrophin, insulin, glucagon, prolactin, thyroid stimulating hormone, adrenocorticotrophic stimulating hormone, and the insulin-like growth factors (and perhaps their binding proteins). Steroid hormones such as oestrogen, progesterone, testosterone and glucocorticoids, have effects on enzyme activities and rates of metabolic conversions important to both 'productive' and 'reproductive' processes. The sympathetic and adrenal neurotransmitters norepinephrine and epinephrine regulate lipolysis and perhaps proteolysis. Readers interested in the endocrine and physiological mechanisms of these processes are referred to recent reviews (Oddy and Lindsay, 1986; Boyd and Bauman, 1989; Mersmann, 1989; Landsberg, 1990; Bauman and Vernon, 1993; Pettigrew *et al.*, 1993; Bell and Bauman, 1994; Wade *et al.*, 1996).

The road to a better quantitative picture was started (as is usual) with an evolution of the qualitative understanding of hormonal control in complex systems. The decades of the 1950s and 1960s were occupied with definitions of functions of homeostatic systems exemplified by insulin and the catecholamines. This led to a tremendous amount of detail on the scope of effects of these two homeostatic controllers on glucose, lipid and amino acid metabolism in the liver, adipose tissues, muscle and mammary glands. Hormonal binding curves, enzyme activation curves, and dose–responses for a variety of metabolic pathways were generated. However, as experiments were done in various stages of growth, pregnancy and lactation, it became clear that 'homeostasis' itself was not a constant. That is, tissue responsiveness differs in various situations such that homeostasis of blood concentrations of nutrients is still maintained, but the net effects of a hormone on accretion of muscle protein, adipose lipid, fetal tissues, or milk components are markedly different. Although this was hypothesized quite early (Hammond, 1944), the concept of 'variable homeostasis' has come to be known as homeorhetic regulation since the original paper of Bauman and Currie (1980).

What are some specific examples, for research in endocrine regulation, of 'quantitative' effects? For qualitative purposes the questions we ask are of the type: 'Does a hormone have an effect on process X in cell type Y in physiological state Z?' or 'What is the chemical mechanism of response to hormone action (binding, uptake) on X process in Y cell type in physiological state Z?' Building upon this beginning, a quantitative description requires questions such as:

**1.** What is the mathematical relationship between the concentration of a hormone, the binding of hormone to receptor, the concentration of second messenger, the activation of intermediary intracellular regulators, the rate of final metabolic activity of one or more affected pathways, and the net integrated effect on growth or other production?
**2.** What are the mathematical forms and parameters which describe the curve, for an exogenously delivered hormone upon the rate of chemical reactions at subcellular or organ level or integrated to whole animal inputs and outputs?
**3.** What is the mathematical description of the effect of one hormone on the kinetic parameters of a response to a different hormone at the enzyme, pathway, whole tissue and whole body levels?
**4.** What are mathematical equations describing the integrated effects among tissues and over time, of the set of critical regulatory hormones on the key chemical interconversions in the major body organs during growth, pregnancy and lactation?

Formulating questions in this manner forces the investigation and experimental design to be of a different nature from that used to answer qualitative questions. Often these experiments are more complex and expensive but this is not always the case. In addition, if the problems are more complex, meaningful experiments by necessity must also be. This is our job as biologists working on practical problems: define the system in a systematic way using basic principles validated with practical experiments.

Figure 9.2 follows from equation 3 in Table 9.1 and exemplifies the relationship of hormone concentration with intermediate fluxes, rates or concentrations such as signal molecules or metabolic activities, which lead ultimately to a net effect on a rate or state (Loeb and Strickland, 1987). The different curves represent intermediate fluxes (abcissa) after receptor binding to one hormone (ordinate); or varied responses to a hormone in different physiological or metabolic states. By substituting time on the abcissa and letting the ordinate be metabolic state or rate, the various curves may also be viewed as a variable response to hormone in different situations. The arrows signify that the responses can vary along a continuum. The major concept is that although such relationships can be described in some mathematical form, there is no *a priori* reason to assume one set of parameter values exists which describes a process in all physiological situations. Rather we need to know the form of the equation, and then estimate the population of parameter values in different situations.

Below is a list of processes in the pig for which we need quantitative descriptions, including estimates of those parameters given in Table 9.1. All can be measured in the pig, and for some, measurements in other animal models may be carefully extrapolated to the pig.

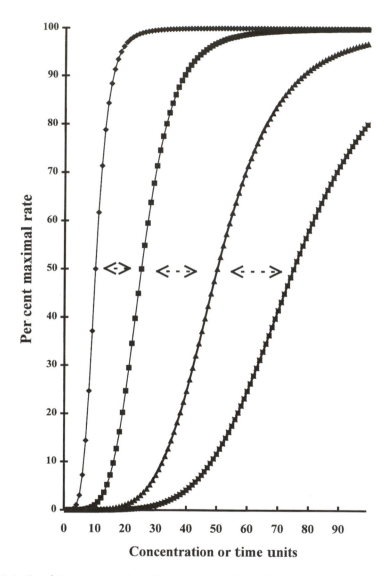

**Fig. 9.2.** Graphic representation of quantitative relationships among hormonally controlled processes. General schematic demonstrating the type of relationships between hormone concentrations, various processes or states, and time. Ordinate may be viewed as concentration or binding of hormone, or per cent of maximal rate or size of state, while abcissa may be viewed as a metabolite concentration or process rate, hormone concentration or unit of time. The equation form and estimates of parameters describing it must be determined for various hormonally controlled processes.

**1.** *The mathematical relationship of hormone concentration, or secretion/ degradation rates, to rates of metabolic processes.* For one enzyme-catalysed reaction, the pathway which contains it, the intracellular pathways connected to it by shared intermediates, and other pathways in other organs which share intermediates, and finally, the net aggregate at the level of animal input or output. The effects of somatotrophin or insulin on muscle protein synthesis and proteolysis and adipose tissue lipogenesis and lipolysis are pertinent examples.

**2.** *Direct evidence on hormonal binding, second messenger generation and destruction, enzyme activation and catalytic activity, metabolic reaction rates and responses, and the kinetic parameters (maximal velocity, substrate sensitivity) within different tissues, organs or cell types.* We lack this basic information for most hormonally stimulated events in the pig. We need a more complete description for hormones in different growth or lactational situations.

**3.** *Responses of cellular systems which are not linearly related to the binding of a hormone, but instead may result in a massive release of a second messenger, such as cyclic AMP.* The purpose of such systems is that the second messenger ensures a total response and not necessarily a graded one. The 'excess' concentration of the second messenger is not easily related to subsequent rates of activation or inhibition. However, often the initial message and net flux result can be described in a mathematical formulation, as has been done for cAMP and lipolysis in adipose tissue during lactation (McNamara *et al.*, 1992). If appropriate, these equations can then be integrated into the whole animal system. Responses to adrenergic agonists and antagonists (endogenous or pharmaceutical) fall into this general category.

**4.** *The opposite problem to that described immediately above is exemplified by systems in which the initial and intermediate steps in response to hormonal binding are very subtle.* These may involve only a slight physical movement of one peptide chain in relation to another, but eventually results in a cascade of events, of which only the final metabolic event is easily measurable. Included in this example are most hormonal responses which include a change in gene transcription or translation. Measurement of the rate of RNA transcription, translation and degradation and the actual concentration is still rather qualitative. It is not clear that the concentration of an enzyme or metabolic rate is a function of the amount or turnover of messenger RNA. We need a better definition of hormonal effects involving transcription, translation and translocation for the pig. This is one example where *in vitro* cell cultures or other animal models may be used to good effect.

**5.** *Quantifying 'permissive' effects of hormones, such that a certain concentration of hormone allows or prevents the response to another hormone, but not in a fully concentration-dependent fashion.* This may also be viewed as a 'threshold' effect. This is a common mechanism of hormonal control, primarily as a long-term strategy, but difficult to define. They may simply lead to a one set of parameter values for one 'hormone-low' state and another set for the 'hormone-high' state. Studies of continuously related systems (such as adipose tissue response to insulin) in different 'permissive' states (such as non-lactating versus lactating females) are relatively straight-forward and useful.

**6.** *The relationship of continuous events (concentration of hormone or nutrient) with discrete events (ovulation or a pulse release of hormone).* Some reproductive events by nature are discrete (either they occur or do not occur) but are regulated by the current rate or state of the overall system. The relationship of glucose and/or insulin status to gonadotrophin releasing hormone secretion and subesequent luteinizing hormone secretion and ovulation is in dire need of more quantitative information (Wade *et al.*, 1996).

**7.** *Integration of subsystems with systems as a whole, a difficulty central to the study of physiology and metabolism.* One technical problem is the integration of results from studies *in vitro* with events occurring *in vivo*. This strikes at the core of reductionist research and we do not pretend nor intend to solve the problem here. A simple example is that of a battery, a piece of wire and an electrical resistor, such as a light bulb. For each one of these subsystems, the characteristics such as voltage, chemical composition, con-ductivity and resistance may be known precisely and calculations made as to exactly what current would be generated when the subsystems are con-nected. Yet, upon connection, heat is generated, which alters the resistance and conductivity, leading to a different result in the system from what was predicted from exact knowledge of all the subsystems. Without prior knowl-edge that heat would be generated, it would be impossible to predict performance of the whole system *only* from knowledge of the subsystems. *In vivo* responses can seldom be predicted *directly* from *in vitro* measure-ments. However, we usually do not (and should not) work in such a vacuum. Rather, in an integrated research approach, observations from whole-animal work lead to specific experiments to test mechanisms and relationships *in vitro*. When the *in vitro* experiments are done, they *must* always be compared against previous, simultaneous or new *in vivo* experiments to provide a framework for validation of the *in vitro* estimates. This is often not done, but we suggest a better experiment is designed to test hypotheses at both animal and subsystem level simultaneously, so such internal validation is made easier.

**8.** *A better quantitative description of homeorhetic regulation.* Working from our improved understanding of homeostatic and homeorhetic systems, we must continue to build a mathematical description of the control of one hormonal system by another. This applies to the differential responses in different physiological states (homeorhesis) as well as differential homeo-static responses induced by exogenous hormones. Included must be an estimate of rates of chemical interconversions among intermediates within and among tissues. For example, what is the dose–response effect of somatotrophin on the parameters describing the response of adipose tissue lipogenesis to insulin, lipolysis to norepinephrine, *and* the resultant aggregate effects on glucose uptake, release of glycerol and fatty acids in this tissue *and* the overall integrated effects on other tissues?

**9.** *A quantitative description of integrated control.* Inherent in 'hormonal control of nutrient metabolism' is 'nutrient control of hormonal systems'. The classic homeostatic systems are a constant loop of that which controls and that which is being controlled. This is central to metabolic regulation, yet we do not really have a quantitative mathematical description of this system, over time, and in the variety of nutritional and physiological states important to pig production. Smaller pieces of this system have been modelled, but these have not yet been fully integrated into whole animal models. The simple answer is a systematic, cooperative approach to define each piece of the system in an iterative and integrative fashion, and then to build the quantitative description of the entire system.

# Quantification of Hormonal Regulation During the Life Cycle

In this section we will point out specific examples in growth and lactation in the pig, working from the general descriptions above. As our efforts in quantitative definition move forward, we continue the iterative question of: Does this new piece of information on a lower level of biological organization help explain or predict the response of a higher level to some effector? For a growing pig, we may ask, would a quantitative description (change in $K_s$ or $V_{max}$) of the effect of insulin and/or somatotrophin on rates of amino acid conversion to muscle protein help to explain the effects noted on muscle accretion, either in normal or hormonally stimulated animals?

Hormonal control of critical reactions by insulin and somatotrophin can be described using the equations in Table 9.1. Kinetic parameters are not constants, but usually vary with age, genetic selection, previous nutritional history or exogenous hormonal treatment. Thus, somatotrophin may alter a $K_s$ or $V_{max}$ directly; it may alter the number of receptors for a different $H$ (insulin in this case); or it may alter the $F$ of the effect of a substrate on $H$ concentration. Additionally, these effects may be different for the same reaction in different organs, providing the quantitative description of homeorhetic control.

## Muscle growth

Muscle growth processes are under hormonal control and are quantifiable, including muscle cell differentiation, myotube formation and muscle fibre hypertrophy, and muscle protein synthesis and degradation. Somatotrophin and insulin cause an increase in proliferation and differentiation of satellite cells to myotubes *in vitro* (Dayton and Hathaway, 1989). These effects are dose-responsive but are not always active within the physiological range of concentrations. Also, mechanisms of hormones on proliferation and differentiation include repression or de-repression of gene sequences. This leads to movement from one phase of the cell cycle to the next and results in increased cell division or differentiation to the subsequent cell type. These processes probably are involved in target cells when hormones such as somatotrophin are given exogenously. However, when hormonal effects on cellular proliferation and differentiation are integrated with effects on enzyme activity and nutrient metabolism within the muscle and among other tissues, often the metabolic effects overwhelm developmental ones. Yet, it is important to understand hormonal effects on cellular proliferation and differentiation, especially in the young animal.

For example, injections of insulin or somatotrophin increase net accretion of muscle protein in a dose-responsive manner. In the younger animal, the total effect may be a combination of the changes in cell proliferation and differentiation and effects on amino acid, glucose and lipid metabolism (Boyd and Bauman, 1989). Insulin stimulates muscle protein synthesis, not simply by an increase in uptake of amino acids, but by an increase in protein synthesis and a decrease in proteolysis (Reeds, 1989; Waterlow, 1995). Somatotrophin acts either directly or through a change in responsiveness to insulin, insulin-like growth factors or changes in insulin-like growth factor binding action (Boyd and Bauman, 1989; McGuire *et al.*, 1995).

Thus, *in vitro* experiments designed to estimate the $K_m$ and $V_{max}$ of amino acid conversion to proteins under insulin stimulation at various concentrations of somatotrophin would answer both a qualitative question (does somatotrophin directly affect response to insulin?) and quantitative ones: the $V_{max}$ of amino acids to protein is $f[\text{Ins}]$; and $(V_{max} f[\text{Ins}])$ is a function of somatotrophin: $\{(V_{max} f[\text{Ins}] f[\text{St}])\}$. A similar set of equations may be generated for sensitivity $(K_m)$ as well. Thus the parameter describing the maximal rate or substrate sensitivity could be adjusted by a parameter (with an experimentally derived value) representing the concentration of insulin; and this coefficient in turn could be adjusted by an experimentally derived coefficient related to somatotrophin concentration.

A similar approach could easily be applied to proteolysis, this would likely be a function of a maximal rate and inhibition by glucose or limiting amino acids (as opposed to a substrate sensitivity function). The maximal rate or inhibition constant may be a function of insulin concentration, and the parameter describing that function in turn would be a function of somatotrophin concentration. These hormonal control parameter values would initially be hypothesized to be different for those controlling protein synthesis. This approach allows testing of appropriate parameter values, and by controlling each reaction (protein synthesis and proteolysis) separately, recognizes that although the rates of these reactions often move in the same direction, they obviously do not change by the same amount and are likely differentially sensitive to hormonal control.

Several pig growth models exist, most base muscle accretion on some allometric function of time, maximal protein accretion, or DNA (Black *et al.*, 1986; Kyriazakis, 1996). This is a valid approach and has great practical utility. However, such models usually have a limited scope, and are not always able to be extrapolated to various genotypes and feeding systems without extensive characterization of each different situation. Describing muscle growth using different equations for protein synthesis and proteolysis allows more flexibility and applicability. However, there is the added cost of determining several different parameter values. The allometric, empirical equation approach has practical value in that various models can be constructed to represent various genotypes with more easily derived parameter values (Black *et al.*, 1986; Kyriazakis, 1996). The various statistics describing the allometric growth equation may be estimated in different genotypes with varying nutrient inputs, for example. Within such a framework one may also design experiments to test the effect of insulin or somatotrophin on maximal protein, efficiency of amino acid use, or the allometric slope as a function of age – some of this work has been done (Boyd and Bauman, 1989; Reeds *et al.*, 1993).

Ideally, a combination of *in vitro* and *in vivo* experiments using, for example, stable isotopes, may be designed to measure both allometric parameters and parameters describing the effects of insulin and somatotrophin on rates of amino acid conversion to protein (and/or protein hydrolysis rates) as given above. These data can be used to begin construction of a model based on a lower level of biological organization, which offers more explanatory power and greater flexibility.

The discussion above on insulin and somatotrophin actions on protein synthesis in the pig also applies to regulation of protein breakdown. Glucagon and the sympathetic neural transmitter norepinephrine cause an increase in proteolysis, as well as decreased insulin concentrations (Reeds, 1989; Waterlow, 1995). Norepinephrine also alters basal muscle contractions or 'tone' and thus may affect both protein synthesis and protein degradation in this manner as well (Mersmann, 1989). For both muscle protein and adipose lipid stores, the synthetic and degradative processes are also linked kinetically by the intermediate substrate concentrations (amino acids and fatty acids).

Although we must recognize and quantify hormonal interactions, certain combinations of hormones are unlikely to occur normally. Biologists should not blindly follow a mathematical recipe and fill in every potential curve or data point. Drawing on past knowledge, we concentrate on the critical areas first, but also recognize that there is tremendous variation in practice. It may be rare to encounter a pig, genetically selected for rapid muscle growth (greater somatotrophin concentration) with high rates of energy intake (leading to elevated insulin) and low rates of amino acid intake (elevated glucagon?) in the cold (elevated norepinephrine). However, it is possible that a genetically selected pig (or a hormonally treated one) may find itself in the cold and consuming a diet not properly balanced or fed for amino acids or energy, thus it is our charge to define that situation.

## Adipose tissue

Early development of adipose tissue consists of hyperplasia and hypertrophy of adipocytes, which occur concurrently at varying rates. Rates also vary among different adipose depots, are under direct hormonal stimulation and much of the pathway biochemistry involved is known (Hausman and Hausman, 1993). Insulin directly stimulates lipogenesis and to a lesser extent esterification: some of this effect comes from increased uptake of glucose; some by enzyme activation, acting both in the short term (such as controlled by phosphorylation states) and the long term (by increased gene expression and protein synthesis). Insulin stimulates activation of lipoprotein lipase for uptake of blood fatty acids from lipoprotein triglycerides. Insulin also decreases triglyceride hydrolysis by hormone-sensitive lipase, probably by a reduction in generation of cyclic AMP. Norepinephrine is a potent stimulator of cyclic AMP production, hormone-sensitive lipase phosphorylation (activation) and lipolysis. There is constant turnover of fatty acids and mono and diglycerides back into triglycerides. In a situation of general carbon and energy balance, esterification and lipolysis occur at equal rates, but this rate of recycling is high even at energy balance. At energy balances even moderately different from zero the rate of recycling is often six to eight times the net rate of fatty acid release (Baldwin, 1995). Also, there is a diurnal pattern to this system, and the metabolic rates and hormonal responses acting over shorter periods of time will be a function of the timing and size of the meals.

A quantitative problem arises because of the nature of control on this system. In large part the lipolytic system developed in order to store and deliver large quantities of energy, often in a short time (survival, 'flight or fight' system). Thus, the response of trigclyceride hydrolysis is more sensitive to hormonal stimulation than lipogenesis, which in turn is sensitive to substrate availability. Coupled with rapid rates of turnover, a very small change in norepinephrine or glucose availability may lead to a doubling or tripling of the net rates of fatty acid release. During normal growth, an inaccurate estimation of rates of recycling becomes an accumulative, quantitative problem for models of longer term effects. For models attempting to describe shorter term (meal-pattern, circadian) regulation, this is a large problem. Thus, the acute and chronic sympathetic nervous system release of norepinephrine effect on the accretion of lipid must be accounted for (Dulloo and Miller, 1985; Landsberg, 1990). Genetic selection, environmental temperature and stress level and the amount of carbohydrate or fat consumed also affect net accretion of lipid and these effects are either carried out hormonally or affect the hormonal systems indirectly. This creates a challenge which only an integrated, systematic, quantitative approach can meet.

The role of somatotrophin in adipose growth continues to evolve but some quantitative data are available. The response of net accretion or loss of total body fat to exogenous somatotrophin varies in different stages of growth, with the effects generally increasing with age (Beermann, 1989; Boyd and Bauman, 1989). These studies should continue, and determine direct effects of somatotrophin on key biochemical pathways of carbon and nitrogen flux through the muscle or adipose depots, such as was begun by Dunshea *et al.* (1992). It is likely that a reduced cellular response to insulin and an increased response to norepinephrine are caused by somatotrophin action (Boyd and Bauman, 1989; Bauman and Vernon, 1993). Differences in the whole-body response to homeostatic signals are characteristic of various exogenous somatotrophin regimens (Bauman and Vernon, 1993). Continued studies will be very useful to estimate the internal parameters of this system.

It would also be helpful if effects of exogenous hormones were determined under various endogenous hormone concentrations, such as occurs during various phases of growth or exposure to different nutritional regimens. Also, there is a relationship between dose of somatotrophin and voluntary feed intake. This forms an interactive loop of dose of hormone, response of adipose tissue fatty acid release, and feed intake. Eventually, feed intake limits amino acid availability to the muscle and energy availability to the adipose tissue. This would not be predicted from either the effect of somatotrophin on adipose tissue or muscle alone. However, recognition of glucose and fatty acid flux effects on nutrient intake would help predict this. Thus, an initial model at a lower (pathway biochemistry) level of organization would contain sufficiently detailed equations on hormonally controlled maximal, sensitivity and inhibition parameters on muscle and adipose tissue metabolism; and effects of nutrient supply on nutrient intake. Such a model would be useful to test hypotheses and describe the interactions of hormonal dose, enzyme (pathway) induction or inhibition, and substrate supply on the net response of the animal. The model of Pettigrew *et al.* (1992) in lactation uses this basic approach to the interaction of hormones, tissue maximal rates and substrate supply (see section on lactation below); however, this model is also limited by a lack of information on control of nutrient intake.

### Ovulation and fetal development

Reproductive processes are also under quantitative hormonal control, though in perhaps a less obvious manner. The reproductive process from growth of the follicle, to ovulation, insemination, fertilization, implantation of the embryo, establishment of the pregnancy, fetal growth (and accompanying tissues); birth (including number and size of live young); lactogenesis, lactation, weaning, and re-initiation of the process, includes a combination of continuous and discrete events. Continuous events such as rates of milk production and the effect of hormones on rate of follicular growth are controlled in a similar fashion to the growth events discussed above and do not present conceptually different quantitative processes.

The timing and number of ovulations and the resulting number of successful implantations; and the timing of re-initiation of luteinizing hormone release pulses after parturition are more discrete than continuous in nature and thus present greater challenges in both biological and mathematical terms. A simple reason that we do not have a better quantitative description is the cost of conducting such studies. It is technically possible to measure the fertility and conception rates, number of viable fetuses at different stages of gestation and the number of pigs born alive, several quantitative studies at the whole animal level describing the relationship of parity, nutrition and repro-duction have been done (Black *et al.*, 1986). It is just a matter of degree to take it to a lower level of biochemical organization.

As a brief example, there has been improvement in our understanding of the relationships of glucose metabolism, insulin, the release of gonado-trophin releasing hormone, luteinizing hormone and ovulation (Wade *et al.*, 1996). Mathematical equations may be developed which relate continuous variables (such as rate of energy intake, or concentration of glucose, insulin or luteinizing hormone) with the rate or pattern of gonadotrophin releasing hormone, luteinizing hormone, the chronological time from initiation of follicular growth to ovulation, the number of ovulations and subsequent fertility rates. As for growth models, this level of description will allow a more precise understanding of the system and a greater range of eventual applica-bility. More cooperation is needed between reproductive endocrinologists, nutritionists and mathematicians to design coordinated experiments and derive such equations whenever appropriate.

For example, a critical area to the breeding herd is the relationship of body fatness, energy balance and similar measures of 'fuel surplus or deficit' with reproductive processes. Several empirical theories have been put forth to explain the observed connections (Wade *et al.*, 1996), and these empirical roads are now converging on some specific biochemical mechanisms. Neu-rons in the forebrain which secrete gonadotrophin releasing hormone are directly affected by blood glucose or insulin, or both, and this in turn affects ovulation timing or rates (Wade *et al.*, 1996).

In lactating pigs, a slower rate of energy intake increases the time after parturition at which the pulsatile secretion of luteinizing hormone resumes (Tokach, 1992a; Koketsu *et al.*, 1996). In pigs consuming increased amounts of energy in the first one or two weeks of lactation, the first luteinizing hormone pulses are delayed and the return to oestrus is shortened, regardless of body fat gain or loss (Tokach, 1992a; Koketsu *et al.*, 1996). In cattle, direct infusions of glucose caused luteinizing hormone release to begin within 24 h, and this mechanism clearly works in rodents (Wade *et al.*, 1996). This has not yet been clearly demonstrated in the lactating pig (Tokach, 1992b) but the empirical evidence suggests a similar mechanism is operating. The new findings from rodents (Wade *et al.*, 1996) can now direct more specific experimental designs in pigs, which will be worth the effort to the swine industry.

## Lactation

The lactating mammary gland exerts a constant kinetic force on the system to remove amino acids, fatty acids and glucose (Fig. 9.1). This is coordinated by lactational hormones acting on the homeostatic regulatory processes. The nutrient demand eventually feeds back to the centres which control feed intake to cause an increase in nutrient influx into the pools. The regulatory loop between nutrient outflow, metabolism in the tissues, changing tissue pool sizes, and nutrient intake changes through lactation as the 'lactation curve' is followed. For an in-depth treatise on quantification of metabolism during lactation, pointing out several areas for which quantitative knowledge on hormonal control is still lacking, see Baldwin (1995).

One example discovered somewhat 'early on' in quantitative lactational endocrinology was an empirical description of a hormone and the process of lactation (Koprowski and Tucker, 1973). They discovered an empirical relationship between prolactin immediately post-milking and the rate of milk production, which followed the general form of the lactation curve. This finding stimulated a tremendous amount of research into hormonal control of metabolism and mammary biology, one eventual result of which was the development of the role of somatotrophin in control of milk production.

Knowledge of lactational endocrinology in the pig is woefully inadequate. We possess little qualitative and no real quantitative evidence for the role of several hormones including prolactin, somatotrophin, insulin, insulin-like growth factors and steroids in lactation of the pigs. From other species, we know prolactin has direct effects on lactogenesis and mammary casein and lipid synthesis and a more quantitative picture of control, especially in early lactation, is clearly warranted. In early lactation, prolactin may be still exerting an effect on differentiation of mammary cells and gene transcription for key enzymes. The interaction of lactational (prolactin, somatotrophin) and homeostatic hormones (insulin, glucagon, norepinephrine) is a 'wide-open' field for study in swine lactation (Boyd *et al.*, 1995). Such studies will require quite frequent sampling of mammary tissue and measurement of metabolic activity, coupled with simultaneous measurements of circulating hormones. It will also require determination of direct *in vitro* effects of hormones on casein, lactose and fat synthesis. Some of this work has been done in mammary culture systems for the cow and rat (Carrington *et al.*, 1983; Collier *et al.*, 1984; Fekry *et al.*, 1989), but this has yet to be reported for the pig. Technical limitations involved in culturing mammary cells and tissues will continue to be a problem. However, some recently revived and improved techniques for measuring mammary nutrient flux *in vivo* in pigs will be useful (Trottier *et al.*, 1996).

One important need for understanding of quantitative homeorhetic regulation in lactation is the coordination of mammary activity with that in other tissues. A major example is the adaptation of metabolism in the adipose tissue (Vernon, 1988; McNamara, 1991, 1994; Bauman and Vernon, 1993; Parmley and McNamara, 1996). It is likely that changes in circulating prolactin and/or somatotrophin alter the sensitivity of the adipose tissue to beta-adrenergic stimulation and diminish that to insulin. The effects of the two hormones vary: it is likely that prolactin is more important in late pregnancy and somatotrophin more important in lactation; however this is far from certain, especially in the pig. Even though one may be 'more important' than the other in different states, that begs the quantitative question of what is the net effect of both.

Mechanisms involved in the action of lactational hormones such as somatotrophin potentially include a greater density of beta-receptors on fat cells, a more tightly regulated G-protein binding, and an increase in cAMP concentration, ultimately leading to more active hormone-sensitive lipase and faster lipolytic rates (McNamara, 1988; McNamara *et al.*, 1992; Bauman and Vernon, 1993). The molecular mechanism of the decreased responsiveness to insulin may include: decrease in insulin receptor density (perhaps to prolactin prepartum); altered insulin signalling and processing; and increased cAMP and intracellular fatty acid concentrations (Vernon, 1988; Marinchenko *et al.*, 1992; Bauman and Vernon, 1993). These adaptations result in changes in net fatty acid uptake and release from adipose tissue to balance the nutrient demand of the mammary gland and energy requirements of maternal tissues with the nutrient and energy intake.

We will provide three pertinent examples of quantitative descriptions of metabolic flux and potential or direct hormonal control from lactating dairy cattle which have application to the pig. The first example relates characteristics at the whole-animal level with metabolism at the tissue level. The curves shown in Fig. 9.3 were constructed from rates determined *in vitro* in adipose tissue samples taken from cows at 60 days of lactation at widely different energy balances (McNamara and Hillers, 1989). Energy balance (a whole-animal integration of energy intake, maintenance energy use and milk energy output) is related to tissue-level, hormonally controlled rates of lipogenesis, esterification and lipolysis. Increasing energy balance above zero has a very large effect on lipogenesis rates, which are negligible at zero energy balance or below. Esterification is not as tightly controlled as lipogenesis, to allow for the increased fatty acid turnover. Lipolysis steadily decreases as energy balance increases, and at close to zero energy balance, lipolysis and esterification effectively balance. These equations demonstrate the extremely sensitive differential hormonal control which occurs, even just within one tissue. Thus, specific measurements must be made for each pathway in each tissue

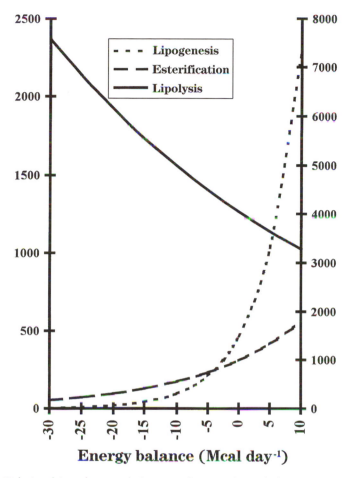

**Fig. 9.3.** Relationships of energy balance with rates of metabolic reactions in adipose tissue of lactating dairy cattle. Data were generated from *in vitro* incubations of adipose tissue biopsied from lactating dairy cattle at day 60 of lactation (McNamara and Hillers, 1989). Equations are the best fit of energy balance the month prior to the biopsy with the rate of lipogenesis, esterification or lipolysis (as glycerol release). Rates are in nanomoles of substrate converted to product (or glycerol release) / 2 h per gram tissue, energy balance is in Mcal day$^{-1}$. These are examples of integrated empirical and mechanistic equations, relating the input–output functions at one level with the underlying subsystem characteristics.

and no one equation with one set of parameter values can describe the full scope of hormonal control of kinetic flux even in one physiological state such as lactation.

An example specific to altered hormonal control of metabolism in

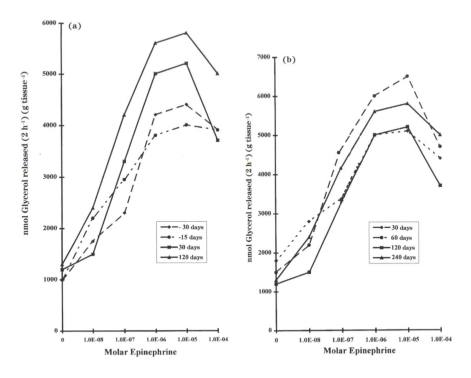

**Fig. 9.4.** Epinephrine stimulation of lipolysis in adipose tissue from dairy cattle in different stages of lactation. These curves demonstrate the variation in sensitivity and maximal responsiveness of a tissue to one hormone in different physiological states. The curves are a specific instance of the general ones shown in Fig. 9.2. (After McNamara, 1988.)

lactation is provided in Fig. 9.4. The response of lipolysis (glycerol release) to epinephrine was measured *in vitro* in adipose tissue, sampled from cows during different lactational states. Immediately postpartum, an increase in both maximal velocity and sensitivity to epinephrine is noted. As lactation continues past 120 days, sensitivity does not increase further, but maximal velocity does. Even though these animals have been in positive energy balance for some time at 240 days of lactation, the adipose tissue still has a very high rate of lipolysis to meet mammary demand. At the same time, adipose tissue lipogenesis is proceeding at fast rates and body reserves are being restored. This seeming contradiction of high rates of lipolysis with positive energy balance had not previously been reported in any other situation other than late lactation. This has now been demonstrated for the sow as well – concurrent high rates of adipose tissue lipogenesis and lipolysis, to meet both the needs of the young and the dam (Parmley and McNamara,

1996). This coordination of opposing pathways to meet simultaneous needs of different organs underscores the ability of the homeorhetic system to fine-tune metabolic control as needed for unique situations.

A third example is the effect of exogenously administered somatotrophin on lipolysis in lactating animals. Using an *in vivo*, whole-body approach, it has been demonstrated that the sensitivity and maximal velocity of lipolysis increases in somatotrophin treated cows in response to epinephrine infused *in vivo* (Bauman and Vernon, 1993). The curves generated follow the general example in Fig 9.2. The control function exerted by exogenous somatotrophin is similar to the normal adaptations in responses to epinephrine in early lactation of dairy cows (Fig. 9.4).

It is these types of hormonal regulation and interactions which must be determined in the lactating sow. Taking the subtleties and complexities discovered in other species into account, we can now design experiments to define control of metabolism in different lactational states by hormones such as insulin, somatotrophin and prolactin in the sow. Again, a combination of *in vivo* and *in vitro* studies will probably be required.

One example of an approach useful for the sow is the recent work of McGuire *et al.* (1995). They used a euglycaemic/hyperinsulinaemic *in vivo* clamp technique in cattle, in which both insulin and glucose are infused to increase both the hormonal signal and maintain the glucose supply. This has advantages over either a simple insulin infusion, which depresses blood glucose, or a glucose infusion, which allows only a fairly short study window. They found that simultaneously elevated levels of insulin and glucose increased milk lactose, fat and protein production in the dairy cow (McGuire *et al.*, 1995), while glucose infusions alone do not increase protein synthesis. This identifies some interesting challenges for defining hormonal mechanisms and determining quantitative effects. It has long been thought that the mammary gland of the dairy cow is not sensitive to insulin. However, the capacity for increased rates of protein synthesis appears to exist in the mammary gland. The effect on protein synthesis may occur as a response to insulin-like growth factor.

This work also emphasizes the interactions between the hormonal systems and the nutrient flux systems. The interactions of nutrient supply with hormonal regulation must be considered in the regulation of milk protein production in the pig. Because the mammary gland of the pig is sensitive to insulin, and protein and energy intake is limiting to piglet growth, it would be useful to utilize this technique to obtain quantitative descriptions for the lactating pig. Following the descriptions in Table 9.1, experiments designed to determine the effects of insulin on the $V_{max}$ of glucose use for lactose synthesis, or of insulin and glucose on amino acid conversion to milk protein, would be straightforward.

For the pig, we need to determine the quantitative descriptions of flux control and hormonal regulation as depicted in Fig. 9.1 and in Table 9.1. These parameters need to be determined in coordinated *in vitro* and *in vivo* studies, for mammary, muscle and adipose issue metabolic pathways during lactation. The effect of litter size (milk demand in the pig) on the $K_s$ and $V_{max}$ parameters and their hormonal regulation must be determined. This can easily be done using tissue biopsies to measure metabolic rates, or with the use of stable isotopes *in vivo*, in dams nursing litters of different sizes. The interaction of endocrine status substrate supply also needs to be known. We already have a tremendous amount of data on the effects of amino acid and energy supply to litter growth and maternal body composition (Pettigrew *et al.*, 1992; see also Chapter 4 of this volume). Estimates are now available on some internal parameters of lipogenesis, esterification and lipolysis in adipose tissue of lactating pigs (Parmley and McNamara, 1996). We are beginning to investigate hormonal control in rat models (McNamara and Murray, 1994). We still need more information on amino acid use by all organs of the lactating pig. Once generated, these parameters can be integrated into the existing model of metabolism in lactating pigs (Pettigrew *et al.*, 1992), as is being done for the metabolic model of the cow (McNamara and Baldwin, 1995).

## Models of Integrative Hormonal Control – Need for Further Research

Our goal is to predict responses to changes in nutritional management, genetic selection, environmental situations or endocrine manipulation in pigs. With this in mind, how do we integrate information from various small experiments into a quantitative description of the metabolism or physiology of the whole animal? How do we define those situations put forth above as critical? The answer is obviously stated, if not easily completed. Following the types of equations and relationships described in Figs 9.1 and 9.2 and in Table 9.1, we must design experiments such that mechanisms of effects are elucidated, and the full range of such an effect is described in biochemical, kinetic and mathematical terms.

Two basic quantitative approaches, generally defined as 'empirical' and 'mechanistic', will be used. Each has its merits and limitations. It is usually easier to define parameter values for empirical models, but mechanistic models will have a greater scope, precision and explanatory power. We think that both types can and should be used, as appropriate, for research and practical purposes which are different but with a tremendous amount of overlap. The research community will be conducting experiments to derive 'internal' or mechanistic parameter values in a context of a practical range of production. Ideally, parameter values for use in empirical models can be derived simultaneously. Determining the effects of different hormonal states *in vivo*, or *in vitro* or *in vivo* exogenous treatments is straightforward following the patterns and equations in Fig. 9.1 and Table 9.1.

This approach is not 'new', it has been applied in the physical, chemical and biological sciences for more than fifty years. It is not new to metabolic biology either (Finklestein and Carlson, 1979; Carson *et al.*, 1981; Baldwin *et al.*, 1987; Cornish-Bowden and Cardenas, 1990; Baldwin, 1995). This quantitative approach has been used with great success in modelling insulin action in a number of systems (Carson *et al.*, 1981), in several physiological processes (Ricard and Cornish-Bowden, 1984), and in several agricultural systems (France and Thornley, 1983) including farm animals (Baldwin *et al.*, 1987; Danfaer and Lescoat, 1995).

The integration of nutrition, reproduction and genetics in the production of pigs has been followed vigorously by several groups (Black *et al.*, 1986; Whittemore and Morgan, 1990; Pettigrew *et al.*, 1992). Recognizing that a quantitative description of metabolic processes would help management of pork production, several model systems have been generated which recognize interactions of metabolism, genetics, the environment and reproduction. Recent developments in practical quantitative systems based on good empirical data are covered in Chapter 4 of this book. This mathematical approach eliminates much of the guesswork and wasted effort which characterize too much of nutritional research and application and stands as a fine example of the integration of basic and applied research using a common language. With each successive iteration of the modelling (hypothesis) and experimentation cycle, the integration of quantitative new knowledge improves our description of reality (see McNamara and Baldwin, 1995).

The examples of hormonally controlled processes given above provide a template for further research. We must continue to improve our methodology to measure more specific molecular events in a quantitative manner, both *in vivo* and *in vitro*. We need to integrate data on hormonal effects in various subsystems studied *in vitro* with the integrated effects measured *in vivo*. For example, one factor may cause cellular proliferation *in vitro*. However, *in vivo*, the *in vitro* conditions may not apply or represent only a part of the developmental process; or the presence of competing factors may inhibit this effect so that it either does not occur, or is balanced out by effects of other factors.

The quantitative, coordinated approach between *in vivo* and *in vitro* research is the only way to eventually know which results from *in vitro* experiments apply *in vivo*, and in what quantitative manner. Examples provided above (McNamara, 1994; Parmley and McNamara, 1996) are relatively easy to do by several scientists or organizations. Nutritional (this could easily be hormonal) treatments are applied to animals, and basic input–output data collected, such as chemical nutrient intake, milk component output, and body composition of the mother. Tissue samples are taken and by standard *in vitro* techniques, metabolic reaction rates are measured (the tissue used was adipose, it could be any of several) and the $K_s$ and $V_{max}$ derived. These are then compared to the actual changes occurring at the whole-animal level. If needed, scaling of parameters estimated *in vitro* is done in recognition of errors systemic to *in vitro* work (McNamara and Baldwin, 1995) and the internal parameter values are integrated into the model. This new iteration now forms the new knowledge base and is used for design of further experiments designed to test more and more refined hypotheses or to make more specific recommendations.

Experiments must be designed to describe effects at different levels of biological organization simultaneously, such as hormone receptor binding, intracellular signalling and flux response. We must investigate these events across several experiments covering various situations of age, physiological state, previous nutritional history, genetic selection, direct manipulation of genetic material, and in various treatments with exogenous agents. Cooperation and coordination among scientists must increase so that individual scientists can continue to design experiments which ask quantitative questions as put forth above. This has been done successfully and needs to continue.

# References

Baldwin, R.L. (1995) *Modeling Ruminant Digestion and Metabolism*. Chapman and Hall; New York.

Baldwin, R.L., France, J. and Gill, M. (1987) Metabolism of the lactating cow. II. Animal elements of a mechanistic model. *Journal of Dairy Research* 54, 77–105.

Bauman, D.E. and Currie, W.B. (1980) Partitioning of nutrients during pregnancy and lactation: a review of mechanisms involving homeostasis and homeorhesis. *Journal of Dairy Science* 63, 1514–1523.

Bauman, D.E. and Vernon, R.G. (1993) Effects of bovine somatotropin on lactation. *Annual Review of Nutrition* 13, 437–462.

Beermann, D.H. (1989) Status of current strategies for growth regulation. In: Campion, D.R., Hausman, G.J. and Martin, R.J. (eds), *Animal Growth Regulation*. Plenum Press, New York, pp. 377–400.

Bell, A.W. and Bauman, D.E. (1994) Animal models for the study of adipose regulation in pregnancy and lactation. In: Allen, L., King, J. and Lonnerdal, B. (eds), *Nutrient Regulation during Pregnancy, Lactation and Infant Growth*. Plenum Press, New York, pp. 71–84.

Black, J.L., Campbell, R.G., Williams, I.H., James, K.J. and Davies, G.T. (1986) Simulation of energy and amino acid utilisation in the pig. *Research and Development in Agriculture* 3, 121–145.

Boyd, R.D. and Bauman, D.E. (1989) Mechanisms of action for somatotropin in growth. In: Campion, D.R., Hausman, G.J. and Martin, R.J. (eds), *Animal Growth Regulation*. Plenum Press, New York, pp. 257–294.

Boyd, R.D., Kensinger, R.S., Harrell, R.J. and Bauman, D.E. (1995) Nutrient uptake and endocrine regulation of milk synthesis by mammary tissue of lactating sows. In: Tucker, H.A., Pettigrew, J.E. and Petitclerc, D. (eds), *Second International Workshop on the Biology of Lactation in Farm Animals. Supply of Precursors and Synthesis of Milk Components in Normal, Extreme and Disease States. Journal of Animal Sciences* (Suppl. 2), 36–56.

Carrington, C.A., Hosick, H.L., Forsyth, I.A. and Dils, R. (1983) Milk-fat synthesis by lobules prepared from rabbit mammary gland: response to insulin, corticosterone, prolactin and progesterone. *Journal of Endocrinology* 97, 157–166.

Carson, E.R., Cobelli, C. and Finkelstein, L. (1981) Modeling and identification of metabolic systems. *American Journal of Physiology* 240, R120–R129.

Collier, R.J., McNamara, J.P., Wallace, C.R. and Dehoff, M.H. (1984) A review of endocrine regulation of metabolism during lactation. *Journal of Animal Science* 59, 498–510.

Cornish-Bowden, A. and Cardenas, M.L. (1990) *Control of Metabolic Processes.* Plenum Press, New York.

Crabtree, B. (1993) Metabolic regulation. In: Forbes, J.M. and France, J. (eds), *Quantitative Aspects of Ruminant Digestion and Metabolism.* CAB International, Wallingford, UK.

Danfaer, A. and Lescoat, P. (eds) (1995) *Proceedings of the IVth International Workshop on Modelling Nutrient Utilisation in Farm Animals.* Denmark Institute of Animal Science, Foulum, Denmark, 199 pp.

Dayton, W.R. and Hathaway, M.R. (1989) Autocrine, paracrine and endocrine regulation of myogenesis. In: Campion, D.R., Hausman, G.J. and Martin, R.J. (eds), *Animal Growth Regulation.* Plenum Press, New York, pp. 183–210.

Dulloo, A.G. and Miller, D.S. (1985) Increased body fat due to elevated energetic efficiency following chronic administration of inhibitors of sympathetic nervous system activity. *Metabolism* 34, 1061–1065.

Dumshea, F.R., Harris, D.M., Bauman, D.E. and Boyd, R.D. and Bell, A.W. (1992) Effect of porcine somatotropin on in vivo glucose kinetics and lipogenesis in growing pigs. *Journal of Animal Science* 70, 141–151.

Fekry, A.E., Keys, J.E., Capuco, A.V., Bitman, J., Wood, D.L. and Miller, R.H. (1989) Effect of bovine growth hormone on incorporation of [14C] acetate into lipids by co-cultures of bovine mammary, liver and adipose tissue explants. *Domestic Animal Endocrinology* 62, 87–94.

Finkelstein, L. and Carson, E.R. (1979) *Mathematical Modeling of Dynamic Biological Systems.* Medical Computing Series, vol. III. Research Studies Press, Forest Grove, Oregon, USA.

France, J. and Thornley, J.H.M. (1983) *Mathematical Models in Agriculture. A Quantitative Approach to Problems in Agriculture and Related Sciences.* Butterworths, London, UK.

Hammond, J. (1944) Physiological factors affecting birth weight. *Proceedings of the Nutrition Society* 2, 8–12.

Hausman, G.J. and Hausman, D.B. (1993) Endocrine regulation of porcine adipose tissue development: cellular and metabolic aspects. In: Hollis, G.R. (ed.), *Growth of the Pig.* CAB International, Wallingford, UK.

Hopkins, J.C. and Leipold, R.J. (1996) On the dangers of adjusting the parameter values of mechanism-based mathematical models. *Journal of Theoretical Biology* 183, 417–427.

Koketsu, Y., Dial, G.D., Pettigrew, J.E., Marsh, W.E. and King, V.L. (1996) Influence of imposed feed intake patterns during lactation on reproductive performance and on circulating levels of glucose, insulin and luteinizing hormone in primiparous sows. *Journal of Animal Science* 74, 1036–1046.

Koprowski, J.A. and Tucker, H.A. (1973) Serum prolactin during various physiological states and its relationship to milk production in the bovine. *Endocrinology* 92, 1480–1488.

Kyriazakis, I. (1996) A solution to the problem of predicting the response of an animal to its diet. *Proceedings of the Nutrition Society* 55, 155–166.

Landsberg, L. (1990) Insulin resistance, energy balance and sympathetic nervous system activity. *Clinical and Experimental Hyperalimentation – Theory and Practice* A12(5), 817–830.

Loeb, J.N. and Strickland, S. (1987) Hormone binding and coupled response relationships in systems dependent on the generation of secondary mediators. *Molecular and Cellular Endocrinology* 1, 75–83.

Marinchenko, G.V., McNamara, J.P., Sun, S. and Becker-Khaleel, B. (1992) Growth hormone alters metabolic effects and proteolysis of insulin in adipose tissue during lactation. *Proceedings of the Society for Experimental Biology and Medicine* 200, 57–66

McGuire, M.A., Dwyer, D.A., Harrell, R.J. and Bauman, D.E. (1995) Role of insulin in the regulation of mammary synthesis of fat and protein. *Journal of Dairy Science* 78, 816–824.

McNamara, J.P. (1988) Regulation of bovine adipose tissue metabolism during lactation 4. Dose-responsiveness to epinephrine as altered by stage of lactation. *Journal of Dairy Science* 71, 643–649.

McNamara, J.P. (1991) Regulation of adipose tissue metabolism in support of lactation. *Journal of Dairy Science* 74, 706–719

McNamara, J.P. (1994) Lipid metabolism in adipose tissue during lactation: a model of a metabolic control system. *Journal of Nutrition* 124, 1383S–1391S.

McNamara, J.P. and Baldwin, R.L. (1995) Modeling metabolism in lactation: estimation of critical parameters in lipid metabolism. In: Danfaer A. and Lescoat, P. (eds), *Proceedings of the IVth International Workshop on Modelling Nutrient Utilisation in Farm Animals*. Denmark Institute of Animal Science, Foulum, Denmark, pp. 63–84.

McNamara, J.P. and Hillers, J.K. (1989) Regulation of bovine adipose tissue metabolism during lactation 5. Relationships of lipid synthesis and lipolysis with energy intake and utilization. *Journal of Dairy Science* 72, 407–418.

McNamara, J.P. and Murray, C.E. (1994) Adaptations in sympathetic nervous system activity in white adipose tissue during pregnancy and lactation: role in regulating adipose tissue metabolism. *Journal of Dairy Science* 77 (Suppl. 1), 226.

McNamara, J.P., Becker-Khaleel, B. and Parmley, K.L. (1992) Quantitative relationships between cAMP and lipolysis in adipose tissue during the peripartum period. *Journal of Dairy Science* 75, 1901–1913.

Mersmann, H.J. (1989) Potential mechanisms for repartitioning of growth by beta-adrenergic agonists. In: Campion, D.R., Hausman, G.J. and Martin, R.J. (eds), *Animal Growth Regulation*. Plenum Press, New York, pp. 337–358.

Oddy, V.H. and Lindsay, D.B. (1986) Metabolic and hormonal interactions and their potential effects on growth. In: Buttery, P.J., Haynes, N.B. and Lindsay, D.B. (eds) *Control and Manipulation of Animal Growth*. Butterworths, London, pp. 231–248.

Parmley, K.L.S. and McNamara, J.P. (1996) Rates of lipid metabolism in adipose tissue of pigs adapt to lactational state and dietary energy restriction. *Journal of Nutrition* 126, 1644–1656.

Pettigrew, J.E., Gill, M., France, J. and Close, W.H. (1992) A mathematical integration of energy and amino acid metabolism. *Journal of Animal Science* 70, 3742–3761.

Pettigrew, J.E., McNamara, J.P., Tokach, M.D., King, R.H. and Crooker, B.A. (1993) Metabolic connections between nutrient intake and lactational performance in the sow. *Livestock Production Science* 35, 137–152.

Reeds, P.J. (1989) Regulation of protein turnover. In: Campion, D.R., Hausman, G.J. and Martin, R.J. (eds), *Animal Growth Regulation*. Plenum Press, New York, pp. 183–210.

Reeds, P.J., Burrin, D.G., Davis, T.A., Fiorotto, M.A., Mersmann, H.J. and Pond, W.G. (1993) Growth regulation with particular reference to the pig. In: Hollis, G.R. (ed.), *Growth of the Pig*. CAB International, Wallingford, UK, pp. 1–32.

Ricard, J. and Cornish-Bowden (1984) *Dynamics of Biochemical Systems. Proceedings of a NATO Advanced Research Workshop on Dynamics of Biochemical Systems*. Plenum Press, New York, USA.

Tokach, M.D., Pettigrew, J.E., Dial, G.D., Wheaton, J.E., Crooker, B.A. and Johnston, L.J. (1992a) Characterization of luteinizing hormone secretion in the primiparous, lactating sow. Relationship to blood metabolites and return-to-estrus interval. *Journal of Animal Science* 70, 2195–2200.

Tokach, M.D., Pettigrew, J.E., Dial, G.D., Wheaton, J.E., Crooker, B.A. and Koketsu, Y. (1992b) Influence of glucose infusions on luteinizing hormone secretion in the energy-restricted, primiparous, lactating sow. *Journal of Animal Science* 70, 2202–2206.

Trottier, N.L., Shipley, C.F. and Easter, R.E. (1996) A technique for the venous cannulation of the mammary gland in the lactating sow. *Journal of Animal Science* 73, 1390–1395.

Vernon, R.G. (1988) The partition of nutrients during the lactation cycle. In: Garnsworthy, P.C. (ed.), *Nutrition and Lactation in the Dairy Cow*. Butterworths, London, UK, pp. 157–170.

Wade, G.N., Schneider, J.E. and Li, H.Y. (1996) Control of fertility by metabolic cues. *American Journal of Physiology* 33, E1–E19.

Waterlow, J.C. (1995) Whole-body protein turnover in humans – past, present and future. *Annual Review of Nutrition* 15, 57–92.

Whittemore, C.T. and Morgan, C.A. (1990) Model components for the determination of energy and protein requirements for breeding sows: a review. *Livestock Production Science* 26, 1–37.

# Food Intake and Metabolism

# Voluntary Food Intake and Diet Selection

**10**

## I. Kyriazakis and G.C. Emmans

*Animal Biology Division, Scottish Agricultural College, West Mains Road, Edinburgh EH9 3JG, UK*

## Introduction

Among the reasons for wanting to describe quantitatively, and to predict, the rate of voluntary food intake in pigs fed *ad libitum* are: (i) the effective simulation of growth and body composition; (ii) efficient least cost diet formulation; and (iii) the determination of optimum feeding strategies (Whittemore, 1994). With feed as the predominant cost of pig meat production (NRC, 1987) these areas are of great economic importance. If we can have the ability to predict, for example, the consequences of genetic selection and environmental effects on food intake, we will be able to make more effective financial and management decisions. While the importance of being able to predict food intake is widely recognized, few effective attempts have been made to do this in pigs. The common method of relating past intake to observed live weight in one, or more, experiments cannot be seen as having any useful generality as will be shown later.

While considerable research effort has been put into the investigation of the possible mechanisms which underlie the control of food intake of pigs (for reviews see NRC, 1987; Forbes and Blundell, 1989; Rayner and Gregory, 1989) there is no indication that this approach will allow the effective prediction of what food intake will be in a given case.

The lack of good, general methods for making quantitative predictions of food intake reflects, at least in part, the fact that pigs in the past, at least in Europe, were often fed restricted, controlled amounts of food (Riley, 1989). As a consequence early, and surprisingly more recent, models of pig growth have treated food intake as an input (e.g. Whittemore and Fawcett, 1974;

© CAB INTERNATIONAL 1999. *A Quantitative Biology of the Pig*
(ed. I. Kyriazakis)

Whittemore, 1976; Pomar *et al.*, 1991; de Lange, 1995) rather than as a consequence to be predicted. Progressive advances in pig breeding over the past twenty years have resulted in leaner pigs which are very often fed *ad libitum* for much, if not all, of their growing period. For this reason an effective method for predicting food intake in the growing pig has become of considerable importance (Emmans, 1995).

In this chapter we will review the existing approaches to food intake prediction in the pig, from simple regression equations, to a more complex and hence more flexible and general approach. We will try to answer the often posed dilemma: Does the pig grow because it eats, or does it eat because it grows? The approach favoured here follows from that of Emmans (1981a) and Emmans and Fisher (1986). In brief the answer to the dilemma is in two parts: the pig seeks to eat because it seeks to grow but, if unable to be successful, it will grow to the extent which depends on what it has managed to eat. We will concentrate almost exclusively on the food intake of growing and non-productive mature pigs. This is because pregnant and dry sows still usually have their food intake restricted (Lawrence *et al.*, 1993). Food intake during lactation is dealt with in Chapter 7 of this book.

We will also attempt, for the first time, to predict the food intake and diet selection when more than one food is offered to the pig as a choice. This seems to be timely now that the method appears to have attracted some increased interest as a potential feeding method (Kyriazakis, 1994).

## Food Intake Descriptions

### Simple regression equations

The first attempts to describe quantitatively the food intake in pigs were in the form of regression equations relating intake to either time or live weight. Often they were derived from a rather narrow range of data. Two such examples, taken from NRC (1987), describe the relationship between digestible energy intake (*DEI*, as the preferred expression of food intake) and either time, *t*, or live weight, *W*:

$$DEI \text{ (from creep feed)} = 46.4 \, (t - 13.5) \qquad \text{kJ day}^{-1} \qquad (10.1)$$
$$DEI = 1.93.W - 0.0407.W^2 - 6.40 \qquad \text{MJ day}^{-1} \qquad (10.2)$$

Equation 10.1 is intended to work only over a very narrow range, i.e. to express creep feed intake of piglets. Intake is predicted to begin at about 13.5 days of age, and the energy intake derived from milk is ignored. It clearly becomes absurd when extrapolated. Equation 10.2 illustrates the dangers of using datasets of small range to estimate the values of parameters of expressions that are not of a sensible form. It predicts that the rate of *DEI* will reach a maximum when the pig weighs less than 25 kg and that it goes to zero when *W* = 44 kg. Such approaches have only very limited value in predicting food intake. They are essentially ways of summarizing and describing data. They have no theoretical basis, can be misleading when extrapolated beyond the bounds of the data from which they were derived, and have no generality even within the range over which they were derived.

## Exponential equations

More progress has been made by the use of exponential equations which relate food intake to the live weight of the pig. The forms which have been used are:

$$DEI = a.W^b \qquad \text{MJ day}^{-1} \text{ (ARC, 1981)} \qquad (10.3)$$
$$DEI = c(1 - e^{-kW}) \qquad \text{MJ day}^{-1} \text{ (ARC, 1981; NRC, 1987)} \qquad (10.4)$$

Equation 10.3 suggests that intake will increase monotonically as weight increases, with no maximum unless $W$ has a maximum; intake tends to zero as $W$ approaches zero. Equation 10.4 predicts very similar changes in intake as live weight increases, but it does so towards a specific value of $c$ which is the food intake at maturity. The forms have been widely used to summarize the intakes observed in growing pigs. Considerable effort has been put into estimating the values of parameters $a$ and $b$ in equation 10.3 (e.g. Cole *et al.*, 1967; Cole and Chadd, 1989; Close, 1994). In some cases the value of $b$ has been forced to be a number such as 0.75 or 0.67 to connect intake with assumptions about the scaling of maintenance.

There are both shortcomings and dangers in adopting either of the above equations:

**1.** They do not have a strong theoretical basis; there is no reason why the allometric form of equation 10.3 should be an appropriate one for relating food intake to live weight. The form of the equation has been criticized by Whittemore *et al.* (1995) on the grounds that a single exponent is unlikely to describe adequately the change in the energy costs of maintenance and growth as the animal increases in weight. That the equation can have no general accuracy can be shown by an example. If the values of the parameters of equation 10.3 are taken, together with those for maintenance in the ARC (1981) system, they predict that pigs have a mature weight of about 800 kg! As this is clearly wrong it means that the two equations cannot both be correct.

**2.** They do not take into account the composition of the food, nor any qualities of the environment and the effects these might have on food intake (Kyriazakis, 1994; see also Chapter 4 of this volume). Hence, they do not allow for differences between different sets of data, nor for predictions to be made under different conditions.

**3.** The sole description of animal state is its live weight. By using this descriptor alone, they ignore the possible effect of body composition and pig genotype on intake at a weight (De Greef, 1992; Kyriazakis and Emmans; 1992a).

**4.** They do not allow intake to have an intermediate maximum value as $W$ increases, before it declines to reach the equilibrium of mature food intake (Emmans, 1997).

### Relationship between intake and time

Parks (1982) suggested a general form to relate the rate of food intake, d$F$/d$t$, and time, $t$, for farm and laboratory animals.

$$\mathrm{d}F/\mathrm{d}t = d + (c - d) \cdot (e^{(-t/t^*)}) \qquad \text{g day}^{-1} \qquad (10.5)$$

where $d$ is the intake at $t = 0$, $c$ is the mature food intake and $t^*$, is the time required for the animal to increase its rate of food intake to 0.63 of its mature food intake. In the system of Parks (1982) growth was seen as the con-

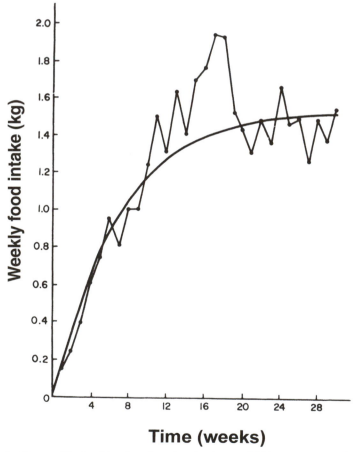

**Fig. 10.1.** The weekly food intake of cockerels given *ad libitum* access to food for 30 weeks from hatching. The fitted function is according to the equation of Parks (1982) which relates the rate of food intake to time. (Adapted from Parks, 1982.)

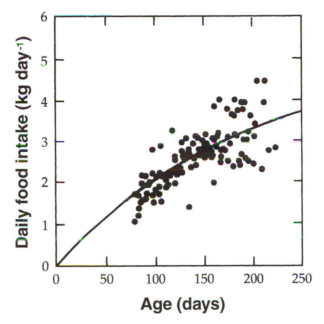

**Fig. 10.2.** Data and fitted function for the daily food intake as a function of age of gilts given *ad libitum* access to food from the experiment of Thompson *et al.* (1996). The fitted function is according to the equation of Parks (1982) which relates the rate of food intake to time.

sequence of food having been eaten. The two non-trivial parameters, $c$ and $t^*$, can be seen as genetically determined variables so that, in principle at least, the food intake of different kinds of pig can be predicted by giving them different values. The equation has recently been applied to data on the food intakes of pigs from different genotypes, so that estimates of $c$ and $t^*$ have been generated (Thompson *et al.*, 1996).

There are three main disadvantages associated with the use of equation 10.5:

**1.** Because intake is made a direct function of time *per se*, it is consequently independent of animal state; given evidence to the contrary (e.g. the intake of pigs following a period of realimentation (Kyriazakis and Emmans, 1991; Stamataris *et al.*, 1991)), this is difficult to accept.
**2.** The equation does not allow food intake to have a maximum value from which it then declines to the equilibrium of mature intake, as observed in the data of Whittemore *et al.* (1988), which is a complete set of food intake data from weaning to maturity. This disadvantage was recognized by Parks (1982) when he attempted to describe the food intake of cockerels (Fig. 10.1); he attempted to overcome the problem by adding extra terms. In fact this

shortcoming of equation 10.5 was also clearly demonstrated in the paper of Thompson *et al.* (1996), in the discrepancy between actual data and predicted food intake (Fig. 10.2).

**3.** As well as being affected by genotype the values of the two main parameters in equation 10.5 will be affected by food composition and environment (see Chapter 4). Incorporating such effects into the equation for the purposes of prediction may not be easy or straightforward.

## An alternative approach to food intake prediction

### Desired food intake

An alternative approach to food prediction of immature farm animals, initially that of poultry, has been proposed by Emmans (1981). The approach has been applied to the prediction and quantitative description of the voluntary food intake of pigs by Emmans and Kyriazakis (1989), Ferguson *et al.* (1994) and Kyriazakis (1994). In this approach it is proposed that the rates at which the animal is seeking to grow protein and lipid are the driving forces, together with maintenance, which lead to the rate at which it will try to eat a given food ($dF/dt$, g day$^{-1}$). The animal is seen as having current goals which it seeks to achieve (Emmans and Kyriazakis, 1995). In the case of the growing animal these are maintenance, the potential rate of protein retention ($dP/dt$, g day$^{-1}$) and the desired rate of lipid growth ($dL/dt$, g day$^{-1}$). These goals, in a thermally neutral environment, lead to a calculation of the rate of food intake needed to meet the energy requirement using an energy system (Emmans, 1994, 1997). Where energy is not the first limiting resource the rate at which the pig will be trying to eat the food is that which it needs to meet its requirement for the first limiting nutrient. This allows for overconsumption of energy on imbalanced foods as seen in the work of Kyriazakis *et al.* (1991), Kyriazakis and Emmans (1991) and Ferguson and Gous (1997).

When energy is the first limiting food resource, then:

$$dF/dt = (dF/dt)_e = (1/FEC) \cdot (MH + k_1 \cdot dP/dt + k_2 \cdot dL/dt) \quad \text{g day}^{-1}$$
$$(10.6)$$

where *FEC*, kJ g$^{-1}$, is the energy content of the food; *MH*, kJ day$^{-1}$, is the maintenance heat; and $k_1$ and $k_2$ are energy constants (Emmans, 1994). The estimation of the potential rate of protein retention, and the desired rate of lipid retention, have been addressed in Chapter 8. These require estimates of three genetically determined variables (the mature protein weight, the fatness at maturity and the Gompertz growth rate parameter), and of a single state variable, the current protein weight. Where conditions have been such that the desired rate of lipid retention has not been met the pig will be either fatter or leaner than it seeks to be. In such cases its desired rate of lipid retention will be either lower or greater than that calculated from its genotype values.

When an amino acid is the first limiting food resource, the desired food intake will be that needed to meet the requirement for the first limiting amino acid:

$$dF/dt = (dF/dt)_a = (1/(FAC)).(M_a + {}^1/e_p .PAC. dP/dt) \quad \text{g day}^{-1} \quad (10.7)$$

where *FAC*, g g$^{-1}$, is the food content of the first limiting amino acid; $M_a$, g day$^{-1}$, is the maintenance requirement for protein; $e_p$ is the net efficiency of using the amino acid for growth (Kyriazakis and Emmans, 1992a) and *PAC* is the content of the amino acid in the retained protein.

The first rule of food intake can now be summarized:

If

$$(dF/dt)_e > (dF/dt)_a, \text{ then } dF/dt = (dFl/dt)_e \tag{10.8}$$

Otherwise

$$dF/dt = (dF/dt)_a \tag{10.9}$$

The above implies that a pig will be prepared to consume: (i) extra energy when it is given access to a food which is low in protein; and (ii) extra protein when its food is relatively low in energy. Both these implications are consistent with experimental observations (Kyriazakis and Emmans, 1992c). On foods marginally deficient in protein the extra energy eaten is deposited as additional fat (Kyriazakis *et al.*, 1990, 1991; Kyriazakis and Emmans, 1991, 1992b; Ferguson and Gous, 1997). Excess protein is deaminated and lost through the urine when excess protein is eaten (Campbell, 1977; Kyriazakis *et al.*, 1990). The extra heat generated by this process may cause a reduction in intake where the environment is already hot (Ferguson and Gous, 1997; see below).

The above approach to food intake prediction has also recently been advocated by Whittemore *et al.* (1995), who suggested that digestible energy intake, *DEI*, can be predicted from:

$$DEI = 0.44.W^{0.75} + 52.dP/dt + 53.dL/dt \quad \text{MJ day}^{-1} \quad (10.10)$$

However, the application of the above equation to adequately predict the rate of intake of pigs of different genotypes has failed. This is because the equation uses past rates of actual protein and lipid gain rather than the *desired* rates of protein and lipid gain. Thus the application of equation 10.10 to any given data set merely results in an approximate test of the energy system used.

Emmans (1995) used the energy system of ARC (1981), purely as an example, to show that different kinds of pig would need to eat very different amounts of food at a particular live weight in order to achieve their *potentials* to grow and fatten in a thermally neutral environment. The examples are in Table 10.1. Within the reasonable range for pig genotypes intake at 50 kg live weight, of the same food, was predicted to vary between 1.30 and 2.37 kg day$^{-1}$.

**Table 10.1.** The predicted food intakes and heat losses of pigs of 50 kg live weight eating to achieve different rates of protein retention (PR, g day$^{-1}$) and different lipid : protein ratios in their gain (LR : PR) on a food with 13 MJ ME kg$^{-1}$. The calculations are based on the ARC (1981) energy system, and the table is modified from Emmans (1995).

| PR (g day$^{-1}$) | LR : PR ratio | Food intake (g day$^{-1}$) | Heat loss (MJ day$^{-1}$) |
|---|---|---|---|
| 120 | 0.5 | 1302 | 11.7 |
|     | 1.0 | 1549 | 12.5 |
|     | 1.5 | 1804 | 13.5 |
| 150 | 0.5 | 1466 | 12.5 |
|     | 1.0 | 1774 | 13.6 |
|     | 1.5 | 2083 | 14.6 |
| 180 | 0.5 | 1629 | 13.3 |
|     | 1.0 | 1999 | 14.6 |
|     | 1.5 | 2369 | 15.8 |

### Constrained food intake

It is possible that the *desired food intake* of a pig may not be able to be met due to constraints arising from the food which it is offered, or to the environment in which it is kept. Such constraints include the bulk of the food, which may limit intake, the hotness of the environment, which limits the pig's ability to lose heat, and toxins in the food. Where one of these constraints operates then the actual food intake will be a constrained rate of *food intake*, d$CF$/d$t$. Two cases of d$CF$/d$t$, the most relevant and likely to arise in modern pig production systems, are considered below.

**1.** It was proposed earlier that pigs will eat at the rate needed to just meet their requirement for energy when the food given is balanced, or has excess of all nutrients unless these are toxic. For balanced foods the prediction is that, as the energy content of the food declines due, for example, to its dilution with an indigestible material, then food intake will increase. This prediction is consistent with ample experimental evidence (for a summary of experimental data see NRC, 1987). However, the food apart from energy will also contribute 'bulk' to the gastrointestinal tract of the pig:

$$\mathrm{d}B/\mathrm{d}t = (\mathrm{d}F/\mathrm{d}t).(D.b_\mathrm{d} + b_\mathrm{u}.(1 - D)) \qquad \text{bulk units day}^{-1} \qquad (10.11)$$

where d$B$/d$t$ is bulk intake, $D$ is the digestibility of the food or its proportion which is digested, and yields $b_\mathrm{d}$ units of bulk; the undigested proportion of the food yields $b_\mathrm{u}$ units of bulk. Kyriazakis and Emmans (1995) have suggested that the water holding capacity of the food (g g$^{-1}$) is an appropriate measure of food bulk.

It is likely that there is an upper limit to the capacity of the animal to accommodate bulk, (d$B$/d$t$)$_\mathrm{max}$, which imposes a constraint on food intake. In these cases the d$CF$/d$t$ will be expressed as:

$$\mathrm{d}CF/\mathrm{d}t = (\mathrm{d}B/\mathrm{d}t)_{max}.(1/(D.b_d + b_u (1 - D))) \qquad \mathrm{g\ day^{-1}} \qquad (10.12)$$

Kyriazakis and Emmans (1995) have suggested that the maximum capacity for bulk is a function of the live weight of the animal, so that:

$$(\mathrm{d}B/\mathrm{d}t)_{max} = z.W \qquad \qquad \mathrm{bulk\ units\ day^{-1}} \qquad (10.13)$$

Suitable ranges for food related variables $b_d$ and $b_u$, and a value for constant $z$ have also been suggested in that paper. It seems likely, however, that live weight might not be an adequate scalar for the capacity for bulk over the whole relevant range. Ferguson *et al.* (1994) suggested the animal's current protein weight, $P$, should be used as a scalar for capacity for bulk. The problems of describing the bulkiness of given foods, and the capacity for bulk of given pigs, remain active areas of research.

**2.** It was also suggested earlier that an animal will be prepared to consume extra energy when it is given access to an 'imbalanced' food, in order to try to meet its requirement for a nutrient. Metabolizable energy, *ME*, is the relevant measure of food energy in this case as it is energy retention and heat loss which is of interest (Emmans, 1994). The extra *ME* eaten can either be deposited as lipid or lost as heat. Emmans and Fisher (1986) showed that intake on an imbalanced food would be able to be maximized if the animal used the rule of maximizing excess lipid retention so that its heat loss would be minimized. This is because it will confront an upper limit to its rate of heat loss which is set by the environment. It must be the case that, in the limit, the excess heat that can be lost, Exd$H$/d$t$, and hence produced, is given by:

$$\mathrm{Exd}H/\mathrm{d}t = (\mathrm{d}HL/\mathrm{d}t)_{max} - \mathrm{d}H/\mathrm{d}t \qquad \mathrm{MJ\ day^{-1}} \qquad (10.14)$$

where, d$H$/d$t$ is the rate of heat production on a perfectly balanced food, in a thermally neutral environment, and $(\mathrm{d}HL/\mathrm{d}t)_{max}$ is the maximum heat loss in the actual environment. For an approach to modelling heat loss in the pig, and hence $(\mathrm{d}HL/\mathrm{d}t)_{max}$, the reader is referred to Chapter 4 in this book, but also to the work of Bruce and Clarke (1979), Emmans (1989) and Ferguson *et al.* (1994).

The above analysis suggests that a pig will be able to grow protein at its potential rate on an imbalanced food (i.e. one that is low in protein in relation to energy), only if it is kept in a suitably cold environment (Table 10.2) and food bulk is not limiting intake. The excess heat production, associated with the extra food intake, can then be dissipated to the environment.

### Food intake following a period of (nutritional) limitation

A special case to be considered is the rate of food intake of pigs following a period of nutritional limitation, during which they have failed to grow as fast as they could, or their gain was of abnormal composition (e.g. too much lipid in relation to protein gain and vice versa) or both. According to the theoretical framework of Kyriazakis and Emmans (1992b) the pig will attempt to do the following three things through its food intake, during rehabilitation: (i) it

**Table 10.2.** The influence of the environmental temperature ($T$) on the daily rates of food intake, heat loss and protein (PR) and fat (LR) deposition in 35 kg pigs fed *ad libitum* on a moderately low protein food. (After Close and Mount, 1978; Close *et al.*, 1978.)

| $T$ (°C) | Food Intake (g kg$^{-0.75}$ day$^{-1}$) | Heat loss (kJ kg$^{-0.75}$ day$^{-1}$) | PR (g kg$^{-0.75}$ day$^{-1}$) | LR (g kg$^{-0.75}$ day$^{-1}$) |
|---|---|---|---|---|
| 10 | 169 | 1009 | 10.5 | 17.9 |
| 15 | 135 | 854 | 7.5 | 14.1 |
| 20 | 140 | 863 | 8.2 | 15.1 |
| 25 | 121 | 714 | 6.5 | 13.6 |
| 30 | 103 | 765 | 5.3 | 7.9 |

will attempt to correct its protein : ash ratio to the normal value (see Chapter 8) and hence may grow protein faster than normal; (ii) it will attempt to correct any abnormalities in its fatness over time; and (iii) the rates at which (i) and (ii) are achieved will always be set by the condition of rehabilitation. *In extremis* it is suggested that the animal will eat at the rate such that it will achieve the above two things immediately (i.e. within a day). So for a pig with a protein deficit of $P_d$, g, relative to its ash weight, and a lipid deficit, $L_d$, g, relative to its protein weight, the predicted rate of food intake on a food first limiting in energy, from equation 10.6 will be:

$$dF/dt = (1/FEC).(MH + k_1.(dP/dt + P_d) + k_2(dL/dt + L_d)) \quad \text{g day}^{-1}$$
$$(10.15)$$

In other words 'the animal does not dally in its attempt to return to normal . . . physiological status' (Parks, 1982). Whether it achieves this, and the consequent rate of food intake, will depend, according to the arguments presented earlier, on the composition of the food used (and in particular its protein : energy content), its bulkiness and the hotness of the environment in which the animal is kept. The paper of Kyriazakis and Emmans (1992b) presents predictions of the above framework, which are consistent with the experimental evidence on the food intake and gain of pigs during rehabilitation.

The system of prediction described here is theoretical in its approach. It connects the pig's ability to grow, its maintenance and growth needs for energy and nutrients, and an energy system in a way that allows a rational method for predicting food intake. Where *ad libitum* feeding is used it is unrealistic to expect that any one simple equation will allow intake to be predicted accurately across a range of conditions. The approach allows, when parameter values are used which are suitable for pigs, for a maximum in the rate of food intake before maturity (Emmans, 1997). By assuming that the rates at which the animal is seeking to grow protein and lipid are the driving forces for the rate of food intake, it also allows for the food intakes of different kinds of pig and at different states to be predicted or adequately described. Both these qualities are unique among the systems of food intake description that have been considered here.

## Prediction of Diet Selection

An interesting feeding system to be considered is when pigs are given free and continuous access to two or more foods and hence are given a choice. Since there is the possibility that the system has some practical importance (Kyriazakis, 1994), and has recently aroused interest in its use as a method of improving our understanding of what is the animal trying to achieve through its feeding behaviour (for a review see Kyriazakis, 1997), it certainly deserves some emphasis in quantitative terms.

For the system where the pig is given access to more than one food as a choice, the problem is to predict how much of each food it will eat, and hence, what the composition of its diet will be. The complexity of the problem increases as the number of foods offered to the pig increases, but for the sake of simplicity here we will consider the case where the animal is given access to two foods only. This particular case seems to be the one of practical relevance (Kyriazakis, 1994) and raises most of the issues of theoretical importance. The assumption is that the animal uses a rule, or rules, to 'decide' its diet composition; if there are no rules it is clearly impossible to predict the outcome of such a feeding system. These rules will also have to be related to quantitative qualities of the foods on offer. Differences in food qualities, such as in their 'palatability', which do not possess this quantitative characteristic, cannot lead to a quantitative description of the outcome of diet selection (Emmans, 1991; Kyriazakis, 1994). The underlying framework here is that the pig is seeking to achieve its output goals through its diet selection, in a manner similar to the one described on p. 235.

The first, and simplest, case to be considered is when pigs are given a choice between two foods which differ in a single nutritional dimension (e.g. the content of one food resource), but a combination of which is such that the pig can meet its requirement for the resource. The first expectation is that the composition of the diet will consist almost entirely of the food which is higher in the relevant resource. This expectation seems to be a reasonable one when two foods offered as a choice differ in their energy content, although the very limited experimental evidence from animals offered such a choice suggests that it is perhaps naive to consider 'energy content' as a single dietary dimension in such cases (Pitries *et al.*, 1980).

When the two foods on offer differ in their protein content – but are equal and abundant in their energy and other nutrient contents – the above suggestion implies that the pig will select a diet which consists predominantly of the high protein food. The diet selected will not be expected to

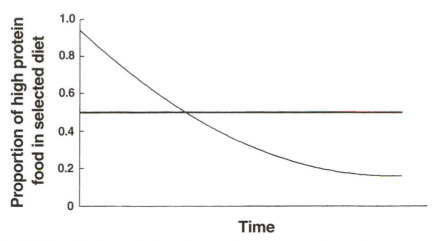

**Fig. 10.3.** Possible changes in diet selection of pigs over time. The crude protein selected over time (g kg$^{-1}$ feed) changes as the requirements of the pig change over time.

change as the requirements of the animal for protein change over time. However, ample experimental evidence suggests that pigs at the same time select a diet which avoids excess protein intake (Bradford and Gous, 1991; Kyriazakis and Emmans, 1992c, 1993; Kyriazakis *et al.*, 1991a,b). Therefore, it is proposed that the diet selected will be such that the pig meets precisely its requirements for protein when in a given state, and that it will change as requirements change over time (Fig. 10.3). In this case the composition of the diet selected is expected to be such that the protein requirement is just met. Where '$p$' is the proportion of the food with the higher protein content, $h$ g g$^{-1}$, and $(1 - p)$ is the proportion of food with the lower protein content, $l$ g g$^{-1}$, it is expected that:

$$(p.h) + (1 - p).l = \text{PRRQ}/(\text{d}F/\text{d}t)_e \qquad (10.16)$$

where PRRQ is the animal's protein requirement, g d$^{-1}$, and $(\text{d}F/\text{d}t)_e$ is the food intake needed to meet the energy requirement. For the sake of simplicity it was assumed that the two foods on offer have the same protein quality and the same energy contents.

A slightly more complex situation is when the two foods on offer differ in two dimensions, e.g. in their contents of protein and a macromineral. The expectation then is that the pig will be prepared to eat an excess of one nutrient in order to satisfy its requirement for the other (Emmans and Fisher, 1986) up to some maximum rate. The quantitative arguments become similar to the ones presented on p. 237 on constrained food intake. This is because the animal will attempt to meet its goals, but there might be constraints on the ability to cope with the excess intake of any nutrient. We are not aware of any experiments which have approached the problem of diet selection of pigs offered such choices this way. There is a quantitative approach to the problem in other kinds of animals (e.g. invertebrates; Raubenheimer and Simpson, 1993).

The complexity of the problem can be substantially increased by offering pigs access to two foods which differ in many nutritional dimensions. Emmans (1991) suggested that in these cases it would be useful if the foods offered to the animals are seen as mixtures. A mixture of three dietary components (for example fat, carbohydrate and protein) can be seen as a geometrical space defined by an equilateral triangle; a mixture of four components would be represented by a regular tetrahedron. Each of the two foods offered as a choice would occupy a point in such a diet composition space, and the animal can be seen as using some rules to decide which mixture of these will come closest to meeting the diet composition that it can be seen as seeking. The main problem then is to define accurately the relevant diet composition space of the foods offered as a choice, and then develop hypotheses on the rates that the animal uses for selecting its diet. It is possible then that the problem of predicting the diet selection of pigs offered a choice between a cereal (e.g. wheat) and a 'balancer' (a high protein supplement feed) – one that has potential practical relevance (Dalby *et al.*, 1996) – can be seen this way. Currently, however, there is no theoretical framework which could predict the diet selection of pigs offered such complex choices.

From the above it is implied that in diet selection experiments it is, if anything, even more critical than in experiments where animals are given access to a single food, to have sufficient description of the foods used. It may even be that such descriptions need to invoke dimensions which are not relevant in single feeding. We, for example, have shown that the food intake of pigs is unaffected by the relatively high levels of a toxin (glucosinolates). However, when given a choice, pigs avoid very strongly, but not absolutely, the foods that contain glucosinolates (Kyriazakis and Emmans, 1993). Explanations for the different pig strategies have been offered in that paper. We are therefore tempted to speculate that in past experiments, in which the diet selection of pigs does not seem to follow specific rules (e.g. Rose and Fuller, 1995; Cameron, 1997), the results can be accounted for by the inadequate or insufficient description of the foods offered as a choice, and the failure to place them in an appropriate dietary space.

## Discussion and Conclusions

We have reviewed here the approaches taken over the past twenty years on the prediction and description of the voluntary food intake of pigs. We have suggested that the majority of the approaches taken constitute a very limited way of describing the rate of food intake in the pig, since they seem to lack the important quality of generality. We have singled out two approaches, those of Parks (1982) and the one which stems from the idea of the 'Desired Food Intake' (Emmans, 1986), since they appear to possess this character-istic. We have concluded that the latter seems to carry more advantages, since it connects the idea of what the animal is trying to achieve with the problem of predicting food intake through the conventional idea of requirement.

Any system which seeks to predict the rate of food intake of a food by an animal at a particular time should be able to be extended to predict the way in which food intake will change over time. Park's (1982) equation is already dynamic in its nature, but fails to describe adequately the change in the rate of intake of pigs (Thompson *et al.*, 1996). It also suggests that intake is some direct function of time *per se*, rather than the state of the animal. Emmans (1997), on the other hand, has recently shown how the idea of 'Desired Food Intake' can be extended to become dynamic, since within its system: (i) the rate of food intake is a function of the current state of the animal; and (ii) the rate of change in the animal's state can be predicted from its initial state and the predicted rate of intake.

Although it was not stated explicitly, the approaches to the prediction of food intake, and by extension to diet selection, concerned an individual animal. It is of course obvious that most nutritional experiments in research, and all experiments in commercial pig production, are carried out on groups or populations of animals, and therefore one is most interested in the prediction of the food intake of groups of pigs. There is the need therefore to include stochastic elements in our predictions of food intake. However, this has not received any substantial attention, although the need to perform stochastic modelling is gradually being recognized in other areas of pig production (i.e. in predicting growth: Knap, 1996; Kyriazakis, 1996). The framework of the Desired Food Intake requires a small number of genetic variables as inputs, which are associated with the estimates of the potential rate of protein and desired rate of lipid (see Chapter 8). Providing that the distributions of these variables are known, i.e. their variances and covar-iances, then the approach can be adapted to predict the food intakes of the individual pigs which comprise a population of pigs. It is of course inherent in this approach that the efficiencies of nutrient utilization, the elements which constrain food intake, and the recovery rules are assumed to have zero variation (i.e. not to differ between pigs). This assumption is not necessarily true as Emmans and Fisher (1986) have suggested, but it constitutes the first step in introducing stochasticity in the prediction of food intake.

An issue which is associated with the above, stems from the observation that the rate of food intake of pigs kept individually can be higher than that of similar pigs kept in groups (Chapple, 1993). An approach to this problem is presented in Chapter 5 of this book, but it is possible that the problem can be approached through the framework applied on the 'constrained food intake' on p. 236.

All descriptions and predictions presented in this chapter dealt with the rate of voluntary food intake of pigs over a period of a day. This seemed reasonable to us since all animals including pigs (Nielsen *et al.*, 1995) exhibit a daily rhythm of food intake which is associated with specific events (such as a daily offering of fresh food, period of darkness, etc.). The question, however, is whether one can predict food intake and diet selection within shorter time intervals and of course what the benefits of such an approach will be. Forbes (1993) suggested that short-term prediction of food intake (e.g. in terms of meals) could enhance our understanding of the factors that are likely to control food intake of animals. However, the view taken here is the one proposed by Kyriazakis (1997), that while longer-term feeding behaviour (e.g. daily) is related very closely to longer-term changes in the

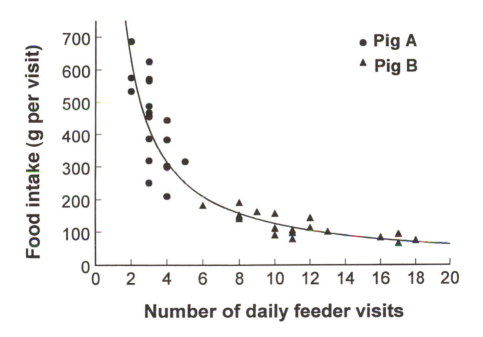

**Fig. 10.4.** Number of daily feeder visits (NDV) plotted against mean food intake per visit (FIV) for two pigs of similar size, housed within a group of 20; both pigs had received identical treatment throughout. Both pigs achieved similar levels of daily food intake (pig A (●) 1255 g day$^{-1}$ and pig B (▲) 1264 g day$^{-1}$) and the drawn isoline represents the combination of NDV and FIV resulting in a daily intake of 1260 g day$^{-1}$. (Data from Nielsen *et al.*, 1995.)

animal's internal state, short-term feeding behaviour (meals) is largely unrelated to short-term fluctuations in the animal's internal state. Short-term feeding behaviour is merely a device to exploit the feeding environment effectively. The data of Nielsen *et al.* (1995) in Fig. 10.4 strongly support this suggestion, since very similar pigs achieve the same rate of daily food intake through very different feeding strategies. It would be very difficult to construct a system which would predict adequately the short-term feeding behaviour of both pigs in Fig. 10.4.

Lastly, we touched all too briefly on the subject of the prediction of diet selection of pigs. We concluded that currently there is no appropriate system to predict the diet selection of pigs offered complex choices (i.e. foods which differ in many nutritional dimensions). We have offered an approach to the problem by suggesting that the first step towards this is by viewing the foods offered to the animals as mixtures, which occupy a specific point within a geometrical space. We considered this as the first step to developing hypotheses on the rules that the animal uses in selecting its diet. Given this, it is then possible to evaluate the system of offering pigs access to more than one food as a choice, for its practical relevance.

# References

ARC (Agricultural Research Council) (1981) The nutrient requirements of pigs. *Technical Review by an Agricultural Research Working Party*. Commonwealth Agricultural Bureau, Farnham Royal, UK.

Bradford, M.M.V. and Gous, R.M. (1991) The response of growing pigs to a choice of diets differing in protein content. *Animal Production* 52, 185–192.

Bruce, J.M. and Clark, J.J. (1979) Models of heat production and critical temperature for growing pig. *Animal Production* 28, 353–369.

Cameron, N.D. (1997) Preferential choice of diets differing in protein contents by pigs selected for high or low lean growth rate. In: Forbes, J.M., Lawrence, T.L.J., Rodway, R.G. and Varley, M.A. (eds), *Animal Choices*, Occasional Publication No. 20. British Society of Animal Science, Edinburgh, pp. 111–112.

Campbell, R.G. (1977) The response of early-weaned pigs to various protein levels in a high energy diet. *Animal Production* 24, 69–75.

Chapple, R.P. (1993) Effect of stocking arrangement on pig performance. In: Batterham, E.S. (ed.), *Manipulating Pig Production IV*, Australian Pig Science Association, Victoria, pp. 87–97.

Close, W.H. (1994) Feeding new genotypes: Establishing amino acid/energy requirements. In: Wiseman, J., Cole, D.J.A. and Varley, M.A. (eds), *Principles of Pig Science*. Nottingham University Press, Nottingham, pp. 123–140.

Close, W.H. and Mount, L.E. (1978) The effects of nutrition and environmental temperature on the energy metabolism of the growing pig. 1. Heat loss and critical temperature. *British Journal of Nutrition*, pp. 413–421.

Close, W.H., Mount, L.E. and Brown, D. (1978) The effects of nutrition and environmental temperature on the energy metabolism of the growing pig. 2. Growth rate, including protein and fat deposition. *British Journal of Nutrition* 40, 423–431.

Cole, D.J.A. and Chadd, S.A. (1989) Voluntary food intake of growing pig. In: Forbes, J.M., Varley, M.A. and Lawrence, T.L.J. (eds) *The Voluntary Food Intake of Pigs*, Occasional Publication No. 13. British Society of Animal Science, Edinburgh, pp. 61–70.

Cole, D.J.A., Duckworth, J.E. and Holmes, W. (1967) Factors affecting voluntary intake in pigs. I. The effect of digestible energy content of the diet on the intake of castrated male pigs housed in holiday pens and in metabolism crates. *Animal Production* 9, 141–148.

Dalby, J.A., Varley, M.A., Forbes, J.M. and Jagger, S. (1996) The diet selection of weaned piglets in different group sizes. *Animal Science* 62, 623.

de Greef, K.H. (1992) Prediction of production. Nutrition induced tissue partitioning in growing pigs. PhD thesis, Wageningen Agricultural University, The Netherlands.

de Lange, C.F.M. (1995) Framework for a simplified model to demonstrate principles of nutrient partitioning for growth in the pig. In: Moughan, P.J., Verstegen, M W A., Visser-Reyneveld, M I. (eds), *Modelling Growth in the Pig. EAAP Publication* No. 78. Wageningen Pers, Wageningen, The Netherlands, pp. 71–85.

Emmans, G.C. (1981) A model of the growth and feed intake of *ad libitum* fed animals, particularly poultry. In: Hillyer; G.M., Whittemore, C.T., Gunn, R.G. (eds), *Computers in Animal Production*, Occasional Publication No. 5. British Society of Animal Production.

Emmans, G.C. (1986) A model of the food intake, growth and body composition of pigs fed *ad libitum*. *Animal Production* 42, 471.

Emmans, G.C. (1989) The growth of turkeys. In: Nixey, C. and Grey, T.C. (eds), *Recent Advances in Turkey Science*. Butterworth, London, pp. 135–166.

Emmans, G.C. (1991) Diet selection by animals: theory and experimental design. *Proceedings of the Nutrition Society* 50, 59–64.

Emmans, G.C. (1994) Effective energy: a concept applied across species. *British Journal of Nutrition* 71, 801–821.

Emmans, G.C. (1995) Energy systems and the prediction of energy and feed intakes. In: Moughan, P.J., Verstegen, M.W.A. and Visser-Reyneveld, M.I. (eds), *Modelling Growth in the Pig*. EAAP Publication No. 78. Wageningen Pers, Wageningen, The Netherlands, pp. 71–85.

Emmans, G.C. (1997) A method to predict the food intake of domestic animals from birth to maturity as a function of time. *Journal of Theoretical Biology* (in press).

Emmans, G.C. and Fisher, C. (1986) Problems in nutritional theory. In: Fisher, C. and Boorman, N. (eds), *Nutrient Requirements of Poultry and Nutritional Research*. Butterworths, London, pp. 9–39.

Emmans, G.C. and Kyriazakis, I. (1989) The prediction of the rate of food intake in growing pigs. In: Forbes, J.M., Varley, M.A., and Lawrence, T.L.J. (eds), *The Voluntary Food Intake of Pigs*, Occasional Publication No. 13. British Society of Animal Production, 110 pp.

Emmans, G.C. and Kyriazakis, I. (1995) The idea of optimisation in animals: uses and dangers. *Livestock Production Science* 44, 189–197.

Ferguson, N.S. and Gous, R.M. (1997) The influence of heat production on voluntary food intake in growing pigs given protein-deficient diets. *Animal Science* 64, 365–378.

Ferguson, N.S. Gous, R.M. and Emmans, G.C. (1994) Preferred components for the construction of a new simulation model of growth, feed intake and nutrient requirements of growing pigs. *South African Journal of Animal Science* 24, 10-17.

Forbes, J.M. (1993) Voluntary feed intake. In: Forbes, J.M. and France, J. (eds), *Quantitative Aspects of Ruminant Digestion and Metabolism*. CAB International, Wallingford, Oxford, pp. 479-494.

Forbes, J.M. and Blundell, J.E. (1989) Central nervous control of voluntary food intake. In: Forbes, J.M., Varley, M.A. and Lawrence, T.L.J. (eds), *The Voluntary Food Intake of Pigs*, Occasional Publication No. 13. British Society of Animal Science, Edinburgh, pp. 7-26.

Knapp, P.W. (1996) Stochastic simulation of growth in pigs: protein turn-over dependent relations between body composition and maintenance requirements. *Animal Science* 63, 549-561.

Kyriazakis, I. (1994) The voluntary food intake and diet selection of pigs. In: Wiseman, J., Cole, D.J.A and Varley, M.A. (eds), *Principles of Pig Science*. Nottingham University Press, Nottingham, pp. 85-105.

Kyriazakis, I. (1996) A solution to the problem of predicting the response of an animal to its diet. *Proceedings of the Nutrition Society* 55, 155-166.

Kyriazakis, I. (1997) The nutritional choices of farm animals: to eat or what to eat? In: Forbes, J.M., Lawrence, T.L.J., Rodway, R.G. and Varley, M.A. (eds), *Animal Choice*. Occasional Publication No. 20. British Society of Animal Science, Edinburgh, pp. 55-65.

Kyriazakis, I. and Emmans, G.C. (1991) Diet selection in pigs: dietary choices made by growing pigs following a period of underfeeding with protein. *Animal Production* 52, 337-346.

Kyriazakis, I. and Emmans, G.C. (1992a) The effects of varying protein and energy intakes on the growth and body composition of pigs. 2. The effects of varying both energy and protein intake. *British Journal of Nutrition* 68, 615-625.

Kyriazakis, I. and Emmans, G.C. (1992b) The growth of mammals following a period of nutritional limitation. *Journal of Theoretical Biology* 156, 485-498.

Kyriazakis, I. and Emmans, G.C. (1992c) The selection of a diet by growing pigs given choices between feeds different in their contents of protein and rapeseed meal. *Appetite*, 19; 121-132.

Kyriazakis, I. and Emmans, G.C. (1993) The effect of protein source on the diets selected by pigs given a choice between a low and high protein food. *Physiology and Behaviour* 53, 683-688.

Kyriazakis, I. and Emmans, G.C. (1995) The voluntary feed intake of pigs given feeds based on wheat bran, dried citrus pulp and grass meal in relation to measurements of feed bulk. *British Journal of Nutrition* 73, 191-207.

Kyriazakis, I., Emmans, G.C. and Whittemore, C.T. (1990). Diet selection in pigs: choices made by growing pigs given foods of different protein concentrations. *Animal Production* 51, 189-199.

Kyriazakis, I., Emmans, G.C. and Whittemore, C.T. (1991a) The ability of pigs to control their protein intake when fed in three different ways. *Physiology and Behavior* 50, 1197-1203.

Kyriazakis, I., Stamataris, C., Emmans, G.C., Whittemore, C.T. (1991b) The effects of food protein content on the performance of pigs previously given foods with low or moderate protein contents. *Animal Production* 52, 165-173.

Kyriazakis, I., Leus, K., Emmans, G.C., Haley, C.S. and Oldham, J.D. (1993) The effect of breed (Large White x Landrace vs purebred Meishan) on the diets selected by pigs given a choice between two foods that differ in their crude protein contents. *Animal Production* 56, 121-128.

Lawrence, A.B., Terlouw, E.M.C. and Kyriazakis, I. (1993) The behavioural effects of undernutrition in confined farm animals. *Proceedings of the Nutrition Society* 52, 219-229.

Nielsen, B.L., Lawrence, A.B. and Whittemore, C.T. (1995) Effect of group-size on feeding behaviour, social behaviour, and, performance of growing pigs using single-space feeders. *Livestock Production Science* 44, 73-85.

NRC (National Research Council) (1987) *Predicting Feed Intake of Food-producing Animals*. National Academy Press, Washington DC.

Parks, J.R. (1982) *A Theory of Feeding and Growth of Animals*. Springer-Verlag, Berlin.

Pitries, Y.Y., Emmans, G.C. and Dun, P. (1980) The response of hens to feeds with different proportion of their energy from fat, protein and carbohydrate and the choices that they make between these. *Proceedings of the Nutrition Society* 39, 59A.

Pomar, C., Harris, D.L. and Minvielle, F. (1991) Computer simulation model of swine production systems: I. Modelling the growth of young pigs. *Journal of Animal Science* 69, 1468-1488.

Raubenheimer, D. and Simpson, S.J. (1993) Rails and arcs: the geometry of feeding. *Animal Behaviour* 45, 953-964.

Rayner, D.V. and Gregory, P.C. (1989) The role of the gastrointestinal tract in the control of voluntary food intake. In: Forbes, J.M., Varley, M.A. and Lawrence, T.L.J. (eds), *The Voluntary Food Intake of Pigs*. Occasional Publication No. 13. British Society of Animal Science, Edinburgh, pp. 27-39.

Riley, J.E. (1989) Recent trends in pig production: The importance of intake. In: Forbes, J.M., Varley, M.A. and Lawrence, T.L.J. (eds), *The Voluntary Food Intake of Pigs*. Occasional Publication No. 13. British Society of Animal Science, Edinburgh, pp. 1-5.

Rose, S.P. and Fuller, M.F. (1995) Choice feeding systems for pigs. In: Garnsworthy, P.C. and Cole, D.J.A. (eds), *Recent Advances in Animal Nutrition*. Nottingham University Press, pp. 211-222.

Stamataris, C., Kyriazakis, I. and Emmans, G.C. (1991) The performance and body composition of young pigs following a period of growth retardation by food restriction. *Animal Production* 53, 373-381.

Thompson, J.M., Sun, F., Kuczek, T. Schinckel, A.P. and Stewart, T.S. (1996) The effect of genotype and sex on the patterns of protein accretion in pigs. *Animal Science* 63, 265-276.

Whittemore, C.T. (1976) A study of growth responses to nutrient inputs by modelling. *Proceedings of the Nutrition Society* 35, 383-391.

Whittemore, C.T. (1994) Growth and the simulation of animal responses. In: Cole, D.J.A., Wiseman, J. and Varley, M A. (eds), *Principles of Pig Science*. Nottingham University Press, UK, pp. 55-74.

Whittemore, C.T. and Fawcett, R.H. (1974) Model responses of the growing pig to the dietary intake of energy and protein. *Animal Production* 19, 221-231.

Whittemore, C.T., Tullis, J.B. and Emmans, G.C. (1988) Protein growth in pigs. *Animal Production* 45, 437-445.

Whittemore, C.T., Kerr, J.C. and Cameron, N.D. (1995) An approach to prediction of feed intake in growing pigs using simple body measurements. *Agricultural Systems* 47, 235–244.

# Digestion, Absorption and Excretion

# 11

## D. Bastianelli[1,2] and D. Sauvant[2]

*[1]CIRAD-EMVT, Campus de Baillarguet, BP 5035, 34032 Montpellier Cedex 1, France; [2]INRA Laboratoire de Nutrition et Alimentation, INAPG, 16 rue Claude Bernard, 75231 Paris Cedex 5, France*

## Introduction

In applied animal nutrition, and pig nutrition in particular, the concepts linked to feed formulation evolve rapidly. The traditional approach, which consists of formulating diets to meet the nutrient requirements driven by production potential, is progressively replaced by an approach which integrates the laws of response of the animal to variations of its diet (Sauvant, 1992; Sauvant *et al.*, 1995). These responses concern not only the performance and efficiency of diet transformation, but also the quality of products, the flows of excretion of organic and mineral matter, the animal welfare, etc. In order to formulate diets which take into account these responses, the digestive and metabolic phenomena, and flows associated with the transformation of feed into animal products need to be taken into account in more detail than in the classical approach. The basic reason for this is the necessity to integrate further the underlying explicative events of responses to the diet. In practice this evolution requires one to be able to quantify the main nutrient flows between the major digestive, tissue and metabolic compartments. This concerns in particular the flows of nutrients absorbed through the gut wall. During the last decades, the formulation of the amino acid (AA) supply has already developed in this direction with the consideration of their digestibility at the ileal level (Chapter 13). It now becomes necessary to quantify not only the flows of absorbed AA, but also those of the other major nutrients such as glucose, fatty acids and volatile fatty acids. This objective leads us to seek a quantitative representation, i.e. a model, which integrates the digestive processes that determine the transformation of ingested feed into absorbed nutrients.

Digestion includes several major processes: transit, degradation, exchanges within the organism (absorption, endogenous secretions), fermentation. Each of these aspects has already been addressed by a great deal of experimental work, and some quantitative approaches have already been applied to them. However, most of the digestive phenomena are very difficult to quantify or to follow accurately in time. In addition, the interactions between the different functions (transit, absorption) are very difficult and costly to investigate experimentally. The relevance of modelling such a system is now obvious, but until now only few attempts have been performed to achieve a consistent and quantitative description of the major digestive events and their effects on feed degradation. In this chapter the various attempts to model pig digestion are reviewed. The major physiological digestive phenomena are briefly discussed with the aim to clarify the basic principles of the proposed models, and their degree of mechanisticity. More 'physiological' and detailed descriptions of digestion phenomena can be found elsewhere.

## Modelling Transit in the Digestive Tract

From the ingestion of feed to the excretion of wastes, digesta are moving in the digestive tract in a unique direction, even if some more or less marked mixing phases (stomach, caecum) or occasional processes of retroactive flux (stomach–duodenum and ileo-caecal junctions) can occur. The stomach has a volume of 5–10 l in the adult; therefore the animal can eat quickly (1.25 g DM $min^{-1}$ $kg^{-0.75}$; Nienaber *et al.*, 1991) a great volume of feed. The small intestine is approximately 20 m long in the adult, which represents almost three-quarters of the total length of the digestive tract. The large intestine is around 5 m long with a maximum volume of 10 l, including the caecum. The dynamic characteristics of the progression of the digesta are not the same in these successive parts of the digestive tract, therefore they cannot be treated *a priori* in the same manner in models.

The most common approach consists of describing transit as the mean retention time (*MRT*) of digesta in a specific compartment or in the whole tract. The duration and the kinetics of the transit can be assessed by various methods, which do not necessarily lead to the same outcome (Warner, 1981). *MRT* in the whole tract is the easiest variable to measure because it does not require animal modification (canulation, surgery). The major contribution to *MRT* is from the large intestine (more than 20 h versus approximately 4 h and 3 h in the stomach and small intestine respectively). *MRT* depends on several factors, and particularly on the level of feed intake (Roth and Kirchgessner, 1985) and the nature of digesta (e.g. proportion of dietary fibre: Cherbut *et al.*, 1988). It is also very variable between individuals. The effect of dietary fibre (described as *NDF*) is obvious, and clearly non-linear (Fig. 11.1). A statistical analysis of existing data provides a prediction of *MRT* as a function

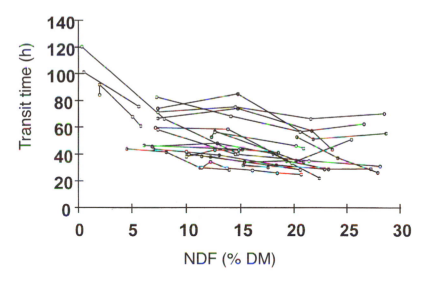

**Fig. 11.1.** Total DM transit time (h) as a function of *NDF* content (% DM) of the diet. Data from literature: Castle and Castle (1956); Castle and Castle (1957); Furuya and Takahashi (1975); Canguilhem and Labie (1977); Furuya *et al.* (1978); Fioramonti and Bueno (1980); Ehle *et al.* (1982); Kuan *et al.* (1983); Roth and Kirchgessner (1985); Stanogias and Pearce (1985); Pond *et al.* (1986); Sandoval *et al.* (1987); Cherbut *et al.* (1988).

of the characteristics of the experimental diets. The following equation is an example of the results that can be obtained:

$$MRT = 108.2 - 0.37 \times FL - 12.69 \times \ln (NDF) \qquad h$$
$$(R^2{=}0.57, RSD{=}12.6, n{=}88) \tag{11.1}$$

where *FL* is the feed level (DM intake/metabolic weight) in g g$^{-0.75}$ and *NDF* is the diet *NDF* content in per cent of feed DM.

Knowledge and prediction of *MRT* cannot be ignored in digestion modelling because it is a major cause of the difference in digestive capacity across ages, and also a major contributor to the efficiency of digestion of certain nutrients, for example of cell wall. Nevertheless, the *MRT* approach has a limited explicative power because it represents an empirical outcome which is not based on the existing knowledge of the underlying phenomena. Therefore it is useful to try to integrate more precisely the components of the transit time, particularly by considering separately the events occurring in the successive digestive compartments.

## Gastric emptying

The stomach accumulates the feed immediately after a very coarse mastication and a rapid passage in the oesophagus. During the gastric phase a large quantity ($10 \, l \, day^{-1}$; Juste, 1982) of gastric juice is added and mixed with the feed. Strong muscular contractions then mash and homogenize digesta. Only the liquid phase and the small particles can be evacuated by the pylorus. The outflow is discontinuous and consists of successive waves. There is a feedback control of gastric emptying by chemical factors acting in the duodenum or further in the small intestine. Some authors (Hunt et al., 1985) state that the regulation of this outflow is homeostatic, based on the energy flow leaving the stomach. A precise modelling of gastric emptying as a function of time has been performed by Laplace and Tomassone (1970). Their approach was empirical relationships expressed as third degree polynomials, and thus did not describe the mechanisms involved. Therefore their equations cannot be used when feeding conditions vary (i.e. quantity and composition of diet).

The simplest models of gastric emptying ignore the short-term discontinuity of the process and are thus first-order processes, also called mass-action laws: i.e. it is stated that the quantity of matter leaving the stomach per unit of time is directly proportional to the total quantity present in the stomach at that time (Usry et al., 1991; Bastianelli et al., 1996). As it is difficult to estimate precisely the quantity of water secreted in saliva and gastric juice, the calculations are based on dry matter rather than on fresh matter. Data from literature confirm the validity of this simple principle when feed is ground and homogeneous (Fig. 11.2). When solid aggregates (several millimetres) are present in the feed, it is possible to improve the model by adopting a structure with two (or more) compartments. There is thus a compartment of small particles that can leave the stomach, and one of large particles that cannot leave the stomach. The flow rate between these two compartments corresponds to the crushing activity of stomach muscle. This process is likely to be facilitated by the liquid impregnation of the large particles. This approach was proposed in Bernier et al. (1988) for humans, and was also used in most rumen models (Sauvant and Ramangasoavina, 1991). The approach of Oko (1979, quoted in Usry et al., 1991) also considered two subcompartments in the stomach: the first one (S1) received and crushed the feed, and the second (S2) accumulated the digesta before providing them to the duodenum. The flow from the latter was proportional to the square root of the size of S2. This representation is complicated and does not seem to represent physiological actions, since the pyloric antrum is considered to have an important role in crushing of particles (Bernier et al., 1988) and thus its action cannot be reduced to that of a delay compartment. Bernier et al. (1988) have pointed out that the rates of gastric emptying and of reduction of particles can vary through time as the outcome of accelerated or slowing down processes. This could be a way to take into account the

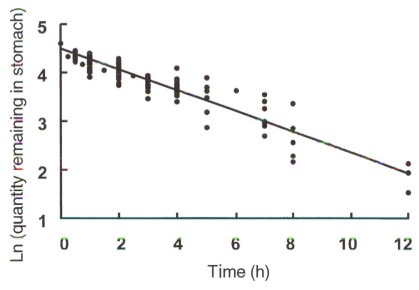

**Fig. 11.2.** Gastric emptying of DM (ln of quantity remaining in the stomach) from homogeneous meals over time in the pig. Data from literature: Rérat and Lougnon (1963); Zebrowska and Horszczaruk (1975); Cuber and Laplace (1979); Cuber *et al.* (1980); Cuber *et al.* (1981); Laplace and Cuber (1984); Low *et al.* (1985); Rainbird and Low (1986).

effects of the regulations of gastric emptying. Corresponding data are unfortunately difficult to find because many factors can affect the rate of gastric emptying.

Suggestions have been made to refine exponential models: Bernier *et al.* (1988) used fresh matter instead of dry matter. This was a more realistic approach regarding distension factors that could regulate outflow. However, it required reliable data on the liquid secretions and kinetics. Usry *et al.* (1991) proposed to use four different coefficients for the rate of gastric emptying, because they considered that this was more rapid during the first hour after the meal than later, and also during the day than during the night. However, the data used came from a single experiment and probably do not have a universal significance.

Some alternative models to describe gastric emptying were further proposed. Turner *et al.* (1987) suggested that there was a maximum rate of DM passage which was modified by a coefficient ($K$) which accounted for the degree of filling (*DF*, in % of a defined maximum) of the stomach according to the equation:

$$K = 3 \times DF^2 - 2 \times DF^3 \tag{11.2}$$

However, even this representation, which is a modification of the mass-action law, does not seem to be able to account more reliably for actual events.

## Transit in the small intestine

The small intestine is a tube of about 15–20 m long in the adult. Digesta are propelled by peristaltic contractions associated with the migration of myoelectric complexes (MC) from duodenum to ileum (Laplace, 1978). These complexes are composed of a quiescent phase whose duration varies according to feeding and time of the day (Rayner and Wenham, 1986), a phase of irregular spiking activity (ISA) lasting 50–80 min, and a phase of regular spiking activity (RSA) lasting 3–5 min. The MC migrate at a decelerating speed from about 20–30 cm min$^{-1}$ in the first parts of the small intestine to about 5 cm min$^{-1}$ in the distal parts (Laplace, 1978). As can be observed by simultaneous electromyographic and radiological observations performed on the pig by Rayner and Wenham (1986), digesta have a rapid progression for a moderate distance during ISA phases associated with peristaltic rushes, and a slow but complete progression with the passage of RSA phases. They literally sweep away digesta and carry them away via the ileum.

Attempts to model transit in the small intestine are rare. Tomassone and Laplace (1973) have already proposed the use of compartmental models to study the physiology of digestion. Nevertheless, their studies have remained at a preliminary stage. Turner *et al.* (1987) used a single compartment representation of the small intestine, which obviously does not correspond to a realistic representation, since part of the digesta leaves the compartment as soon as it enters. Moreover, such a system underestimates 'structurally' the mean retention time in the small intestine. If the outflow rate constant is lowered to get an acceptable retention time then the quantity of matter in the compartment becomes overestimated (D. Bastianelli, unpublished observations). This means that a minimum of multicompartmental structure must be adopted to achieve a minimum of reliability for the small intestinal part of a model of digestion.

The compartmental structure of the model of Bastianelli *et al.* (1996) took these observations into account. The small intestine was divided into two parts: the duodenum and the jejunum + ileum. The former was considered as a simple compartment, whereas the latter included a delay. This delay was equivalent to the passage of digesta through 'waiting compartments' that accounted for the transit time in the small intestine. The outflow from this compartment followed a mass-action law applied to the total DM quantity minus the DM 'in transit' which could not yet be evacuated. The results presented in that paper were realistic, but the representation did not account for the mechanisms of digesta propulsion.

The model proposed by Usry *et al.* (1991) was structured in a different way (Fig. 11.3). The small intestine was divided into a large number of

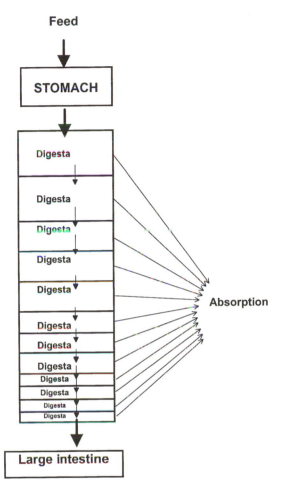

**Fig. 11.3.** Diagrammatic representation of the model of Usry *et al.* (1991). Gastric emptying follows a mass-action law, with a higher emptying rate during the first hour after each meal. Transit in small intestine is represented by 142 sections of 6 cm each. Large intestine is an accumulation compartment (for details see text).

sections (142 sections of 6 cm length). Each of them had a maximum volume. The propulsion of digesta was linked to contractions associated with the slow waves. Every 3 s (or 6 s in the distal part of small intestine), each section had a probability to contract and consequently to transmit a part of its DM content to the following section. The probability followed a Markov process of third order, which meant that it was linked to the events that occurred in the three preceding periods. The validation of the model described in the paper of Usry *et al.* (1991) was not very convincing, which suggests that the representation was not optimal. The effect of the fibre level in the diet was also considered since a different set of parameters was used when *NDF* level was higher or lower than 15%. It is noticeable that the equations describing transit were based on total matter (and not DM), but the result was the same since the DM content of digesta was considered as constant. This model was an actual attempt to integrate dynamic properties of digesta, but it was limited by lack of experimental data: most of the choices (parametization but also structure of the model) relied on the outcomes of a single experiment. This model was rather complicated but not so close to the known underlying mechanisms.

The representation proposed by Rivest (1995) (Fig. 11.4) was more mechanistic. He also considered a succession of small sections, with a DM passage from one to the other occurring every 5 min. The length of the sections decreased all along the small intestine, in order to account for the decreasing speed of digesta. The total retention time was an input of the model, from which the length of sections and consequently the digesta flow were calculated. The resulting description was comparative to the data of Laplace (1978) on the speed of migration of MCs, which validated the calculations. Nevertheless, the transit was regular, without showing the digesta waves associated with RSA phases. This concept of transit was only applied to the protein fractions which were the main interest of the model. It cannot therefore be compared to experiments on DM transit, but the validations made on nitrogen passage through the intestines seem to be sound.

## Transit in the large intestine

The transit processes occurring in the large intestine are less well documented than those in the small intestine. The caecum acts more or less as a mixing reservoir of digesta, and the slow transit in the colon is based on peristaltic waves, the regulation of which does not seem to be precisely known. A simple compartment representation is not recommended because it would lead to a very large compartment with respect to the retention time: more than 1000 g DM for a pig eating 1.5 kg DM day$^{-1}$, which is obviously too high. The only published model including a large intestinal compartment is the one of Bastianelli *et al.* (1996). The representation was compartmental, with a fractional rate of outflow depending on the DM present in the compartment. This accounted for the observation that a greater ileal DM flow decreases the retention time in the large intestine (Warner, 1981). However, because of the lack of experimental data, the authors chose a conceptual approach expressed by the following equation:

$$\mathrm{d}QLI/\mathrm{d}t = FLIE \times \exp\left((QLI - QN)/kn\right) \times QLI \qquad \text{(g min}^{-1}\text{)} \qquad (11.3)$$

where *QLI* is the DM in the large intestine, *FLIE* is the basal fractional rate of outflow (constant), *QN* is a neutral or 'set point' value of *FLIE* for which the transit followed the basal fractional rate and was neither markedly quickened

**Feed proteins**

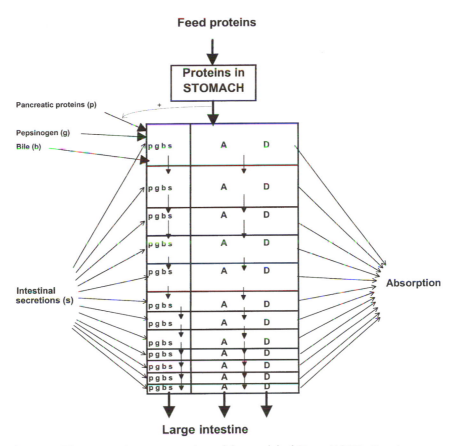

**Large intestine**

**Fig. 11.4.** Diagrammatic representation of the model of Rivest (1995). Gastric emptying follows a mass-action law. Transit in small intestine is represented by approximately 50 sections of increasing length. Digesta pass from one section to the next one at regular intervals. Only the protein components are considered. The dietary protein is originally intact (A) and is progressively degraded into absorbable end products (D). Endogenous protein sources considered are pancreatic protein (p), pepsinogen (g), bile protein (b) and intestinal secretions (s). Absorption of D is limited by a maximum rate of absorption.

nor slowed down, and *kn* is a constant. Although this solution did not represent directly physiological events, it offered a consistent description of the transit kinetics, and represented fairly realistically the other digestive events which occur in this organ (fermentations, etc.).

# Modelling Endogenous Secretions

Endogenous secretions have been extensively studied, but investigations have mainly dealt with the ileal nitrogen, or the qualitative description of individual secretions. In fact very few attempts have been made to construct a quantitative model of endogenous secretions (Bastianelli, 1996). However, some authors present their results in diagrams which could be the basis of relevant mechanistic models (e.g. Krawielitzki *et al.*, 1990; Souffrant *et al.*, 1993). These studies, however, are often limited to nitrogen or amino acids (AA), although it is well documented that there are also significant endogenous flows of minerals and lipids (Juste, 1982).

The model of Usry *et al.* (1991) did not involve any input from endogenous material. Although this model only considered DM flows, such an omission is important since the daily DM flow of endogenous origin can be as high as 100 g (calculated from Juste, 1982). In the model of Rivest (1995) the input from endogenous proteins was only considered. All sources of endogenous proteins were assumed to be of a constant value throughout the day with the exception of pancreatic juice. The latter was represented by a conceptual submodel, based on a theoretical 'stimulating hormone' whose synthesis rate was linked to the duodenal protein flow. In the small intestine, the secretion of AA was distinguished from the secretion of proteins. The proteins originating from endogenous secretions had a different degradation rate than the alimentary ones.

The model of Bastianelli *et al.* (1996) included endogenous secretions for proteins, AA, fat and minerals in all sections of digestive tract. These secretions were considered as proportional to the DM flow. It was further assumed that for a 'standard' intake level, the daily quantity of endogenous material was the mean of literature values. A higher DM flow (higher DM intake or lower digestibility) corresponded to a higher endogenous input. The material from endogenous origin was pooled entirely with the corresponding alimentary one. The degradation and absorption kinetics were therefore the same irrespective of their origin. These assumptions were made for the sake of simplicity, but should be refined in future models since feed and endogenous material do not behave in the same way in the intestines.

This is the reason why a more analytical description of the endogenous protein and AA flow in the stomach and small intestine has been proposed by Bastianelli (1996). The various sources of endogenous nitrogen (saliva, digestive and intestinal secretions) were quantified according to mean values deriving from literature data. Each of these secretory flows presented a constant part and a variable one indexed on the DM flow. The results obtained were consistent with those in the literature, and the model allowed estimation of the quantities of AA reabsorption and their kinetics of absorption and ileal passage. The results of the simulations could be interpreted according to the concepts of 'specific' and 'non-specific' endogenous nitrogen developed by Sève (1994). Nevertheless, some modelling factors, such as the effect of antinutritional factors, should also be taken into account as suggested in the review of Nyachoti *et al.* (1997).

# Modelling Degradation and Absorption Phenomena

Degradation and digestion processes have been classically studied by three methods:

1. Enzymatic degradations and *in vitro* hydrolysis.
2. Digestibility at ileal or faecal level.
3. Kinetics of appearance of nutrients in the blood.

The accuracy of the representation of degradation and absorption in mechanistic models is limited by lack of data on the kinetics of *in vivo* phenomena. The first two methods are not really mechanistic modelling, but they can provide some quantitative information and will therefore be discussed briefly. The attempts for a mechanistic description of degradation kinetics will be examined in more detail.

## Enzymatic hydrolysis

*In vitro* studies allow characterization of feed or its components in terms of potential degradability and degradation kinetics. However, extrapolation to *in vivo* conditions is somewhat risky because experimental conditions are generally very different from the *in vivo* ones. In particular the digesta entering the small intestine have already been markedly modified by the mouth and the stomach activities. Besides, such studies are very often focusing on only one specific constituent (proteins, starch, fibre, … ). In ruminants, *in situ* measurements of rumen degradation kinetics are classical and fairly well standardized (Michalet-Doreau *et al.*, 1987); unfortunately there is no similar methodology in the pig. The mobile nylon bag technique (Sauer *et al.*, 1983) is not as simple and as reliable as the rumen nylon bags technique. Some potential degradability or degradation rates data can be found in human nutrition, but there is no published standard *in vitro* method to evaluate the *in vivo* digestibility of the constituents of the whole feed.

The classical mathematical representation of the enzymatic degradation is a Michaelis–Menten kinetics equation. Thus the flow rate ($F$) of appearance of the products of hydrolysis of a substrate of concentration $C$ follows the relation:

$$F = F_{max} \times C/(C + K_m) \tag{11.4}$$

where $F_{max}$ is the maximum value of $F$ and $K_m$ is the affinity constant which corresponds to the value of $C$ for $F = F_{max}/2$. However, the *in vivo* degradation of feed can also be limited by other aspects such as a limited access of enzyme to substrate (due to cell wall for instance). Moreover, some constituents are aggregated in equations (different starches are gathered under 'starch', different kinds of proteins are gathered under 'proteins') although they include fractions having different degradation kinetics. In such situations the Michaelis representation cannot be effectively applied. The use of such a representation is therefore more aimed at representing an aggregated relation to describe saturable processes than at describing a specific mechanism.

'Digester' systems allow us to follow the kinetics of degradation of a feed from the removal of digestion products by dialysis (e.g. Savoie and Gauthier, 1986). They are in fact 'physical models' which can provide degradation kinetics (e.g. Holm and Björck, 1992) or at least rank the feeds according to their potential degradability. They can be used as experimental tools to identify explicative parameters or to test hypotheses facilitating the elaboration of mechanistic models of digestion.

## Digestibility

Digestibility coefficient values represent the most frequently used aggregated criteria to assess degradation and absorption in practice. They are an estimate of the fraction of the feed that has apparently been absorbed by the animal. As this is the outcome of numerous phenomena, the approach is 'empirical' rather than mechanistic. Feed digestibility is the main source of variation on its nutritive value, and therefore digestibility trials are major experimental methods for feed evaluation and feed unit systems. Within this framework, relationships between digestibility and feed composition have often been studied to take into account the causes of variation in digestive balance trials (e.g. Noblet *et al.*, 1989). These balance trials are and will remain indispensable in validating more explicative models. Simulated results of mechanistic models need to be compared at their highest (aggregated) level, to data of digestibility to check the consequences of underlying relationships. They can also be used as a part of the input parameters in digestion models as they reflect the potential digestibility values of some constituents (cf. Usry *et al.*, 1991).

## Kinetics of absorption or appearance of nutrients in blood

Nutrient absorption can be a simple diffusion process, an active transport or, more generally, a combination of these two mechanisms (Matthews, 1991). The flow of absorption by diffusion is proportional to the difference of concentration between the two considered compartments: the intestinal lumen and the portal blood. The mechanism for active transport is more complicated. In this case, the flow rates of absorption are curvilinear functions of nutrient concentration, as shown by Adibi (1969). This relation can be well represented by a Michaelis–Menten function. This representation of absorption is simple, but its parameterization still remains difficult because actual quantitative data are scarce. On the other hand, absorption mainly concerns the end products of degradation, which are limited in number, and whose characteristics do not systematically depend on specificities of the original substrates. These latter are nevertheless important because they can influence the physicochemical properties of digesta such as their viscosity (Bernier *et al.*, 1988). The main problems concern the estimation of their concentration in the lumen, which brings us back to degradation flows.

The measurement of the kinetics of the difference of concentration of nutrients between portal vein and arterial blood (Rérat, 1977) is a very useful experimental method for building or validating models of absorption. This method allows, when blood flow is simultaneously measured, the kinetics of nutrient appearance in the blood to be followed. This method provides useful patterns of nutrient absorption, if we take into account the bias induced by the metabolism of the intestinal wall. Unfortunately it is technically almost impossible to obtain simultaneously and on the same animal reliable values of disappearance rates of digesta in the lumen. Such a possibility would allow one to validate digestive models on their most useful terminal outputs. The technique of portal flow has been extensively used in the pig, in particular by Rérat and colleagues, and data can be found in the literature. Empirical models of the cumulative absorption as a function of time or diet quantity have been sometimes carried out (Rérat *et al.*, 1984). This has provided us with an indirect quantitative estimation of the metabolization of nutrients in the gut wall.

### Representation in mechanistic models

Degradation and nutrient absorption processes have been addressed in various ways in published models, depending on their aim: very mechanistic and precise descriptions were proposed in specific situations, which can hardly be extrapolated to actual *in vivo* situations, whereas more rough and aggregate models, also further away from basic processes, have a more general application.

Fairly precise models of absorption in simple situations have been proposed in human nutrition. This is for instance the case in Bernier *et al.* (1988) who studied the absorption of glucose in the intestine with a specific model which simulated the liquid flow rate in the duodenum, as well as the water, sodium and glucose movements. The liquid speed was estimated from the flow rate (as an input parameter) by the relationship:

$$Xn = 1 + (LFR / 8) \qquad \text{cm min}^{-1} \qquad (11.5)$$

where $Xn$ is the progression length per minute (cm min$^{-1}$) when flow rate is $LFR$ (ml min$^{-1}$). Glucose absorption was a Michaelis–Menten function of its luminal concentration. This model presented the advantage of considering the liquid fraction and not only the DM one. Water absorption was estimated in order to maintain a fairly stable osmolarity at a value close to the one in plasma. This kind of model is well suited to simple situations. This is for instance the context of simulation of experimental data obtained by liquid perfusion of controlled composition and flow rate. These models seem to be useful because they indicate the shape of the relationships and the range values of key parameters. Nevertheless, it is not recommended to use them directly for simulating more complicated situations where a complete diet is degraded during transit. Indeed, the viscosity properties observed for some digesta do not allow the use of the above-mentioned simple relationships used to determine propulsion rate of digesta. Moreover the determination of flows of absorption of water is complicated since numerous factors can influence digesta, osmolarity and water retention.

At the opposite end of the precision scale, the approach developed by Usry *et al.* (1991) considered a simple 'disappearance rate' for DM moving in the intestine. This disappearance was the sum of the effects of degradation and absorption. At each time step, a quoted quantity of DM, proportional to total DM, left each intestinal portion. The disappearance rate was calculated from the diet ileal DM digestibility which was an input parameter. This approach is highly integrated but it can be useful if the main concern is the DM transit. However, applications of such a model have several limits. Firstly, the model does not take into account diet composition although it is well known that the various fractions of the diet can have significantly different degradation rates (even with similar ileal digestibility). In addition, the absorption rate is assumed to be the same all along the small intestine. This contrasts with experimental results which show that the fractions present in the ileum are less degradable than the duodenal digesta. As a consequence, the predictions made by the model cannot represent well the appearance of nutrients in the blood – all the more so since nutrients are not identified.

In the model of Bastianelli *et al.* (1996), the processes of degradation of the various substrates in the anatomical compartments were assumed to follow a simple mass-action law. This means that enzyme quantities were not supposed to limit degradation. Fractional rates of degradation were therefore considered as feed attributes and should be used as input parameters (Bastianelli and Sauvant, 1995). Michaelis–Menten functions were used for the absorption, because several data, as mentioned above, have shown that this is a saturable phenomenon. Functions were based on the quantity of nutrients available for absorption in each of the compartments, rather than on their concentrations, because the use of concentration is debatable in a model that considers DM only. The choice of Michaelis–Menten functions was more conceptual than mechanistic since nutrients were pooled into categories (e.g. all AA together) and not considered separately.

The model of Rivest (1995) used another type of representation, which was partly linked to the different ways of representing transit. The model dealt only with protein. The fractional rate of dietary protein degradation (*FRD*) decreased with time (and therefore during transit) following an exponential function:

$$FRD = a \times \exp(-b \times t) \qquad \% \; min^{-1} \qquad (11.6)$$

where *a* was the initial value for *FRD*, *b* was the decay parameter and *t* was the time of presence in the intestine (from the entrance). The algebraic integration of this equation is:

$$NDP = 100 \times \exp\{a/b\,[\exp(-b \times t) - 1]\} \qquad \% \; min^{-1} \qquad (11.7)$$

where *NDP* is the fraction of dietary protein which is not degraded at time *t*. The validation exhibited fairly satisfactory results for this model. It must be noted that for low values of *b* (0.009 for casein according to the author) the quantity $\exp(-b \times t)$ is close to $(1 - b \times t)$, and the relationship then becomes almost a mass-action law (of parameter *a*). Protein of endogenous origin was considered separately and its degradation followed a first-order process. Flows of absorption were linked to the level of the compartment corresponding to the degraded protein (free AA and peptides). A maximum value of the absorption rate was calculated from the data of Buraczewska (1981a, b): 8 g $m^{-1}$ $day^{-1}$ for the proximal parts of intestine and 12 g $m^{-1}$ $day^{-1}$ in the distal ones. The authors tried to use a Michaelis–Menten function but as the $K_m$ obtained were very low, a constant absorption rate was preferred.

In conclusion, this model allowed the prediction of the amino nitrogen absorbed and of the protein flow in the small intestine with a higher level of mechanisticity. The values for degradation parameters were specific of the proteins considered. Nevertheless, the calculation of these parameters requires the actual availability of digestible protein at various levels of small intestine as input parameters, which are seldom known. In the present version, this model was limited to the protein fraction, with no consideration of non-protein nitrogen or any other dietary components.

## Modelling Microbial Digestion in the Large Intestine

In the pig, because of the long MRT of digesta and of the physicochemical conditions in the hindgut, there is a significant microbial fermentation of substrates arriving from the small intestine. The energy released by the fermentation of carbohydrates and proteins is partly transformed into ATP which supports maintenance and synthesis of microbial matter (Kirchgessner *et al.*, 1994). The end products of this process are volatile fatty acids (VFA), mainly acetate, propionate and butyrate, ammonia ($NH_3$) and gases ($CO_2$, $CH_4$). VFA are absorbed by the organism and contribute to the energy supply at a probable average level of 10% of the maintenance requirements (Imoto and Namioka, 1978). However, the instantaneous contribution of VFA to energy supplies is likely to be higher within the time range of 5 to 8 h after meals (Giusi, 1986). Ammonia is quickly absorbed and transformed into urea in the liver. Urea is essentially eliminated through urine, however it can also be recycled from the blood into the distal part of the digestive tract (urea–$NH_3$ cycle). This diffusion of urea is in general quantitatively smaller than $NH_3$ uptake so that the ileal digestibility of nitrogen is less than the faecal one. Amino acids cannot be absorbed from the large intestine in significant quantity (Just *et al.*, 1981). It is assumed that processes of reflux of digesta into the small intestine can permit the digestion of some bacterial proteins synthesized in the hindgut. Faecal digestibility of amino acids is not a good predictor of their uptake by the small intestine since the microbial protein synthesis largely alters the AA profile of digesta. This fact is the major reason to measure ileal and not faecal digestibility to evaluate AA availability in pig. Numerous studies have been carried out on the metabolism in the large intestine, particularly in relation to diet level (e.g. Dierick *et al.*, 1989). However, as far as we are aware, the only modelling attempt was the very simple representation in Bastianelli *et al.* (1996). This model was based on the principles of rumen models since there is a strong analogy between rumen and caecum+colon digestive processes (Argenzio and Stevens, 1984). Several rumen models have been published since the initial attempt of Baldwin *et al.* (1970) to the more recent developments of Lescoat and Sauvant (1995). The model of Bastianelli *et al.* (1996) considered a microbial population having a constant composition and a growth indexed on glucose availability with a constant efficiency, in a manner similar to the simple model of the fermentations in RUSITEC (*in vitro* digester simulating ruminal fermentations) of D. Sauvant (1990, unpublished). Nitrogen was not considered as potentially limiting bacterial growth since urea diffusion provides a source of nitrogen when digesta have a low N concentration. The model also took into account other major nutrients (lipids, minerals), but these were not considered as potentially limiting.

## Major Criticisms Applied to Digestion Models

In order to model the whole digestion, it is necessary to choose a representation for each of the processes discussed above: transit, endogenous secretions, degradation and absorption, fermentation. Only three models of those mentioned above have achieved a degree of integration and completion which allows constructive criticism: those of Usry *et al.* (1991), Rivest (1995, and Rivest *et al.*, 1994) and Bastianelli *et al.* (1996).

The model of Usry *et al.* (1991, Fig. 11.3) describes the transit of DM in the stomach and small intestine. The digesta then enter a distal compartment of accumulation. The main limit of this model is the fact that it considers only DM. Therefore, it can hardly be considered as a complete model of digestion. However, it is an interesting approach because it aims at representing the transit more mechanistically than with a simple single compartment. However, the basic mechanism used is rather complicated and unlikely to represent the physiological processes involved in the intestinal transit. The fact that the absorption rate is uniform along the small intestine contradicts literature data for protein (Nixon and Mawer, 1970), as well as for carbohydrate absorption (Keys and de Barthe, 1974). This drawback probably leads to prediction of excessive values for DM flow in distal parts of the small intestine. Moreover, the structure of the model cannot be improved by taking into account diet composition.

The model of Rivest (1995, Fig. 11.4) describes more mechanistically protein transit and absorption in the stomach and small intestine. It accounts for most of the known major phenomena, even if the description of some of them could be further improved, as the author states. This could be done by including other major nutrients and also a large intestine compartment. It can already be used as such to investigate the effect of some factors (transit time, rate of protein degradation), and it is therefore a useful research tool. However, its running requires some parameter values that are difficult to obtain. For that reason the author proposes a method to predict these parameters from ileal digestibility data.

To date, the model of Bastianelli *et al.* (1996; Fig. 11.5) is the only one to include all the major parts of the digestive tract (stomach, small intestine, large intestine), and also the only one to consider the major dietary components (starch, protein, fibre, fat, minerals). The degree of aggregation of digestive phenomena is high compared to the current knowledge in some specific areas of physiology; however this is allowed to be consistent throughout the model. In its current version transit description does not take into account explicitly the mechanisms involved in digesta propulsion, however simulated retention times and transit kinetics are fairly sound. The influence of feeding level and transit duration on digestibility of some diet components, particularly cell wall ones, has not yet been validated. One of the major objectives of digestion models is to simulate the kinetics of nutrient absorption. To achieve this objective with maximum reliability, this model focuses on the 'animal' part of digestion phenomena. It can be enriched by a more detailed description of some feed attributes (Bastianelli and Sauvant, 1995), to take more realistically into account the differences in nutrient digestibility between feeds. Moreover, in order to be comparable to the portal flow rates of absorbed nutrients, the model would have to include a minimum of information on the gut wall metabolism and its consequences. This is particularly true for the absorption rate of nutrients and the delaying role of gut wall on their pattern of entry. However, the results obtained are encouraging since the simulation of nutrient absorption is fairly consistent with experimental data (Fig. 11.6)

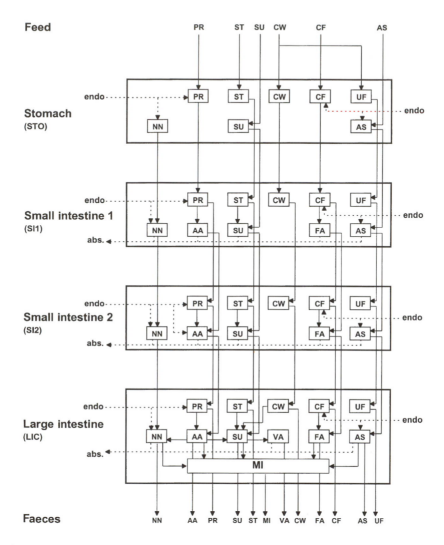

**Fig. 11.5.** Diagrammatic representation of the model of Bastianelli *et al.* (1996). Digestive tract is divided into four anatomical compartments: stomach (STO), small intestine (SI1 and SI2), and large intestine (LIC). Gastric emptying follows a mass-action law. Transit in SI1 and SI2 is based on a mass-action law, with a delay function. Transit in LIC is a non-linear function of digesta quantity. Principal digesta constituents are considered by biochemical subcompartments: non-protein nitrogen (NN); protein (PR); pool of amino acids (AA); starch (ST); sugars (SU); digestible cell walls (CW); lipids (CF); fatty acids (FA); undigestible cell walls (UF); and minerals (AS). In addition, there is a microbial subcompartment in LIC (MI). Degradation flows are based on mass-action laws, whereas absorption is a Michaelis–Menten function of nutrient quantities. Endogenous sources of PR, AA, CF and AS are considered. Flows between compartments are represented by solid lines. Other flows (broken lines) are endogenous secretions (endo) and absorption (abs).

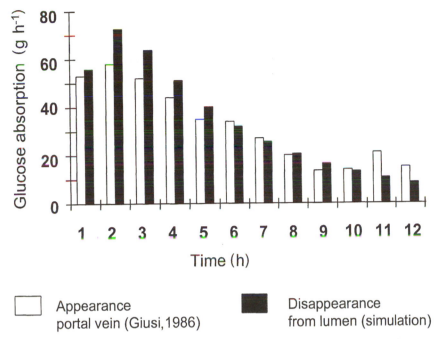

**Fig. 11.6.** Simulation of glucose absorption by Bastianelli *et al.* (1996). The figure shows a representation of glucose absorption (as glucose appearance in portal vein) in the experiments of Giusi (1986, white bars) and as disappearance from the lumen simulated in the model (black bars).

## Modelling Excretion Flow Rates

In industrial countries, one of the most important challenges of pig produc-tion is to predict excretion flows of some nutrients which are associated with deleterious environmental influences (see Chapter 12). This problem cannot be solved by modelling only at single animal level. Effectively, it must take into account higher levels of organization, particularly the herd and the area which can be used to spread the piggery effluents. Moreover, interactions between the various levels of organization cannot be ignored. For instance if there is a global constraint of maximum of N effluents per hectare there is a degree of substitution between the diet N content (lower level) and the number of pigs housed in the farm (higher level).

The most direct way to predict the excretion flows of a piggery is to calculate for each class of animal (sow, piglet, growing/finishing pig) the difference between the levels of intake and the most probable retention of the considered nutrient. Such an approach has already been developed for nitrogen excretion by Dourmad *et al.* (1992). These authors estimate nitro-gen retention from the growth rate (ADG) and lean proportion of the animals, which allows N excretion to be predicted quite accurately since nitrogen intake is easily known. This empirical method is practically operational and provides reliable estimates in standard conditions. However, as the calculations are based on growth performance, the model cannot be used to predict the effects of factors of variation of nitrogen retention, e.g. a deficiency in essential AA or other factors limiting growth and altering composition of weight gain. In these cases, more complete models of prediction, able to account for the influence of these parameters, are needed.

Remaining at the animal level the minimization of the feed/gain ratio, which approximately parallels that of the relative excretion of wastage, has a fairly similar target to minimization of effluents. This principle is particularly valuable for nutrients of economical significance (faecal excretion of organic matter and of phosphorus, urinary and faecal excretion of nitrogen). These imply that environmental considerations are widely accounted for, particularly at the digestive level, with classical nutrition strategies.

Growth models which have been published in the past twenty or thirty years (e.g. Whittemore and Fawcett, 1976; Black *et al.*, 1986) were nutrient partitioning models that predicted the most probable nitrogen and energy retention as a function of feed (namely energy and protein) intake, including AA profile. The potential of the animal, which is a characteristic of the genotype, was also considered. These models could be applied to calculate the nitrogen excretion from the undigested nitrogen and the nitrogen digested but not retained. In this context, the major limitation of these models is that digestibility is an input parameter. Moreover they cannot predict the nature of nitrogen excretion (proteins or urea) since they do not deal with nitrogen metabolism and fermentations in the large intestine. It is possible to use them to study the nitrogen excretion in various conditions, and in particular with limiting AA or energy intake. Some of the models developed with these basic principles integrate further factors of variation of the animal response: genotype, environment (housing and climate), etc. As a consequence, the prediction of nitrogen excretion can *a priori* allow simulation of the effects of some management strategies on nitrogen excretion. Formulation allowances applied to some particular situations or non-standard conditions can also be calculated.

The level of accuracy that is hoped to be achieved has also an influence on the level of aggregation of nutrients which must be involved in the models of excretion. The above-mentioned laws for N assumed that diet protein is well balanced for amino acids. If one desires to be more accurate, it is then necessary to work with laws of responses to supplies of essential amino acid (EAA) or even of ileal digestible EAA (Chapter 13). Such laws remain today scarce, particularly for predicting N outflows. Moreover if one wishes to be more exhaustive and therefore more realistic it would be necessary to integrate other known responses to EAA supply (intake, weight gains of lipids and proteins . . . ); this would require a metabolic submodel connected to a digestive one.

# Conclusion

In the long term, digestion models are likely to be the first step in modelling more accurately animal tissue growth than the simple nutrient partitioning models mentioned above. In the shorter term, more detailed models of digestion would be useful tools to understand digestive interactions between some major feed components. Such interactions include the lower digestibility values obtained for nitrogen or minerals in a feed with higher fibre content, or for energy from a feed with high mineral content. However, such digestion models must include feed degradability characteristics as input parameters. This target requires a large use of *in vivo* data or of digestibility values (ileal or faecal level). Moreover, a precise modelling of fermentations in the large intestine could provide an estimate of the bacterial protein present in the faeces and therefore help to assess the partition between urinary and faecal protein excretion. Lastly, as several authors have already pointed towards the similarities between the digestive processes and digestibility efficiency for monogastrics, it is not an unrealistic target to adapt pig models to make predictions for other monogastrics such as the rat, rabbit, horse and man.

# References

Adibi, S.A. (1969) The influence of molecular structure of neutral amino acids on their absorption kinetics in the jejunum and ileum of human intestine in vivo. *Gastroenterology* 56, 903–913.

Argenzio, R.A. and Stevens, C.E. (1984) The large bowel – a supplementary rumen? *Proceedings of the Nutrition Society* 43, 13–23.

Baldwin, R.L., Lucas, H.L. and Cabrera, R. (1970) Energetic relationships in the formation and utilisation of fermentation end-products. In: Phillipson, A.T. (ed.), *Physiology of Digestion and Metabolism in the Ruminant*. Oriel Press, Newcastle upon Tyne, UK, pp. 319–334.

Bastianelli, D. (1996) Modélisation de la digestion chez le porc en croissance. PhD dissertation. INAPG, Paris, France, 207pp.

Bastianelli, D. and Sauvant, D. (1995) Modelling digestion and absorption in the pig. In: Danfaer, A. and Lescoat, P. (eds), *Proceedings of the IVth International Workshop on Modelling Nutrient Utilisation in Farm Animals*. NIAS, Foulum, Denmarks, pp. 107–115.

Bastianelli, D., Sauvant, D. and Rérat, A. (1996) Mathematical modeling of digestion and nutrient absorption in the pig. *Journal of Animal Science* 74, 1873–1887.

Bernier, J.J., Adrian, J. and Vidon, N. (1988) *Les aliments dans le tube digestif*. Editions Doin, Paris.

Black, J.L., Campbell, R.G., Williams, I.H., James, K.J. and Davies, G.T. (1986) Simulation of energy and amino acid utilisation in the growing pig. *Research and Development in Agriculture* 3, 121–145.

Buraczewska, L. (1981b) Absorption of amino acids in different parts of the small intestine in growing pigs. I- Absorption of free amino acids and water. *Acta Physiologica Polonica* 32, 419–427.

Buraczewska, L. (1981b) Absorption of amino acids in different parts of the small intestine in growing pigs. II- Absorption of constituents of protein hydrolysates. *Acta Physiologica Polonica* 32, 569–583.

Canguilhem, R. and Labie, C. (1977) Variation de la durée du transit intestinal chez le porc en fonction de la teneur de la ration en cellulose. *Revue de Médecine Vétérinaire* 12, 1669.

Castle, E.J. and Castle, M.E. (1956) The rate of passage of food through the alimentary tract of pigs. *Journal of Agricultural Science* 47, 196.

Castle, E.J. and Castle, M.E. (1957) Further studies on the rate of passage of food through the alimentary tract of pigs. *Journal of Agricultural Science* 48, 106.

Cherbut, C., Barry, J.L., Wyers M. and Delort-Laval, J. (1988) Effect of the nature of dietary fibre on transit time and faecal excretion in the growing pig. *Animal Feed Science and Technology* 20, 327–340.

Cuber, J.C. and Laplace J.P. (1979) Evacuation gastrique de la matière sèche d'un régime semi-purifié à base d'amidon de maïs chez le porc. *Annales de Biologie Animale Biochimie Biophysique* 19, 899.

Cuber, J.C., Laplace, J.P. and Villiers, P.A. (1980) Fistulation de l'estomac et contenus gastriques résiduels après ingestion d'un régime semi-purifié à base d'amidon de maïs chez le porc. *Reproduction Nutrition Développement* 20, 1161.

Cuber, J.C., Kabore, C. and Laplace, J.P. (1981) Evacuation gastrique comparés de l'orge et du blé broyés chez le porc. *Reproduction Nutrition Développement* 21, 773.

Dierick, N.A., Vervaeke, I.J., Demeyer, D.I. and Decuypere, J.A. (1989) Approach to the energenic importance of fibre digestion in pigs. I. Importance of fermentation in the overall energy supply. *Animal Feed Science and Technology* 23, 141–167.

Dourmad, J.Y., Guillou, D. and Noblet, J. (1992) Development of a calculation model for predicting the amount of N excreted by the pig: effect of feeding, physiological state and performance. *Livestock Production Science* 31, 95–107.

Ehle, F.R., Jeraci, J.L., Robertson, J.B. and Van Soest, P.J. (1982) The influence of dietary fibre on digestibility, rate of passage and gastrointestinal fermentation in pigs. *Journal of Animal Science* 55, 1071.

Fioramonti, J. and Bueno, L. (1980) Motor activity in the large intestine of the pig related to dietary fibre and retention time. *British Journal of Nutrition* 43, 155.

Furuya, S. and Takahashi, S. (1975) Rate of passage of chromic oxide and polyethylene glycol and digestibility in the digestive tract of pigs. *Japanese Journal of Zootechnical Science* 46, 630.

Furuya, S., Sakamoto, K., Asano, T., Takahashi, S. and Kameoka, K. (1978) Effects of added dietary sodium polyacrylate on passage rate of markers and apparent digestibility by growing swine. *Journal of Animal Science* 47, 159.

Giusi, A. (1986) Influence du niveau alimentaire et de la composition du régime sur la production intestinale et l'absorption des acides gras volatils chez le porc éveillé. (Influence of the level of feeding and of the composition of feed on the production and absorption of volatile fatty acids in the conscious pig.) PhD dissertation, University of Paris VI, Paris, France.

Holm, J. and Björck, I. (1992) Bioavailability of starch in various wheat-based bread products: evaluation of metabolic responses in healthy subjects and rate and extent of in vitro starch digestion. *American Journal of Clinical Nutrition* 55, 420–429.

Hunt, J.N., Smith, J.L. and Jiang, C.L. (1985) Effect of meal volume and energy density on the gastric emptying of carbohydrates. *Gastroenterology* 89, 1326–1330.

Imoto, S. and Namioka, S. (1978) VFA production in the pig large intestine. *Journal of Animal Science* 47, 467.

Just, A., Jorgensen, H. and Fernandez, J.A. (1981) The digestive capacity of the caecum-colon and the value of the nitrogen absorbed from the hindgut for protein synthesis in pigs. *British Journal of Nutrition* 46, 209–219.

Juste, C. (1982) Apports endogènes par les secrétions digestives chez le porc. In: Laplace, J.P., Corring, T. and Rérat, A. (eds), *Digestive Physiology in the Pig*. INRA, Versailles, pp. 155–173.

Keys, J.E. and de Barthe, J.V. (1974) Site and extent of carbohydrate, dry matter, energy and protein digestion and the rate of passage of grain diets in swine. *Journal of Animal Science* 39, 29–52.

Kirchgessner, M., Kreuzer, M., Machmüller, A. and Roth-Maier, D.A. (1994) Evidence of a high efficiency of bacterial protein synthesis in the digestive tract of adult sow fed supplements of fibrous feedstuffs. *Animal Feed Science and Technology* 46, 293–306.

Krawielitzki, K., Zebrowska, T., Schadereit, R., Kowalczyk, J., Hennig, U., Wünsche, J. and Herrmann, U. (1990) Determining of nitrogen absorption and nitrogen secretion in different sections of the pig's small intestine by digesta exchange between 15N labelled and unlabelled animals. *Archives of Animal Nutrition* 40, 25–37.

Kuan, K.K., Stanogias, G. and Dunkin, A.C. (1983) The effect of proportion of cell wall material from lucerne leaf meal on apparent digestibility, rate of passage and gut characteristics in pigs. *Animal Production* 36, 201–209.

Laplace, J.P. (1978) Organisation de la motricité de l'intestin grêle chez le porc et influence de l'alimentation. *Annales de Zootechnie* 27, 377–408.

Laplace, J.P. and Cuber, J.C. (1984) Déafférentation vagale totale et évacuation gastrique chez le porc. *Reproduction Nutrition Développement* 24, 655.

Laplace, J.P. and Tomassone, R. (1970) Evacuation gastro-duodénale chez le porc: fistulation chronique par voie thoracique extra-pleurale: Recherche d'une technique d'analyse mathématique de l'évacuation. *Annales de Zootechnie* 19, 303–332.

Lescoat, P. and Sauvant, D. (1995) Development of a mechanistic model for rumen digestion validated using the duodenal flux of amino acids. *Reproduction Nutrition Développement* 35, 45–70.

Low, A.G., Pittman, R.J. and Elliott, R.J. (1985) Gastric emptying of barley-soyabean diets in the pig: effects of feeding leve, supplementary maize oil, sucrose or cellulose, and water intake. *British Journal of Nutrition* 54, 437–447.

Matthews, D.M. (1991) *Protein Absorption: Development and Present State of the Subject*. Library of Congress, New York.

Michalet-Doreau, B., Verité R. and Chapoutot, P. (1987) Méthodologie de mesure de la dégradabilité in sacco de l'azote des aliments dans le rumen. *Bulletin Technique CRZV Theix* 69, 5–7.

Nienaber, J.A., McDonald, T.P., Hahn, G.L. and Chen, Y.R. (1991) Group feeding behavior of swine. *Transactions ASAE* 34, 289–294.

Nixon, S.E. and Mawer, G.E. (1970) The digestion and absorption of protein in man, *British Journal of Nutrition* 24, 227–240.

Noblet, J., Fortune, H., Dubois, S. and Henry, Y. (1989) Nouvelles bases d'estimation des teneurs en énergie digestibles, métabolisable et nette des aliments pour le porc. INRA, Paris, 106pp.

Nyachoti, C.M., de Lange, C.F.M., McBride, B.W. and Schulze, H. (1997) Significance of endogenous gut nitrogen losses in the nutrition of growing pigs: A review, *Canadian Journal of Animal Science* 77, 149–163.

Pond, W.G., Pond, K.R., Ellis, W.C. and Matis, J.H. (1986) Markers for estimating digesta flow in pigs and the effects of dietary fiber. *Journal of Animal Science* 63, 1140.

Potkins, Z.V., Lawrence, T.L.J. and Thomlinson, J.R. (1991) Effects of structural and non-structural polysaccharides in the diet of the growing pig on gastric emptying rate and rate of passage of digesta to the terminal ileum and through the total gastrointestinal tract. *British Journal of Nutrition* 65, 391.

Rainbird, A.L. and Low, A.G. (1986) Effect of various types of dietary fibre on gastric emptying in growing pigs. *British Journal of Nutrition* 55, 111–121.

Rayner, V. and Wenham, G. (1986) Small intestine motility and transit by electromyography and radiology in the fasted and fed pig. *Journal of Physiology* 379, 245–256.

Rérat, A. (1977) Mise au point d'une méthode quantitative d'étude de l'absorption digestive chez le porc: application à quelques problèmes nutritionnels concrets. *Bulletin de l'Académie Vétérinaire de France* 50, 93–107.

Rérat, A. and Lougnon, J. (1963) Etudes sur le transit digestif chez le porc. *Annales de Biologie Animale Biochimie Biophysique* 3, 21.

Rérat, A, Vaissade, P. and Vaugelade, P. (1984) Absorption kinetics of some carbohydrates in conscious pigs. 2-Quantitative aspects. *British Journal of Nutrition*, 51, 517–529.

Rivest, J. (1995) Modèle mathématique de la digestion de protéines chez le porc. Mémoire de MSc, Faculté des sciences de l'Agriculture et de l'Alimentation, Université Laval, Québec, Canada, 150pp.

Rivest, J., Pomar, C. and Bernier, J.F. (1994) Simulation of digestion of proteins in the small intestine of growing-finishing pigs. *Journal of Animal Science* 72 (Suppl.1), 264 (Abstr.).

Roth, F.X. and Kirchgessner, M. (1985) Verdaulichkeit und intestinale Passagerate beim Schwein in Abhängigkeit vom Fütterungsniveau und Rohfasergehalt des Futters. *Zeitschrift für Tierphysiology Tierernährung und Futtermittelkunde* 53, 254–264.

Sandoval, R.A., Nielsen, T.K. and Sorensen, P.H. (1987) Effects of fibre on nutrient digestion and time of passage in growing pigs. *Acta Agricultura Scandinavica* 37, 367.

Sauer, W.C., Jorgensen, H. and Berzins, R. (1983) A modified nylon bag technique for determining apparent digestibilities of protein in feedstuffs for pigs. *Canadian Journal of Animal Science* 63, 233–237.

Sauvant, D. (1992) La modélisation systémique en nutrition. *Reproduction Nutrition Developpement* 32, 217–230.

Sauvant, D. and Ramangasoavina, B. (1991) Rumen modelling. In: Jouanny J.P. (ed.), *Rumen Microbial Metabolism and Ruminant Digestion*. INRA, Paris, France, pp. 283–296.

Sauvant, D., Bastianelli, D. and van Milgen, J. (1995) Systèmes d'unités d'alimentation et lois de réponse du porc aux régimes alimentaires. *Journées de la Recherche Porcine en France* 27, 237-244.

Savoie, L. and Gauthier, S.F. (1986) Dialysis cell for the in vitro measurement of protein digestibility. *Journal Food Science* 51, 494-498.

Sève, B. (1994) Alimentation du porc en croissance: intégration des concepts de protéine iléale, de disponibilité digestive des acides aminés et d'énergie nette. *Productions Animales* 7, 275-291.

Souffrant, W.B., Rérat, A., Laplace, J.P., Darcy-Vrillon, B., Köhler, R., Corring, T. and Gebhardt, G. (1993) Exogenous and endogenous contributions to nitrogen fluxes in the digestive tract of pigs fed a casein diet. III- Recycling of endogenous nitrogen. *Reproduction Nutrition Developpement* 33, 373-382.

Stanogias, G. and Pearce, G.R. (1985) The digestion of fibre by pigs. 1- The effects of amount and type of fibre on apparent digestibility, nitrogen balance and rate of passage. *British Journal of Nutrition* 53, 513.

Tomassone, R. and Laplace, J.P. (1973) Essai d'utilisation des systèmes compartimentés dans une étude sur la physiologie de la digestion. *Annales de Biologie Animale Biochimie Biophysique* 13, 747-753.

Turner, L.W., Bridges, T.C., Stahly, T.S., Usry, J.L., Bark, L.J. and Loever, O.J. (1987) A physiological model for growing pig: simulating the digestive system. *ASAE paper* 872548, 322-331.

Usry, J.L., Turner, L.W., Stahly, T.S., Bridges, T.C. and Gates, R.S. (1991) GI tract simulation model of the growing pig. *Transactions ASAE* 34, 1879-1890.

Warner, A.C.I. (1981) Rate of passage of digesta through the gut of mammals and birds. *Nutrition Abstracts and Reviews series B* 51, 789-826.

Whittemore, C.T. and Fawcett, R.H. (1976) Theoretical aspects of a flexible model to simulate protein and lipid growth in pigs. *Animal Production* 22, 87.

Zebrowska, T. and Horszczaruk, F. (1975) Wplyw jedno- i dwukrotnego karmienia na ilosc i sklad tresci jelita cienkiego swin (Effect of feeding frequency on the amount and composition of digesta in the small intestine of pigs). *Rocz. Nauk Roln.b seria B*, 96, 91.

# Quantification of Absorbability and Requirements of Macroelements

12

## A.W. Jongbloed[1], H. Everts[2], P.A. Kemme[1] and Z. Mroz[1]

[1]Department of Pigs and Poultry, ID-DLO, PO Box 65, 8200 AB Lelystad, The Netherlands; [2]Department of Large Animal Medicine and Nutrition, Faculty of Veterinary Medicine, Utrecht University, PO Box 80.152, 3508 TD Utrecht, The Netherlands

## Introduction

In the body of pigs there are approximately twenty inorganic elements that are essential for maintenance and normal functioning. Lack or insufficient amounts of these elements results in deficiency symptoms leading to reduced performance. Minerals are present in all kinds of organs and tissues in the body. The functions of the inorganic elements are extremely diverse. They range from structural functions in some tissues, to a wide variety of regulatory functions in other tissues (NRC, 1980; Underwood, 1981; McDowell, 1992). The minerals Ca, Mg, Na, K, P, S and Cl are present in the body in relatively larger amounts (macroelements) than the microelements, like Fe, Cu, Zn, Co, Mo, Mn, I, Cr, F, Ni, Si and Se.

Absorption and utilization of macroelements may be mediated by hormonal control, which is primarily based on their concentration in the extracellular fluid. The pig is able to maintain a homeostasis in the extracellular and intracellular fluids by means of several regulatory mechanisms. As a consequence, large differences in absorption and utilization of minerals can be found, depending on the nutritional status of the animal. Also, several dietary factors may affect absorption and utilization of minerals.

Requirements for macroelements should therefore be based on absorbable or utilizable/available minerals rather than on total amounts. However, quantitative data on absorbable or utilizable/available amounts of several minerals are still lacking; phosphorus is the sole exception. Lately, we have become more aware that a surplus of minerals excreted in pig manure may be harmful to the environment, especially in regions with intensive pig production, like The Netherlands. This surplus may result in undesirable accumulation of pollutants in the soil and eutrophication of groundwater and freshwater. Minerals, like phosphorus, potassium, copper and zinc, several heavy metals, like cadmium, but also lead and mercury that are present as contaminants in several dietary mineral additives and raw materials are of concern (NRC, 1994). In a few countries legislation has been imposed to limit the amount of P which can be applied on each hectare of land. This implies that, apart from the price of the minerals and their nutritional value, the environmental impact is another important factor that should be taken into account when discussing minerals in pig feeding (Jongbloed and Henkens, 1996).

The aim of this chapter is to provide quantitative information on several aspects related to minerals in pig feeding. First, definitions regarding the nutritive value of minerals are explained, followed by information on digestibility of several macroelements and factors affecting this. As phosphorus is important in relation to environmental pollution, a large part will be devoted to this element. Next, the requirements for macroelements of pigs are described wherever possible by the factorial approach. Finally, some important issues in relation to the nutritive value of minerals and estimation of the requirements are discussed. Microelements are not discussed in this chapter because sufficient quantitative data are still lacking.

## Nutritive Value of Minerals

In the literature, different terms are used to determine nutritive values of minerals for pigs, e.g. digestibility, absorbability or (bio-)availability (Partridge, 1980). Digestibility and absorbability relate to the gastrointestinal tract (feed – faeces). The term (bio-)availability, however, is used in different contexts, and can be misleading. ARC (1981) defines (bio-)availability as the fraction that is retained in the body [feed − (faeces + urine)]. With regard to P in pig feeding, the term availability of P is used in certain cases for total P minus phytate P, because it is assumed that phytate P is not available, whereas non-phytate P is regarded as 100% available. In addition, DLG (1987) uses the term intestinal availability of P in feedstuffs of plant origin, which is calculated as $0.4 \times$ phytate P $+ 0.8 \times$ non-phytate P. The term availability is also used in studies assessing the nutritive value of mineral sources, whereby the outcome is compared with a reference that is assumed to be 100% available (NRC, 1988). In defining availability in the latter situation, various response parameters, not directly associated with absorption, are used. From the scientific point of view, the term absorbability should be used as the net fraction that is absorbed from the gastrointestinal tract. As in most balance techniques for pigs the net absorption of minerals is measured, we do not recommend using the term availability, because this term may lead to great confusion.

Before absorption by the enterocytes of the gastrointestinal tract can take place, the elements must become available in ionic form (as cations and anions), which is suitable for uptake and transport. In principle, the trans-epithelial transport consists of an active transcellular component which can be regulated and/or a passive paracellular component which depends on chemical and electrical gradients existing across the intestinal wall (Schröder *et al.*, 1996). Suitability for transport is not problematic for the well-soluble monovalent elements, like Na, K and Cl, but the solubility of various other elements is often low at a neutral pH, and moreover, it is dependent on the presence of other compounds since it can relatively easily precipitate or form non-absorbable complexes. Well-known complexing food components are the phytic and oxalic acids (Harland, 1989). In addition, there are several interactions among various minerals (e.g. Ca and P) or with the enzyme phytase, which complicate our understanding of absorption. Also, viscosity of the chyme of the intestinal tract may negatively affect absorption of minerals (Van der Klis, 1993). It is obvious that the absorbability of P and other minerals in feedstuffs from plant origin can be substantially enhanced by using extrinsic phytases, but this aspect will not be discussed further in this chapter (Jongbloed *et al.*, 1996).

In the next section, we will use the term digestibility to strengthen a practical understanding like in feed formulation, and uniform interpretation or comparison of literature data. The term (apparent) digestibility is, there-fore, equal to (apparent) absorbability.

## Apparent digestibility of phosphorus

Nowadays, P is regarded as one of the most important elements with regard to environmental pollution (Jongbloed and Kemme, 1990). This is especially the case in regions where pig density is high, and application of P in pig manure per hectare of land exceeds withdrawal of P by the crop. This may lead to accumulation of P in the soil, together with leaching and runoff. The effect is eutrophication of freshwater resources (Roland *et al.*, 1993). This has led to considerable research efforts to obtain tabulated data on the nutritive value of P in feedstuffs for pigs in some countries.

So far tabulated values for P are the only ones provided for practice which are based on availability, absorbability or apparent digestibility in raw materi-als for pigs (NRC, 1988; CVB, 1996; Tuori *et al.*, 1996). In judging these values, it is important to keep in mind the procedure which has been used for the assessment of the amount of digestible P in a feed. Furthermore, one should appreciate that there are a lot of factors that affect mineral digestibility (Jongbloed, 1987). The most well known is the interaction between Ca and P (Pointillart *et al.*, 1987; Jongbloed *et al.*, 1995). Dietary Ca content should be standardized for assessing the P digestibility of feeds. This factor is also necessary to take into account when comparing literature data. The supply of vitamin D should be adequate as the absorption process of P is mediated by metabolites of vitamin D (Schröder *et al.*, 1996). Another important aspect is that due to hormonal regulation, the rate of absorption depends on the level of P supply in relation to the requirement. This means that only a maximal absorption rate of P is obtained when the pig is fed below its P requirement. Finally, an appropriate adaptation period is necessary before collection of faeces can be done (Jongbloed, 1987; Düngelhoef *et al.*, 1994). For pigs, we have developed two procedures to measure the apparent digestibility of P: first, in feedstuffs originating from plants we use a quantitative collection of faeces for 7 to 10 days (Jongbloed and Kemme, 1990); and, second, in feedstuffs from animal origin or in feed phosphates we use qualitative grab sampling of faeces and evaluation by the slope-ratio technique (Dellaert *et al.*, 1990).

Results on P digestibility in feedstuffs from plant origin, animal origin and feed phosphates for pigs are listed in Table 12.1. Additional values on P digestibility of feed ingredients for pigs can be found in CVB (1996). It can be concluded from Table 12.1 that relatively large differences in P digestibility can be observed among feedstuffs. The lowest P digestibility was noted for tapioca meal and rice bran, whereas the highest values were obtained for lupins and wheat. The difference in P digestibility between conventional wheat and wheat with inactivated intrinsic phytase shows that wheat phytase has a large effect. This was also confirmed in the experiments of Fourdin *et al.* (1986) and Eeckhout and DePaepe (1992). The large variation within a feedstuff is attributed to differences in phytate P content, phytase activity and processing (Jongbloed and Kemme, 1990). Results in Table 12.1 also show that the apparent digestibility of P between various types of feed phosphates and animal products differs substantially, and it even differs between the same type of feed phosphate, such as monocalcium phosphates (type A or B). Differences in P digestibility between various products of animal origin may be attributed to differences in technological treatments and physicochemical structure of the products.

## Apparent digestibility of Ca, Mg, Na, K and Cl

Data on the apparent digestibility of Ca, Mg, Na, K and Cl are scarce. In the case of Na, K and Cl, it is suggested that these monovalent ions are highly soluble and absorbed to a high degree ( > 80%). This was confirmed in recent studies at our Institute on both growing pigs and breeding sows (Mroz *et al.*, 1996a,b).

There is a hormonal regulation for the absorption of Ca through metabolites of vitamin D and parathyroid hormone (PTH). In the pig, however, PTH does not seem to play an important role (Pointillart *et al.*, 1978). Therefore,

**Table 12.1.** P digestibility coefficients (% of intake) of some feedstuffs of plant origin, animal origin and feed phosphates for pigs.

| Feedstuff | Number of trials | Mean | SD | Range | Feedstuff | Number of trials | Mean | SD | Range |
|---|---|---|---|---|---|---|---|---|---|
| Barley | 5 | 39 | 4 | 34–44 | Hominy feed | 8 | 21 | 8 | 13–34 |
| Maize | 10 | 19 | 6 | 12–26 | Rapeseed extr. | 3 | 27 | 4 | 22–33 |
| Wheat | 4 | 48 | 2 | 46–51 | Rice bran | 6 | 14 | 4 | 9–20 |
| Wheat inact.* | 2 | 26 | 1 | 26–27 | Soybean meal extr. | 9 | 39 | 4 | 33–46 |
| Peas | 4 | 45 | 4 | 42–51 | Sunflower meal extr. | 10 | 15 | 4 | 9–20 |
| Beans | 3 | 37 | 8 | 29–48 | Wheat bran | 3 | 30 | 5 | 24–35 |
| Lupins | 3 | 50 | 4 | 47–56 | Tapioca | 4 | 6 | 5 | 1–13 |
| Meat+bone meal | 2 | 81 | 1 | 80–81 | DCP.OH$_2$O (A,B) | 4 | 64 | 2 | 63–66 |
| Bone precipit. A | 1 | 87 | – | – | DCP.2H$_2$O (A,B) | 3 | 69 | 1 | 69–71 |
| Bone precipit. B | 1 | 61 | – | – | MCP.1H$_2$O (A) | 3 | 75 | 3 | 72–78 |
| Skim milk dry | 1 | 90 | – | – | MCP.1H$_2$O(B) | 4 | 82 | 2 | 80–84 |
| Fish meal | 2 | 72 | 17 | 61–84 | MSP.2H$_2$O(ref.) | 3 | 90 | 3 | 88–93 |

SD = standard deviation.
* Intrinsic phytase inactivated by thermal treatment.
A and B are from different production processes.
ref. = reference chemically pure monosodium phosphate.

absorption coefficients of Ca can be influenced to a large extent by the supplied amount of Ca. However, adaptation of the absorption to the changed supply of Ca takes more time than for P (Pointillart *et al.*, 1986). Furthermore, both Mg and P are antagonistic to Ca. This implies that the ratios between these minerals should also be between certain levels. Furthermore, Ca levels of several raw materials are very low (between 0.1 and 3.0 g kg$^{-1}$), so that it is very complicated to assess reliable digestibility figures of Ca in such feedstuffs. Another important factor is the presence of phytase. Phytase may largely affect the digestibility of Ca as was demonstrated by Jongbloed *et al.* (1995). In their studies on growing pigs, the apparent digestibility of Ca ranged from 40% without phytase to 70% with phytase. Nevertheless, there are some (limited) data on the digestibility of Ca from various Ca sources. Ross *et al.* (1984) showed that in pigs the relative availability of Ca in some Ca sources was high (compared to $CaCO_3$ precipitate), and there were only slight differences among the sources tested. Only two sources of dolomitic limestone were less available. Mroz *et al.* (1996a) found that Ca digestibility and retention from $CaSO_4$, $CaCl_2$ and Ca-benzoate were similar to $CaCO_3$. For breeding sows, Everts *et al.* (1998a) observed values of about 30%, but the level of Ca supply was rather high.

ARC (1981) suggest that the availability of Mg for pigs is 80%. This figure seems to be extremely high for compound feeds in the light of data presented by Jongbloed (1987), and other studies in our Institute (e.g. Kemme *et al.*, 1997b). We have shown that the apparent digestibility of this mineral never exceeded 30%. Perhaps the Mg source used in the ARC studies could explain the difference. Similarly, in balance trials with breeding sows (Everts *et al.*, 1998a), digestibility of Mg ranged between 20 and 30%.

It can be concluded that the assessment of digestibility coefficients for Ca is rather complicated due to its hormonal regulation and interactions with dietary factors like phytase, and level of supply of Ca, Mg and P. For growing pigs a digestibility coefficient between 0.5 and 0.6 can be adopted, while for Mg 0.25 may be assumed.

## Digestibility as affected by age and physiological status

Little is known concerning the effect of age or physiological status of the pig on the absorbability of minerals. In general, the literature points out that adult animals have a lower absorbability than growing animals, and this may also be the case for pigs (Armbrecht, 1987; McElroy *et al.*, 1991). Recently, Kemme *et al.* (1997a) performed an experiment with piglets, growing-finishing pigs, and pregnant and lactating sows to test the hypothesis that apparent P and Ca digestibilities depend on their physiological status. Pigs of the four categories were fed diets with identical feedstuff composition. The digestibility of P increased in piglets and growing-finishing pigs while their LW increased from 30 to 60 kg and then remained stable until 100 kg LW (Table 12.2). The

**Table 12.2.** Apparent digestibility of P and Ca (% of intake) in pigs of different categories, fed the same diet.

|  | Piglet | Growing-finishing pig | Breeding sow | | |
|---|---|---|---|---|---|
|  |  |  | Pregnancy | | |
|  | 34 kg BW | 70 kg BW | 60 days | 100 days | Lactation |
| P | 23.0 | 26.0 | 13.7 | 18.3 | 19.4 |
| Ca | 39.3 | 39.2 | 13.4 | 23.8 | 30.6 |

pregnant sows had a lower P digestibility than the piglets and growing-finishing pigs. During lactation, the P digestibility was higher than during pregnancy. For Ca even larger differences were observed in digestibility between growing pigs and breeding sows. The P digestibility figures of feeds recommended for practice in the Dutch feedstuff table (CVB, 1996) are obtained from experiments with growing pigs. The use of these values for breeding sows is, therefore, questionable. Everts *et al.* (1998a) showed that the P digestibility of pregnant sows was lower than expected on the basis of tabulated values. One of the explanations is that the sows were fed above their digestible P requirements, but it is also possible that there is indeed an effect of age on the P digestibility of feeds.

To test this hypothesis we have performed a second study in which the digestibility of identical feeds for growing pigs and lactating sows was measured (Kemme *et al.*, 1997b). The conclusion was that the P digestibility of the feeds for the lactating sows was not significantly lower than for growing pigs. However, we have again observed that the Ca digestibility was 10 to 15% lower than in growing pigs. No effect of parity of the sows on the digestibility of both Ca and P could be demonstrated.

## Requirements for Minerals

Estimation of the exact requirement of pigs for minerals is a difficult task. Two methods are most often used for this purpose: the empirical and the factorial. The latter is used for macroelements with increased frequency (ARC, 1981; Guéguen and Perez, 1981; NRC, 1988; Jongbloed *et al.*, 1994).

The requirement for minerals depends on several factors, such as the animal (physiological status, production level and type of production), the diet and feeding strategy (amount of feed and chemical composition, chemical binding form and several interactions), the environment (temperature, health status, management, aim of production) and the criterion used (minimal or safe, and the evaluation method). This means that the estimated requirements can be variable according to differences in assumptions. For both scientific and clarity reasons, we prefer the factorial approach in which

an estimate is given for maintenance and production. This will be outlined
below only for P, since sufficient data for the other minerals are still lacking.
Estimates of the requirements for Ca will be derived from those of P.

## Requirement for phosphorus for maintenance

The maintenance requirement for P is determined by the loss of endogenous
P in faeces and the loss of P in urine. Above, we have outlined the nutritive
value of P as expressed in terms of apparent digestible P. Therefore, one has
to appreciate that the contribution of endogenous faecal excretion has
already been taken into account in the digestibility coefficient. So in this case,
maintenance requirement for apparent digestible P equals only urinary loss
of P.

   Literature shows that faecal endogenous loss of P is 9 and 3 mg kg$^{-1}$ LW
per day for pigs that are fed a sufficient and insufficient amount of P,
respectively (Jongbloed, 1987; Jongbloed and Everts, 1992). It has been
described previously that in assessing the nutritive value of P, pigs are fed
below their P requirement, which does not lead to maximal animal performance. In practice, this is not wanted. Therefore, we have proposed a
'maintenance requirement' for P for all categories of pigs at a level equal to
6 mg kg$^{-1}$ LW per day, to compensate for the difference between a sufficient
and insufficient supply of P. In addition to the 6 mg kg$^{-1}$ LW per day, 1 mg
kg$^{-1}$ LW per day is adopted for endogenous loss of P in urine. In total 7 mg P
kg$^{-1}$ LW per day is used for 'maintenance' for all categories of pigs. However,
Fernández (1995) concluded that faecal endogenous losses of this element
are not related to LW. It is our view that his observation needs to be
confirmed.

## Requirement for phosphorus for production

*Requirement for phosphorus for production of piglets and growing pigs*
Piglets and growing pigs deposit P in lean tissue, organs and bones. For
estimation of the P requirement for growth we use an allometric function
developed by Jongbloed and Everts (1992). This function is based on the
results of 53 slaughter experiments in literature and on our own experiments,
in which the amount of P had been determined pigs in the LW range of 1 to
110 kg (Jongbloed, 1987). Only those data which allowed maximal bone
mineralization were used. The function with a quadratic term for live weight
(LW) was:

$$\ln P = 1.494 + 1.108 \ln LW - 0.018 \, (\ln LW)^2, \quad g$$
$$R^2 = 0.995, RSD = 0.081) \qquad (12.1)$$

Requirements for pigs with a LW below 5 kg were not described adequately
by the above equation. This could be attributed to the amount of minerals in

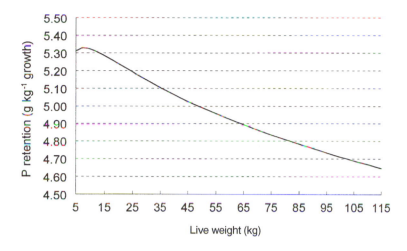

**Fig. 12.1.** Course of P retention (g kg⁻¹ weight increase) from 5 to 115 kg live weight in pigs.

newly born piglets. The amount of P deposited at a certain LW, the first derivative of equation 12.1, is illustrated in Fig. 12.1 and in Table 12.3. As has been mentioned earlier, data used for equation 12.1 were from pigs slaughtered from 1960 to 1985. However, it can be questioned whether these data are representative of current pig genotypes. Therefore, it is shown in Table 12.3 that higher requirements (about 0.25 g P kg⁻¹ LW gain) can be considered for very lean pig types as suggested by Jongbloed (1987). More data, however, are required to support this.

The total requirement for digestible P of piglets and growing pigs can now be calculated. Therefore, one should know the LW of the pig (for maintenance requirements), the daily growth at that LW, and the type of growth (very lean or normal). As an example, we give the estimated requirement for digestible P of a gilt with an average growth rate of 783 g day⁻¹ and an energy conversion ratio of 36.0 MJ ME kg⁻¹ growth (Table 12.4). Feed

**Table 12.3.** Calculated requirement of P for growth (g kg⁻¹ live weight gain) in two types of growing pigs (Jongbloed and Everts, 1992).

| Pig type | Live weight (kg) | | | | | | | | | | |
|---|---|---|---|---|---|---|---|---|---|---|---|
| | 10 | 20 | 30 | 40 | 50 | 60 | 70 | 80 | 90 | 100 | 110 |
| Normal | 5.35 | 5.20 | 5.10 | 5.02 | 4.94 | 4.87 | 4.80 | 4.75 | 4.70 | 4.65 | 4.60 |
| Very lean | 5.45 | 5.45 | 5.35 | 5.27 | 5.19 | 5.12 | 5.05 | 5.00 | 4.95 | 4.90 | 4.85 |

**Table 12.4.** Recommended requirement for digestible P (dP) of a gilt with an average growth rate of 783 g day$^{-1}$ and an average energy conversion ratio of 2.87 EW kg$^{-1}$ growth (1 EW = 12.55 MJ ME).

| Day | LW (kg) | Growth (kg day$^{-1}$) | Digestible P (g day$^{-1}$) Maint. | Growth | Total | Intake (EW day$^{-1}$) | dP (g EW$^{-1}$) |
|---|---|---|---|---|---|---|---|
| 1 | 25.4 | 0.44 | 0.18 | 2.27 | 2.44 | 1.05 | 2.33 |
| 8 | 28.5 | 0.50 | 0.20 | 2.56 | 2.76 | 1.20 | 2.30 |
| 15 | 32.1 | 0.60 | 0.22 | 3.05 | 3.28 | 1.40 | 2.34 |
| 22 | 36.3 | 0.69 | 0.25 | 3.48 | 3.74 | 1.60 | 2.34 |
| 29 | 41.0 | 0.72 | 0.29 | 3.61 | 3.89 | 1.77 | 2.20 |
| 36 | 45.9 | 0.72 | 0.32 | 3.58 | 3.90 | 1.95 | 2.00 |
| 43 | 50.9 | 0.77 | 0.36 | 3.80 | 4.15 | 2.10 | 1.98 |
| 50 | 56.3 | 0.84 | 0.39 | 4.11 | 4.51 | 2.30 | 1.96 |
| 57 | 62.2 | 0.91 | 0.44 | 4.42 | 4.85 | 2.50 | 1.94 |
| 64 | 68.5 | 0.94 | 0.48 | 4.53 | 5.01 | 2.70 | 1.85 |
| 71 | 75.0 | 0.94 | 0.53 | 4.49 | 5.02 | 2.80 | 1.79 |
| 78 | 81.5 | 0.95 | 0.57 | 4.50 | 5.07 | 2.90 | 1.75 |
| 85 | 88.1 | 0.96 | 0.62 | 4.52 | 5.13 | 3.00 | 1.71 |
| 92 | 94.8 | 0.94 | 0.66 | 4.39 | 5.06 | 3.00 | 1.69 |
| 99 | 101.3 | 0.92 | 0.71 | 4.27 | 4.98 | 3.00 | 1.66 |
| 106 | 107.6 | 0.90 | 0.75 | 4.15 | 4.91 | 3.00 | 1.64 |

allowances are also presented to estimate the digestible P content per energy value (EW; 12.55 MJ ME). It can be seen from Table 12.4 that the daily requirement for digestible P increases up to 88 kg LW after which there is only a slight decrease. The content of digestible P in terms of g EW$^{-1}$ decreases gradually from 2.3 at 25 kg LW to 1.6 at 108 kg LW. It is obvious that the required amounts of digestible P are different at other growth rates.

### Requirement for phosphorus for production of breeding sows

PREGNANT SOWS. Estimation of the requirement of digestible P during pregnancy is also made on the basis of the factorial approach which has three components:

1. Maintenance (see above).
2. Development of the products of conception.
3. Maternal growth.

*Development of the products of conception.* During pregnancy, the different products of conception are growing according to specific patterns. The growth of a fetus, the placenta, the empty uterus and the udder can be described by a Gompertz equation, while the dynamic changes of the intrauterine fluids are described by a polynomial equation (Noblet *et al.*, 1985). The prediction of the growth of the products of conception at different stages of pregnancy requires additional information about the litter size and the level of ME supply. Noblet *et al.* (1985) have also presented equations to predict the rates of protein and energy accretion in gilts. Although there are no objections to using these equations, the development of the empty uterus and the udder in older sows seems to differ significantly from that of gilts. However, due to the lack of data from older sows, we have used the equations derived from gilts also for older sows.

For the calculation of the P requirement the rate of mineral accretion is needed. However, data about the development of the mineral content of the fetuses during the whole pregnancy period are scarce in the literature. Den Hartog *et al.* (1988) analysed fetuses at several stages of gestation. The mean birth weight of these piglets was 1326 g. The original data of Den Hartog *et al.* (1988) were used to derive a Gompertz equation:

$$\ln P \text{ (g)} = 4.591 - 6.389 \times \exp(-0.02398 \times (t-45))$$
$$+ 0.0897 \times n, R^2 = 0.99 \tag{12.2}$$

where $P$ = the amount of P in the fetuses (g), $t$ = days of pregnancy and $n$ = the number of fetuses. For a litter size of 11 piglets, this equation estimated about 80 g P at day 115 of pregnancy. The value is considerably lower than expected on the basis of a litter of 11 piglets of 1500 g and a P concentration in newborn piglets of $6.25 \text{ g kg}^{-1}$ estimated by Jongbloed (1987) ($11 \times 1.5 \times 6.25 = 103$ g). To prevent such an underestimation, we have adopted a birth weight of 1500 g and the relatively high concentration of P, and therefore, we introduced a correction factor of $103/80 = 1.28$ to equation 12.2. This correction factor can be changed according to the expected birth weight and P content of the piglets. The course of the daily retention of P in 11 fetuses during pregnancy is shown in Fig 12.2.

Since the amounts of P in the placenta, the uterine fluids, empty uterus and udder are not cited in the literature, we estimated their amount according to Jongbloed (1987) as $0.0096 \times$ amount of protein contained as energy. The amounts of protein in the placenta (equation 12.3), empty uterus (equation 12.4) and udder (equation 12.5) were adopted from Noblet *et al.* (1985), assuming an energy supply of 30 MJ ME day$^{-1}$.

Amount of protein energy ($Pr$, kJ) in placenta:

$$\ln Pr = 7.34264 - 1.40598 \exp(-0.0625 (t-45)) + 0.00759t$$
$$+ 0.06339n \tag{12.3}$$

Amount of protein energy (kJ) in empty uterus:

$$\ln Pr = 7.02748 + 0.33484 \times \exp(+0.01489 \times (t-45))$$
$$+ 0.00609 \times t + 0.05826n \tag{12.4}$$

Amount of protein energy (kJ) in udder:

$$\ln Pr = 1.43401 + 3.32153 \times \exp(+0.0099 \times (t-45)) + 1.4409 \tag{12.5}$$

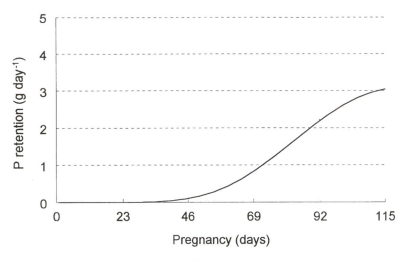

**Fig. 12.2.** Course of P retention (g day$^{-1}$) in 11 pig fetuses during pregnancy.

where $t$ = days of pregnancy and $n$ is the number of fetuses.

The amount of P in uterine fluids is not included. The development of the amounts of P in the products of conception of gilts are given in Table 12.5. The fetuses contribute most to the total amount of required P in the products of conception. At day 115 of pregnancy nearly 90% of the amount of P is present in the fetuses. The contribution from the placenta, empty uterus and udder is relatively small. Their course of daily retention of P during pregnancy is illustrated in Figs 12.3, 12.4 and 12.5, respectively.

**Table 12.5** The amounts of P in the products of conception for gilts (litter size 11) during pregnancy.

| Days of pregnancy | 0 | 28 | 56 | 84 | 98 | 105 | 115 |
|---|---|---|---|---|---|---|---|
| Amount of P (g) | | | | | | | |
| Fetus[a] | 0 | 0.02 | 2.51 | 27.68 | 56.59 | 74.65 | 103.13 |
| Placenta[b] | 0 | 0.02 | 0.94 | 2.09 | 2.50 | 2.68 | 2.94 |
| Intrauterine fluid[c] | — | — | — | — | — | — | — |
| Empty uterus[b] | 0 | 1.33 | 1.80 | 2.62 | 3.28 | 3.71 | 4.49 |
| Udder[b] | 0.10 | 0.12 | 0.29 | 0.95 | 1.96 | 2.94 | 5.51 |
| Total | 0.10 | 1.49 | 5.54 | 33.34 | 64.33 | 83.98 | 116.07 |

[a] According to equation 12.2 with a correction to reach a level of 9.37 g P per piglet.
[b] Amount of nitrogen (according to Noblet *et al.*, 1985) × 0.06.
[c] Not included.

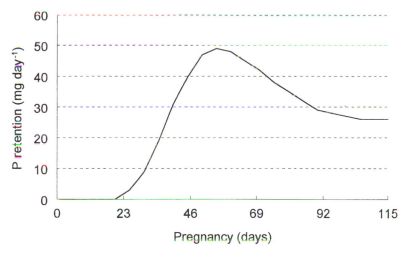

**Fig. 12.3.** Course of P retention (mg day$^{-1}$) in pig placenta during pregnancy (11 fetuses).

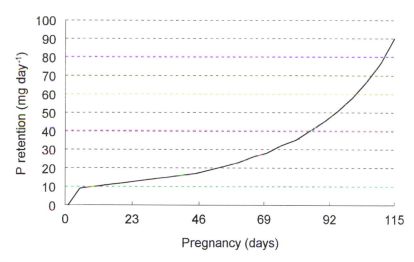

**Fig. 12.4.** Course of P retention (mg day$^{-1}$) in the uterus during pregnancy (11 fetuses).

*Maternal growth.* Maternal growth in the gilt is strongly related to the feed allowance (Walach-Janiak *et al.*, 1986) and the dietary composition (Shields *et al.*, 1985). Data on the maternal growth of older sows are rather scarce.

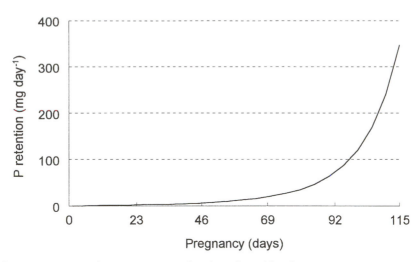

**Fig. 12.5.** Course of P retention (mg day$^{-1}$) in the udder during pregnancy (11 fetuses).

Phosphorus retention in sows can be separated into P retained in protein (0.0096 × g protein; Jongbloed, 1987) and P in bones. The maternal protein gain during pregnancy depends on parity number, feeding history and genetic potential of the sow. Due to lack of a definitive information, we have assumed maternal protein gain in the body to be constant during the whole pregnancy period. When the maternal protein gain during pregnancy is known, then the daily amount of P retained in maternal protein can be calculated as:

$$P_{maternal\ protein} = (total\ maternal\ protein\ gain\ (kg) \times 0.0096)/115$$
$$g\ P\ day^{-1} \tag{12.6}$$

On the basis of many data from different large breeding farms in The Netherlands, we have estimated that the maternal protein gain of sows during the first pregnancy is 7 kg (Everts *et al.*, 1994).

Data about mineral accretion in the maternal bones of sows in relation to parity number are not available. Therefore, according to the results of a slaughter experiment with sows at mating, end of first pregnancy and after the third lactation (Everts and Dekker, 1991) we propose to use as daily accretion of P in maternal bones the following values: for the first parity 1.5 g P day$^{-1}$, for the second 0.8 g P, the third 0.4 g P day$^{-1}$, the fourth 0.2 g P day$^{-1}$ and for the fifth 0.1 g P day$^{-1}$. Although it is known that demineralization of bones can occur during preceding lactation, the rate to which this is restored during pregnancy remains to be investigated.

*Digestible phosphorus requirement during pregnancy.* By knowing the weight of the products of conception, the LW at mating and the daily growth, the actual LW of the sow can be calculated at different stages of gestation. This LW is used to calculate the maintenance requirement. From the equations used to calculate the amount of P in the products of conception, it is also possible to estimate the daily accretion of P in the products of conception. The P required for the development of the bones is already indicated and the P retained in the maternal protein can be calculated from equation 12.6.

The requirement for digestible P during pregnancy for 11 piglets is given in Table 12.5 for a first parity sow. The present requirements are lower than earlier calculations published by Jongbloed and Everts (1992). This difference is mainly due to a difference in the retention of P in the fetus. In the earlier calculations we used the growth curve of piglets suggested by Den Hartog *et al.* (1988), which is very steep in the last phase of pregnancy. The Gompertz curve in the present calculations has a moderate increase in the last phase of pregnancy.

Everts *et al.* (1998a) concluded, on the basis of balance trials with first to third parity sows, that a digestible P supply of 4.2 g day$^{-1}$ was sufficient during mid-pregnancy (days 50 to 60) to guarantee that P retention in the maternal bones was more than 2.0 g day$^{-1}$. During late pregnancy (days 105 to 112) they indicated that a digestible P supply of 6 g day$^{-1}$ was sufficient to guarantee a P retention in the maternal bones of more than 1.4 g day$^{-1}$. These findings are well in line with the requirements shown in Table 12.5. However, it remains uncertain if these levels of P retention in the maternal bones guarantee sound legs and longevity in the sow. The figures for maternal gain can vary considerably depending on the (applied) feeding strategy. Maternal gains presented here are a reflection of the feeding practice on some large Dutch breeding farms. However, differences in maternal protein gain of the range reported in the literature (e.g. Walach-Janiak *et al.*, 1986; Whittemore and Yang, 1989) will have a limited impact on the P requirement.

LACTATING SOWS. The mineral requirement of sows during lactation is closely related to the amounts of minerals excreted in the milk. Due to the high P digestibility of milk (0.91; CVB, 1996), the P requirement is also closely related to the amount of P retained in the piglets. Because the estimate of the P retention in the piglets is more reliable than measuring milk production and milk composition, we derive the digestible P requirement of lactating sows from the retention in the suckling piglets, using a factorial approach with three components:

1. Maintenance (see above).
2. Phosphorus excreted in milk.
3. Mobilization of maternal body stores.

*Phosphorus excreted in milk.* It is shown in Table 12.3 that 1 kg of piglet gain contains about 5.45 g P. For example, assuming a mean daily piglet gain of 250 g day$^{-1}$, a mean LW of 5 kg, P maintenance requirement of 0.007 g kg$^{-1}$ LW and a P digestibility of 0.91 (CVB, 1996), we can calculate a P requirement

**Table 12.6.** Digestible P requirement of lactating sows during a 28 day lactation period (g day$^{-1}$).

| Parity | 1 | 2 | 5 |
|---|---|---|---|
| Mean live weight sow | 160 | 185 | 225 |
| Litter size | 10 | 11 | 11 |
| Daily piglet gain (g day$^{-1}$) | 225 | 250 | 250 |
| Mean piglet weight (kg) | 4.6 | 5.0 | 5.0 |
| Phosphorus for sow maintenance | 1.12 | 1.30 | 1.58 |
| Phosphorus in milk | 13.83 | 16.94 | 16.94 |
| Phosphorus retention in sow | −0.8 | −0.8 | −0.8 |
| Digestible P requirement | 14.15 | 17.44 | 17.72 |

(expressed as P in the milk) for one piglet of 1.54 g day$^{-1}$ (= (5.45 × 0.25 + 5 × 0.007)/0.91). The calculation of the amount of P excretion in milk can, therefore, be summarized by the following equation:

$$P_{\text{in milk}} = (6.096 \times DG + 0.0115) \times n \qquad \text{g} \qquad (12.7)$$

where $DG$ = mean daily LW gain of the piglets (kg) and $n$ = number of suckling piglets.

*Mobilization of maternal body stores.* In practice, sows are mobilizing body stores during the lactation period. The degree of mobilization can differ considerably due to the concentration of P in the body stores and the applied feeding regime. Phosphorus can be mobilized through demineralization of bones, but also some P can be mobilized by protein catabolism.

Reliable data about bone demineralization during lactation are not available and, therefore, we have not included this factor in our calculations. In balance trials with lactating sows, fed according to their P requirements, a mobilization between 0 and 3 g of P from the maternal body was observed (Everts and Dekker, 1991).

Whittemore and Yang (1989) found that protein mobilization during lactation was about 3 kg, whereas Everts and Dekker (1994) observed values between 0.7 and 2.9 kg. When a mean protein mobilization of 2.3 kg during lactation is assumed, then about 0.8 g P are mobilized per day.

*Digestible phosphorus requirement during lactation.* With a known LW, litter size, litter gain and the assumed P mobilization from protein, we can calculate the digestible P requirement as shown in Table 12.6.

The results from balance trials with lactating sows (Everts *et al.*, 1998b) showed a linear relationship between ingested digestible P and the amount of P retained in the suckling piglets. The amount of P per kg piglet gain was

**Table 12.7.** Estimated optimal Ca : digestible P ratio in pig feeds.

|  | Ca : P retention | DC Ca | Ca : dP |
|---|---|---|---|
| Piglets and growing pigs | 1.6 | 60–50 | 2.9 (2.7–3.2) |
| 1st–3rd Parity pregnant sows | 1.6 | 50–40 | 3.6 (3.2–4.0) |
| > 3rd Parity sows pregnant | 1.6–1.8 | 50–40 | 3.8 (3.4–4.2) |
| All parity sows lactating | 1.3–1.5 | 50–40 | 3.2 (2.8–3.5) |

DC = digestibility coefficient.
dP = digestible P.

lower than assumed in the present calculations. This can be due to a change in the genetic potential of piglets during the last decade. Therefore, additional research in this field is needed. To prevent underestimation of the requirement, we used a high level of P retention for the suckling piglets.

## Requirement for calcium

As already stated earlier, the recommended level of Ca should be in a close relationship with the level of P. In the past, an optimal ratio was recommended between total Ca and total P of about 1.2–1.4:1 (Jongbloed, 1987). Nowadays, it is recognized that a ratio which is expressed as total Ca to digestible P, or even digestible Ca to digestible P, is more accurate. Because there are insufficient data available with regard to digestible Ca, the ratio between total Ca and digestible P is increasingly used. It is difficult to advise an optimal Ca : digestible P ratio as it depends on the digestibility of the Ca source, phytase activity of the feed, the concentration of digestible P and the type of production and pig category.

### Growing pigs

At our Institute, we have recently performed an experiment with growing pigs in which the effect of microbial phytase (0 or 800 FTU kg$^{-1}$), four levels of Ca (2, 4, 7 and 10 g kg$^{-1}$) in two diets differing in phytate P content (2.0 and 3.0 g kg$^{-1}$) on performance and on the apparent digestibility of Ca, Mg and P was studied (Jongbloed *et al.*, 1995). We have found that there was a negative linear Ca effect on the digestibility of P. The differences between the apparent digestibility of both Ca and Mg were larger at the lower Ca levels than at the higher Ca levels, irrespective of the diet. Ca, Mg and P contents in urine were significantly affected by the interaction of dietary Ca level and microbial phytase. It was also shown that the dietary Ca level, at a marginal supply of digestible P and Zn, exerted a large effect on performance of the pigs. Based on several assumptions we estimated an optimal Ca : digestible P ratio in the feed for growing pigs of 2.9 : 1 (Table 12.7).

**Table 12.8.** Survey of recommendations of Mg for pigs (g kg$^{-1}$ feed).

| Growing pigs (kg) | | | | Breeding sows | | |
|------|------|------|------|----------|-----------|------------|
| 5 | 25 | 45 | 90 | Pregnant | Lactating | Reference |
| 0.37 | 0.25 | 0.21 | 0.15 | — | — | ARC (1981); 80% availability |
| 0.95 | 0.65 | 0.53 | 0.36 | 1.7 | 1.7 | Own calcul.; 25% digestibility |
| 0.4 | 0.4 | 0.4 | 0.4 | 0.4 | 0.4 | NRC (1988) |
| 0.8 | 0.8 | 0.8 | 0.8 | 0.8 | 0.8 | INRA (1984) |
| < ------------------------ 0.3 to 0.5 ------------------------ > | | | | | | DLG (1987) |

## Breeding sows

The retention in pregnant sows can vary considerably in Ca : P ratio. This variation is due to the differences in amounts of maternal protein and bone retention. Everts *et al.* (1998a) indicated that the Ca : P ratio ranged between 1.35 and 1.7. In lactating sows this ratio is more or less equal to the Ca : P ratio in the suckling piglets or in milk. Everts *et al.* (1998b) found a mean value of 1.45 for lactating sows, when the mobilization from the maternal body was not included. With these figures we have calculated the required ratio between calcium and digestible phosphorus for pregnant and lactating sows assuming a digestibility coefficient for Ca of 0.4 to 0.5 (Table 12.7). In this table the supply of Ca from drinking water is not included. There is still uncertainty about the digestibility coefficient of Ca for breeding sows.

## Requirements for magnesium, sodium, potassium and chlorine

### Magnesium

There are large differences in recommendations regarding the requirement for Mg (ARC, 1981; INRA, 1984; DLG, 1987; NRC, 1988), as can be concluded from Table 12.8. Especially the estimates by ARC (1981) are substantially lower, which could be due to high availability (see above). So far, the highest digestibility of Mg in common feeds has never exceeded 30% in our own experiments. Our conclusion is that for growing pigs a concentration of 1.0 g Mg kg$^{-1}$ diet will be sufficient.

There are only a few references on the requirement for Mg of breeding sows. During late pregnancy about 1 g Mg is retained (Everts *et al.*, 1998a). With a mean digestibility of 25%, 4 g Mg is needed above a maintenance requirement. This suggests that a daily supply of 4.5 g Mg should be enough (at a feeding level of 3 kg, equal to 1.5 g Mg per kg diet). During lactation Mg is excreted in milk. Assuming a milk production of 10 kg with a Mg content of 0.25 g kg$^{-1}$, about 10 g Mg are needed above maintenance. A total Mg requirement of 10.5 g is thus enough (at a feeding level of 6 kg, equal to

**Table 12.9.** Survey of the recommendations of Na for pigs (g kg$^{-1}$ feed).

| Growing pigs (kg) | | | | Breeding sows | | |
|---|---|---|---|---|---|---|
| 5 | 25 | 45 | 90 | Pregnant | Lactation | Reference |
| 1.2 | 0.9 | 0.7 | 0.5 | 0.3 | 0.8 | ARC (1981) |
| 1.0 | 1.0 | 1.0 | 1.0 | 1.3 | 1.6 | NRC (1988) |
| 1.3 | 1.3 | 1.3 | 1.3 | 1.3 | 1.3 | INRA (1984) |
| 1.4 | 1.3 | 1.0 | 0.8 | 1.5 | 1.7 | DLG (1987) |

1.75 g Mg kg$^{-1}$ diet). This value is much higher than the values recommended on Table 12.8.

### Sodium

A survey of the literature concerning the recommendations for Na of pigs is listed in Table 12.9. Our own research showed that levels exceeding 0.8 g Na kg$^{-1}$ feed did not result in better performance of growing pigs from 25 to 110 kg LW (Van Diepen and Lenis, 1989). For pigs up to 25 kg a level of 1.3 g Na kg$^{-1}$ feed seems to be sufficient, and a level of 1.0 g Na kg$^{-1}$ feed for growing pigs from 25 to 110 kg. Table 12.9 shows that ARC recommendations for breeding sows are substantially lower than the others. We recommend a level of 1.5 g Na kg$^{-1}$ feed for breeding sows during pregnancy and 1.7 g Na kg$^{-1}$ during lactation.

### Potassium and chlorine

For an extensive survey of the requirements for K the reader is referred to Vreman (1993). Most handbooks recommend levels between 2 and 4 g K kg$^{-1}$ feed. Vreman (1993) concluded that 3.0 g K kg$^{-1}$ feed will be sufficient for all categories of pigs. Most feeds, however, contain an excess of K that may impair the environment. In contrast to K, there are fewer data from experiments reported in which the requirement for Cl has been assessed accurately. DLG (1987) suggests that the concentration of Cl should be 1.5 times the Na content, although no scientific evidence is provided for this. Based on available literature data we recommend 1.5 g Cl kg$^{-1}$ feed for all categories of pigs.

### Electrolyte balance

Lately, an increased interest has been shown regarding the dietary electrolyte balance (dEB) in pig feeds. This is related to a possible positive effect on digestion and utilization of organic matter and minerals. In this respect, also the buffering capacity of the feed should be mentioned to achieve a good environment in the digestive tract (Roth and Kirchgessner, 1989). Mostly, dEB is defined as Na + K − Cl, in which the concentrations are expressed as meq kg$^{-1}$ feed. In addition, there are more sophisticated formulas, like dUA (dietary undetermined anion), as has been proposed by Patience *et al.* (1986).

Based on American literature (Crenshaw, 1991), it can be concluded that dEB should be in the range of 100–400 meq kg$^{-1}$ feed, whereby between 200 and 250 meq kg$^{-1}$ feed is likely to be preferred using performance as a response parameter.

More attention to the dEB (we prefer the term dietary cation anion difference; dCAD) is expected with regard to animal health, a possible positive effect on meat quality, and the role of dEB in relation to ammonia emission (Mroz *et al.*, 1996a). A lower dEB is indicated to be more effective in reducing indoor ammonia volatilization.

# Discussion

During the last decade, minerals have received more attention due to the possible environmental threat they impose in some areas. The production of minerals in these areas exceeds the amount that can be taken up by the crops. Therefore, to reduce excretion of minerals by pigs, the supply of minerals should be in close balance with the requirement of the animal at any stage of production. This requires a detailed knowledge on the digestibility of minerals of feeds and all kinds of raw materials that are used for pigs. It has been shown that there is a lot of confusion about the terminology used to express the nutritive value of minerals for pigs. Especially the term availability can have at least four different meanings. Therefore, there is an urgent needed to standardize the terms and to define them. For several practical reasons we are in favour of using the term absorbability or digestibility for the methods we apply to assess the nutritive value of the minerals in the feeds. Different terms in expressing the nutritive value of minerals also result in different mineral evaluating systems.

We have developed two techniques to evaluate the digestibility of various raw materials. It has been noted that techniques applied in different research stations are not exactly the same, and therefore, different results may be expected. To enhance the exchange of data, standardization of the evaluation techniques among the different research centres is a prerequisite. The acceptance of a single system all over the world should be encouraged. Efforts to achieve this have so far failed.

The other aspect is that the mineral requirements should be known in more detail, in order to match supply and requirements as closely as possible. In the review, it was shown that there are large differences among countries in mineral recommendations. Because there are quite a lot of factors that affect the requirements, a factorial approach is the best way to get a more precise estimate of the requirements. A more sophisticated model should be developed to achieve this. One should also consider the criteria for the recommendations: are they for maximizing bone mineralization in growing pigs or not? Furthermore, what are the right criteria for reproducing animals regarding their longevity? In addition, do we know the requirements of the current and future breeds, which may have a different course of mineral retention during their growth?

# References

ARC (1981) Agricultural Research Council. *The Nutrient Requirements of Pigs*. Common Agricultural Bureaux, Slough, UK, 307 pp.

Armbrecht, H.J. (1987) Age and the effects on calcium and phosphorus uptake by rat small intestine. *Nutrition Research* 7, 1169–1177.

Crenshaw, T.D. (1991) Sodium, potassium, magnesium and chloride in swine nutrition. In: Miller, E.R., Ullrey, D.E., Lewis, A.J. (eds), *Swine Nutrition*. Butterworth-Heinemann, Stoneham, UK, pp. 183–191.

C.V.B. (1996) *Veevoedertabel: Gegevens over chemische samenstelling, verteerbaarheid en voederwaarde van voedermiddelen*. Centraal Veevoederbureau, Lelystad, 450 pp.

Dellaert, B.M., Van der Peet, G.F.V., Jongbloed, A.W. and Beers, S. (1990) A comparison of different techniques to assess the biological availability of feed phosphates in pig feeding. *Netherlands Journal of Agricultural Science* 38, 555–566.

Den Hartog, L.A., Zandstra, T., Kemp, B. and Verstegen, M.W.A. (1988) Chemical composition of intra uterine tissue and mammary tissue in pigs as related to the stage of pregnancy. *Journal of Animal Physiology and Animal Nutrition* 60, 4–7.

DLG (1987) *Energie und Nährstoffbedarf landwirtschaftlicher Nutztiere no. 4 Schweine*. DLG Verlag Frankfurt (Main), 159 pp.

Düngelhoef, M., Rodehutscord, M., Spiekers, H. and Pfeffer, E. (1994) Effects of supplemental microbial phytase on availability of phosphorus contained in maize, wheat and triticale to pigs. *Animal Feed Science and Technology* 49, 1–10.

Eeckhout, W. and DePaepe, M. (1992) [Wheat phytase and a microbial phytase and the apparent digestibility of a single piglet diet.] *Landbouwtijdschrift* 45, 205–212.

Everts, H. and Dekker, R.A. (1991) Vermindering van de uitscheiding aan stikstof en fosfor bij fokzeugen door het gebruik van twee verschillende voeders voor dracht en lactatie: resultaten van balans-metingen en vergelijkende slachtproef. *IVVO-DLO rapport* no. 230, Lelystad, 110 pp.

Everts, H. and Dekker, R.A. (1994) Effect of nitrogen supply on nitrogen and energy metabolism in lactating sows. *Animal Production* 59, 445–454.

Everts, H., Blok, M.C., Kemp, B., Peet-Schwering, C.M.C. van der and Smits, C.H.M. (1994) [Requirements for pregnant sows. Starting points and factorial approach of the requirement of energy and ileal digestible amino acids for pregnant sows](in Dutch). *CVB documentation report* no. 9, Lelystad, The Netherlands, 51 pp.

Everts, H., Jongbloed, A.W. and Dekker, R.A. (1998a) Calcium, phosphorus and magnesium retention in pregnant sows during three parities. *Livestock Production Science* 53, 113–121.

Everts, H., Jongbloed, A.W. and Dekker, R.A. (1998b) Calcium, magnesium and phosphorus balance of sows during lactation for three parities. *Livestock Production Science* 54 (in press).

Fernández, J. (1995) Calcium and phosphorus metabolism in growing pigs. II. Simultaneous radio-calcium and radio-phosphorus kinetics. *Livestock Production Science* 41, 243-254.

Fourdin, A., Fontaine, N. and Pointillart, A. (1986) [Importance of cereal phytase activity for phytate phosphorus utilization by growing pigs fed triticale or maize diets.] *Journées de la Recherche Porcine en France* 18, 83-90.

Guéguen, L. and Perez, J.M. (1981) A re-evaluation of recommended dietary allowances of calcium and phosphorus for pigs. *Proceedings Nutrition Society* 40, 273-278.

Harland, B. (1989) Dietary fibre and mineral bioavailability. *Nutrition Research Reviews* 2, 133-147.

INRA (1984) *L'alimentation des animaux monogastriques: porc, lapin, volailles.* INRA, Paris, 282 pp.

Jongbloed, A.W. (1987) Phosphorus in the feeding of pigs: Effect of diet on the absorption and retention of phosphorus by growing pigs. PhD thesis, Landbouwuniversiteit Wageningen, Rapport IVVO-DLO no. 179, Lelystad, The Netherlands, 343 pp.

Jongbloed, A.W. and Everts, H. (1992) Apparent digestible phosphorus in the feeding of pigs in relation to availability, requirement and environment. 2. The requirement of digestible phosphorus for piglets, growing-finishing pigs and breeding sows. *Netherlands Journal of Agricultural Science* 40, 123-136.

Jongbloed, A.W. and Henkens, C.H. (1996) Environmental concerns of using animal manure – the Dutch case. In: E.T. Kornegay (ed.), *Proceedings Symposium on Nutrient Management of Food Animals to Enhance the Environment.* Lewis Publishers/CRC Press, Boca Raton, Florida, pp. 317-333.

Jongbloed, A.W. and Kemme, P.A. (1990) Apparent digestible phosphorus in the feeding of pigs in relation to availability, requirement and environment. 1. Digestible phosphorus in feedstuffs from plant and animal origin. *Netherlands Journal of Agricultural Science* 38, 367-575.

Jongbloed, A.W., Everts, H. and Kemme, P.A. (1994) Verteerbaar fosfornormen voor varkens. *CVB-documentatierapport* no. 10. Centraal Veevoederbureau, Lelystad, The Netherlands, 38 pp.

Jongbloed, A.W., Kemme, P.A., Mroz, Z. and Bruggencate, R. ten (1995) Apparent total tract digestibility of organic matter, N, Ca, Mg and P in growing pigs as affected by levels of Ca, microbial phytase and phytate. In: van Hartingsveldt, W., Hessing, M., van der Lugt, J.P. and Somers, W.A.C. (eds), *Proceedings of Second Symposium on Feed Enzymes (ESFE2).* TNO Nutrition and Food Research Institute, Zeist, pp. 198-204.

Jongbloed, A.W., Kemme, P.A., Mroz, Z. and Diepen, J.Th.M. van (1996) Phytase in swine rations: impact on nutrition and environment. A review. *Proceedings BASF Technical Symposia*, Des Moines, Iowa, pp. 44-69.

Kemme, P.A., Jongbloed, A.W., Mroz, Z. and Beynen, A.C. (1997a) The efficacy of *Aspergillus niger* phytase in rendering phytate phosphorus available for absorption in pigs is influenced by their physiological status. *Journal of Animal Science* 75, 2129-2139.

Kemme, P.A., Radcliffe, J.S., Jongbloed, A.W. and Mroz, Z. (1997b) The effects of sow parity on digestibility of proximate components and minerals during lactation as influenced by diet and microbial phytase supplementation. *Journal of Animal Science* 75, 2147-2153.

McDowell, L.R. (1992) *Minerals in Animal and Human Nutrition*. Academic Press, San Diego, 524 pp.

McElroy, S.T., Link, J.E., Dowdy, R.P., Zinn, K.R. and Ellersieck, M.R. (1991) Influence of age and magnesium on calcium metabolism in rats. *Journal of Nutrition* 121, 492–497.

Mroz, Z., Jongbloed, A.W., Canh, T. and Vreman, K. (1996a) Environmental implications of dietary acidogenic and alkalogenic additives in pigs. Paper P1.10 *European Association of Animal Production*, Lillehammer, Norway, p. 245.

Mroz, Z., Jongbloed, A.W., Dekker, R.A., Vreman, K., Van Diepen, J.Th.M., Kemme, P.A. and Kogut, J. (1996b) The effect of different water and urea-nitrogen supplies on manure volume and nitrogen excretion by pregnant sows. *Report ID-DLO no. 96.002*, Lelystad, The Netherlands, 59 pp.

Noblet, J., Close, W.H., Heavens, R.P. and Brown, D. (1985) Studies on the energy metabolism of the pregnant sow. 1. Uterus and mammary tissue development. *British Journal of Nutrition* 53, 251–265.

NRC (1980) *Mineral Tolerance of Domestic Animals*. National Academic Press, Washington (DC), USA, 577 pp.

NRC (1988) *Nutrient Requirements of Swine*. National Academy of Sciences, National Academy Press, Washington (DC), USA, 93 pp.

NRC (1994) *Nutrient Requirements of Poultry*. National Research Council, National Academy Press, Washington (DC), USA, 155 pp.

Partridge, I.G. (1980) Mineral nutrition of the pig. *Proceedings Nutrition Society* 39, 185–192.

Patience, J.F., Austic, R.E. and Boyd, R.D. (1986) The effect of sodium bicarbonate or potassium bicarbonate on acid–base status and protein and energy digestibility in swine. *Nutrition Research* 6, 263–273.

Pointillart, A.N., Garel, J.M. and Guéguen, L. (1978) Plasma calcitonin and parathyroid hormone levels in growing pigs on different diets I. High phosphorus diet. *Annales Biologique Animale Biochimique Biophysique* 18, 699–709.

Pointillart, A.N., Fontaine, N, and Tomasset, M. (1986) Effects of vitamin D on calcium regulation in vitamin-D-deficient pigs given a phytate-phosphorus diet. *British Journal of Nutrition* 56, 661–669.

Pointillart, A., Fourdin, N. and Delmas, A. (1987) Consequences de l'exces de calcium chez des porcs non supplementés en phosphore mineral. *Journées de la Recherche Porcine en France* 19, 281–287.

Roland, D.A., Gordon, R.W. and Rao, S.K. (1993) Phosphorus solubilization and its effect on the environment. *Proceedings Maryland Nutrition Conference*, 138–146.

Ross, R.D., Cromwell, G.L. and Stahly, T.S. (1984) Effects of source and particle size on the biological availability of calcium supplements for growing pigs. *Journal of Animal Science* 59, 125–134.

Roth, F.X. and Kirchgessner, M. (1989) Bedeutung von pH-Wert und Pufferkapazität des Futters für die Ferkelfütterung. 1. Mitteilung: pH-Wert und Pufferkapazität, von Futtermischungen bei Einsatz von Säuren. *Landwirtschaftliche Forschung* 42, 157–167.

Schröder, B., Breves, G. and Rodehutscord, M. (1996) Mechanisms of intestinal phosphorus absorption and availability of dietary phosphorus in pigs. *Deutsch Tierärztliche Wochenschrift* 103, 209–214.

Shields Jr, R.G., Mahan, D.C. and Maxson, P.F. (1985) Effect of dietary gestation and lactation protein levels on reproductive performance and body composition of first-litter female swine. *Journal of Animal Science* 60, 179–189.

Tuori, M., Kaustell, K., Valaja, J., Aimonen, E., Saarisalo, E. and Huhtanen, P. (1996) *Feed Tables and Requirements: Ruminants, Pigs, Poultry, Fur Animals, Horses.* Yliopistopaino, Helsinki, Finland, 112 pp.

Underwood, E.J. (1981) *The Mineral Nutrition of Livestock,* 2nd edn. Commonwealth Agricultural Bureaux, Slough, UK, 237 pp.

Van der Klis, J.D. (1993) Physico-chemical chyme conditions and mineral absorption in broilers. PhD thesis, Agricultural University Wageningen, The Netherlands, 159 pp.

Van Diepen, J.Th. M. and Lenis, N.P. (1989) Effect van het zoutgehalte in mestvarkensvoer op de wateropname en mestresultaten. *Rapport IVVO-DLO* no. 208, Lelystad, 30 pp.

Vreman, K. (1993) Potassium in nutrition of growing pigs. *IVVO-DLO report no. 258,* Lelystad, the Netherlands, 34 pp.

Walach-Janiak, M., Raj, St. and Fandrejewski, H. (1986) Protein and energy balance in pregnant gilts. *Livestock Production Science* 15, 249–260.

Whittemore, C.T. and Yang, H. (1989) Physical and chemical composition of the body of breeding sows with differing body subcutaneous fat depth at parturition, differing nutrition during lactation and differing litter size. *Animal Production* 48, 203–212.

# Protein Metabolism in the Growing Pig

## 13

## P.J. Moughan

*Institute of Food, Nutrition and Human Health, Massey University, Palmerston North, New Zealand*

## Introduction

The absorption of amino acids, subsequent metabolism of the absorbed dietary amino acids and the metabolism of body protein and amino acids of body origin constitute a complex biological system. To gain insight into such a system it is helpful to construct mathematical models. Such quantitation of amino acid transactions in the animal's body is essential if predictions are to be made of an animal's physical performance and ultimately its economic performance. It is also becoming increasingly important to be able to predict, with accuracy, the amounts of elements and especially nitrogen, excreted from the animal into the environment. Dijkstra and France (1995) view a model as an 'idea, hypothesis or relation expressed in mathematics'. This highlights a further basic aspect of quantitation. Very importantly, modelling allows concepts or theories, often complex and involving extensive interaction, to be couched as hypotheses and compared with the real world, via experimentation.

The modelling of amino acid and protein metabolism makes an important contribution to the development of whole-animal growth models as the deposition of body protein has a disproportionate effect on the prediction of body growth rate. Protein has a central role in whole-body animal metabolism. The aim of this chapter is to provide a quantitative overview of body protein metabolism in the growing pig.

# Approaches to a Quantitative Description of Metabolism

A great deal is now known, at least qualitatively, about the digestion of food, the uptake of nutrients over time, their subsequent metabolism and the rate of accumulation and turnover of body tissues. Digestion and metabolism are complex processes involving numerous controls and influencing factors. Factors such as nutrient transport across cell walls, rates of blood flow, organ uptake and rates of enzyme activity associated with different biochemical pathways all interact to control metabolism and these are influenced by complex and sensitive humoral and neural systems. A completely mechanistic model of animal growth would attempt to simulate directly all these physiological and biochemical processes. There is a problem, however, in attempting to do this, in that a comprehensive quantitative description of the control of metabolism is just not possible given our current state of knowledge (Fox and Barry, 1995; Pettigrew, 1995). There are still technical difficulties in measuring certain metabolic processes. For example, there is considerable debate (Riis, 1983a; Baldwin *et al.*, 1994) as to the accuracy of *in vivo* measurement of body protein synthesis and breakdown, two fundamental processes driving growth. The same can be said for the measurement of amino acid oxidation using stable isotopes (Millward and Rivers, 1988). It is also unclear as to what are the rate-limiting steps in protein synthesis and protein breakdown (Madsen, 1983a). Moreover, the rate of catabolism of amino acids in the liver is affected not only by the activity of the amino acid transferases but also by cross-membrane transport which is influenced by amino acid supply and hormonal status. The physiological control of liver transport systems and regulation of the amino transferases for the different amino acids are not fully elucidated (Madsen, 1983b). Lobley (1988) has discussed control of the catabolism of leucine, in which it is believed that the activity of the enzyme branched-chain oxo-acid dehydrogenase is rate limiting. This enzyme is known to exist in two forms: one phosphorylated (inactive) and one dephosphorylated (active). The balance between the two forms of the enzyme is regulated by the competing actions of a protein kinase and a protein phosphatase. The kinase activity is influenced by several factors, including the concentration of the branched-chain oxo-acid, ATP, glucocorticoids and aceto-acetyl-CoA as well as the extracellular or intracellular concentration of leucine. A fully mechanistic model would need to describe explicitly the latter regulation, and this is just one example of many.

A further difficulty with a fully mechanistic model would be its sheer size and complexity, and thus the time required to simulate growth. An explicit representation of biochemical pathways often requires an iteration interval of less than 1 min for the model to remain stable, which also leads to unacceptably long simulation periods for practical application (Black and de Lange, 1995).

It is possible, however, to emphasize only major pathways and to aggregate biochemical reactions within these and thus to develop models at a biochemical level (Schulz, 1978; Pettigrew *et al.*, 1992). Such models have made a significant contribution to our understanding of metabolism and relative to the objectives for which these models were constructed, they have been successful. However, it remains that such models are often simplistic and possibly naïve in their description of the control of growth. These models are not expected to be effective for accurately predicting growth of body tissues, for practical application in agriculture.

It is for this reason that several models of pig growth have been developed which describe growth at an organ/tissue/whole-animal level and attempt to encapsulate the underlying biological processes into implicit concepts of growth causation. An example of the latter is that in this type of model, rather than describe an animal's capacity for body protein retention in terms of the processes of protein synthesis (DNA, messenger RNA per unit DNA, activity of mRNA) and protein degradation, it has been customary to assume that an upper limit to daily body protein retention exists (which is actually the outcome of the underlying mechanisms). This facilitates quantitation and allows the influence of factors known to affect protein retention (e.g. gender, breed, strain, age) to be described.

Modelling at this level, in which concepts concerning the control of a process are represented mathematically, has been described (Black, 1995a) as deductive or semi-mechanistic. Models of this type, instead of describing the growth processes explicitly, do so implicitly by attempting to represent the underlying biological processes, using biologically-interpretable functions and parameters. This is the type of quantitation addressed in the present chapter. The approach has the advantage that model parameters and factors influencing them can be determined with accuracy in the whole animal. Moreover, the model parameters are biologically interpretable and hypotheses concerning their function can be tested rigorously. It is likely that models will become more causal, as understanding develops and thus will become more general in their predictions. It is important, however, that these models do not become more complex simply for the sake of it. There is no point in a more mechanistic description, unless it leads to greater clarity and a higher accuracy of prediction.

A strength of deductive models is that they provide a cohesive framework for testing 'ideas' and 'concepts' about growth and its control. Models and their subcomponents should be continually tested and when found not to perform adequately, they should be revised. There is a real danger that modellers can become so possessed with their belief in how a system works that discrepancies with real-world data can be ignored. At least the deductive model and its components are readably testable, the concepts within are falsifiable. This allows for a proper, orderly progression of knowledge.

**Table 13.1.** Biological processes underlying amino acid (AA) metabolism in the pig.

| Process |
| --- |
| 1. Amino acid absorption |
| 2. Maintenance:[a] |
| Turnover of body protein |
| Integumental amino acid loss |
| Gut endogenous amino acid loss |
| Synthesis of 'other' compounds |
| Urinary amino acid losses |
| 3. Growth: |
| Protein accretion |
| Support costs: Inevitable amino acid catabolism |
| Gut endogenous amino acid loss |
| Turnover of body protein |
| Synthesis of 'other' compounds |
| Preferential amino acid catabolism |

[a] A distinction is made between basal or maintenance processes (i.e. those occurring in the hypothetical state whereby body tissue is neither being gained nor lost) and those processes associated with the accretion of new body tissue. The rate of a process at 'maintenance' is defined as that rate commensurate with a daily food intake under which body weight is neither being gained nor lost. Rates of the processes during growth are variable. It should be noted that for most of the metabolic processes there is actually a natural continuum between maintenance and growth and that the distinction between states is arbitrary and reliant upon definition.

# A Quantitative Description of Protein Metabolism

Amino acid absorption and metabolism is a continuum of interactive processes which are subject to complex multidimensional control and regulation. Thus the flows of absorbed compounds and their rates and sites of accumulation in the body are continually changing over time. However, and with deductive modelling, in which nutrient partitioning is described relative to several biological controls, it is considered useful to regard metabolism as comprising a number of distinct processes (Table 13.1).

In Table 13.1 a distinction has been made between the 'maintenance' or 'basal' processes and those associated with growth. However, the so-called maintenance processes and growth processes are interrelated. For example, gut endogenous amino acid losses are often considered part of the maintenance cost, but increase substantially with food dry-matter intake and are thus correlated with growth. Another example is that the body protein turnover and its associated amino acid losses increase with rate of body protein deposition. For the growing pig, 'maintenance' is a concept rather than a reality and by definition never exists for a *growing* animal. If an animal is forced to a state of zero dietary nitrogen balance, it is no longer growing and, at least in the short term, it is in a highly catabolic state with respect to body lipid reserves.

Nevertheless, it is traditional and is considered here to be helpful to conceptualize overall metabolism in two parts: maintenance and growth. At zero nitrogen retention there are still costs associated with body protein metabolism and these are the classical 'basal' or 'maintenance' costs. Above zero nitrogen retention, whereby the pig is depositing body protein, costs associated with maintaining the proteinaceous body tissues remain, but these may be better classified as 'support costs for growth' (Table 13.1), rather than 'maintenance'. This is a departure from a previous approach (Moughan, 1989), whereby these support costs were included in the overall maintenance term. It is relatively easy and clear cut to describe the 'basal' amino acid costs, but it is not easy, nor necessary, to distinguish between all the different support costs. The partitioning of overall amino acid metabolism into a relatively constant 'basal' or 'maintenance' component and a quite separate, more variable component associated with production (protein deposition and support costs) is consistent with the early arguments of Folin (1905) reiterated by Mitchell (1959), relating to endogenous and exogenous catabolism. It is usually assumed that the basal processes take priority over growth in relation to amino acid supply (Reeds *et al.*, 1993).

## Amino acid absorption

Mechanistically, the uptake of an amino acid from food depends upon the integrated and highly regulated action of numerous processes such as the rate and frequency of food ingestion, the rate of stomach emptying and rate of digesta transit through the small intestine, the activity of several digestive (including bacterial) enzymes, the degree of contact between enzymes and the food protein, the rate of absorption of amino acids and the degree of competition for transport sites. Several of these phenomena are influenced by the physical and chemical composition of the food and the physical environment (e.g. temperature). In fact, the degree of amino acid digestion and absorption is as much a food as it is an animal attribute and numerous food related factors (e.g. presence of cross-linkages) affect the animal's ability to digest and absorb.

It is not common for pig growth models to represent the process of digestion explicitly, but rather experimentally determined coefficients of digestibility are used to represent the overall effect. Obviously, such an approach does not give any indication of the kinetics of amino acid absorption following a meal, and this is a limitation. In pursuit of a greater generality of prediction, some modellers have attempted to model the mechanisms known to underlie digestion and absorption (Bastianelli and Sauvant, 1995).

However, the use of an empirically derived single digestibility coefficient to describe the overall absorption of an amino acid does appear to lead to reasonably accurate predictions of uptake and subsequent tissue deposition (Moughan, 1995a). 'True' as opposed to 'apparent' ileal digestibility is the best means of predicting amino acid uptake, except when the effects of plant antinutritional factors and fibre are represented explicitly in the model, whereby 'real' ileal digestibility coefficients should be used (Boisen and Moughan, 1996a). When true ileal digestibility is used in a growth model, the basal endogenous loss associated with dry-matter flow in the intestines must be represented in the model directly. For feedstuffs which have undergone damage during processing, fluorodinitrobenzene (FDNB) lysine should be used to predict the lysine concentration of the feedstuff and digestibility coefficients for reactive lysine (Moughan and Rutherfurd, 1996) or total nitrogen should be used, rather than conventional lysine digestibility coefficients. Minor effects of food intake and age of animal on nutrient digestibility have been reported (Kidder and Manners, 1978; Noblet and Shi, 1994; Black, 1995b). Generally there appears to be little influence of animal genotype or gender. Porcine somatotrophin (PST) has been shown not to influence nutrient digestibility (Beerman, 1993) though dietary inclusion of enzymes and antibiotics may lead to increases in amino acid digestibility.

## Maintenance

If an animal is fed a protein-free diet, over time it loses body protein, in spite of a considerable intake of energy. Moreover, it can be observed that if an animal is to be maintained at zero nitrogen balance (neither gaining nor losing body protein) it must be supplied with some dietary protein. What gives rise to this 'maintenance' amino acid requirement?

Firstly, skin and hair are continually being synthesized and are ultimately lost from the body surface. These losses need to be replaced. Also, considerable quantities of amino acids are lost via the digestive tract associated with the food dry-matter flow and some amino acids released from body protein during body protein turnover will be oxidized and the nitrogen will be lost in the urine, mainly as urea. Some of the urinary urea will also arise from bacterial catabolism of gut endogenous amino acids to ammonia in the large intestine and there may also be a direct loss of unmodified amino acids in the urine. Finally, some amino acids may be specifically used for the irreversible synthesis of non-amino-acid nitrogen-containing compounds, or may be irrevocably altered during metabolism.

## Basal turnover of body protein

It is recognized (Schoenheimer, 1946) that in a day of growth an animal breaks down and resynthesizes considerable amounts of body protein. Some protein depots are turned over faster than others and rate of turnover appears to be affected by age and rate of nutrient intake (rate of production). Non-producing adult animals still turn over considerable amounts of their body protein and young animals on 'maintenance' or 'below maintenance' food intakes also turn over body protein. Thus animals appear to have a 'basal' rate of body protein turnover, which confers adaptational advantages to the organism (Riis, 1983a).

It appears, at least in terms of nitrogen economy, that body protein degradation/resynthesis is a highly efficient process. Baldwin *et al.* (1994) have discussed evidence which indicates that amino acids arising from protein degradation may be channelled directly to protein synthesis without mixing with intracellular amino acids and without being exposed to enzymes involved in amino acid oxidation. The extent of this channelling (a structurally organized multi-enzyme complex) may be an important determinant of the fate of amino acids derived from body protein turnover. However, and although the extent of reutilization does appear to be high, which is understandable biologically, the process is unlikely to be completely efficient (Simon, 1989). Estimates of efficiency given in the literature range from 80 to 95% (Oldham, 1988; Reeds, 1988; Simon, 1989; Linder, 1991). The simplest explanation for the loss of amino acids during turnover is that some amino acids are inevitably lost to catabolic processes, which exhibit activity even at low substrate concentrations. It has been argued (Riis, 1983b; Simon, 1989) that certain amino acids (often dietary limiting, e.g. lysine) may be preferentially retained within the cell during protein turnover, and thus do not become available for catabolism. The reasoning of Riis (1983b) is based on observed relatively low extracellular flux rates for some amino acids in relation to their concentration in body protein. However, Reeds (1988) has concluded that there is insufficient evidence to substantiate such a claim. If such preferential retention and reutilization does occur to a significant extent, then this will affect modelled estimates of amino acid requirements.

An explanation for the loss of body amino acids associated with turnover, in spite of seemingly highly efficient mechanisms of amino acid recapture, might lie in the diurnal pattern of body protein metabolism. For a large part of the day for the meal-fed animal (postabsorptive period), in which there is no direct supply of absorbed dietary amino acids to the body cells, the loss of any amino acid (indispensable), presumably leads to an imbalance of amino acids for resynthesis with a consequent loss of all other amino acids proportional to their concentration in body protein. A net loss of indispensable amino acids such as lysine, threonine, histidine and valine from muscle during the postabsorptive period with a concomitant uptake by the splanchnic bed has been demonstrated (Felig, 1975) as has a substantial loss of liver amino acids, postabsorption (Waterlow *et al.*, 1978).

The extent of amino acid reutilization is difficult to measure, but can be estimated by comparing the rate of amino acid liberation by protein breakdown with the net rate of loss from the body by oxidation and excretion (Simon, 1989). Several workers have attempted to estimate the loss of amino acid nitrogen from the body due to basal turnover, by measuring the urinary nitrogen excretion of pigs on a protein-free diet. However, these losses may overestimate the actual basal loss as some indispensable amino acids (eg. methionine) are used in the synthesis of essential non-protein nitrogen-containing compounds. This use will in itself lead to an imbalancing of turned-over amino acids, with a consequent overall catabolic loss. For an animal fed just sufficient amino acids to maintain body protein, the requirement for the indispensable amino acid would be met from the diet, and thus the additional imbalance and catabolism would not occur. The role of methionine in relation to the latter phenomenon is implied from the observation that adding methionine to a protein-free ration significantly reduces urinary nitrogen excretion in the growing pig. Also, part of the urinary urea and total nitrogen of animals fed a protein-free diet will have been derived from the hindgut breakdown of gut endogenous protein, not intracellular body protein turnover. Thus the protein-free approach will lead to overestimation. A better estimate of the basal losses may be found by extrapolating urinary urea excretion (arising directly from amino acid catabolism) against dietary protein intake, to give the output associated with zero dietary protein intake. Using this approach, Moughan *et al.* (1987a) found a urinary urea nitrogen excretion in the 30 kg live weight pig of 667 mg day$^{-1}$.

Accepting the latter value as an estimate of the basal urinary urea nitrogen excretion, related to the inefficiency of intracellular body protein turnover, and assuming an efficiency of utilization of dietary nitrogen for maintenance of 0.9 (based on observed values for oxidation of the dietary first limiting amino acid at low amino acid intakes), this corresponds to a dietary protein requirement in the pig of 361 mg protein kg$^{-0.75}$ day$^{-1}$. This indicates an efficiency for intracellular protein turnover of around 98%, assuming a body protein turnover at nitrogen equilibrium of $15.1 \times 10^3$ mg kg$^{-0.75}$ day$^{-1}$ (Fuller, 1980) which is similar to the generalized value for mammals at maintenance of $16 \times 10^3$ mg kg$^{-0.75}$ day$^{-1}$ given by Buttery and D'Mello (1994). Based on these calculations, it does appear as though basal intracellular turnover is a highly efficient process. For a 50 kg live weight pig, and assuming that the degraded amino acids are sourced from the breakdown of body protein (Millward and Rivers, 1988), it can be further calculated that the maintenance lysine cost related to this biological process is 450 mg day$^{-1}$, which is about 2% of the total daily dietary requirement for lysine for a rapidly growing pig of this live weight.

Thus, the dietary maintenance protein requirement (due to inefficiency in protein turnover) for the growing pig (PM$_1$) can be given as:

**Table 13.2.** Amino acid composition (g 100 g$^{-1}$ protein) of skin and hair protein in the pig.

| Amino acid | Composition |
|---|---|
| Lysine | 4.3 |
| Methionine | 1.0 |
| Cysteine | 4.5 |
| Tryptophan | 0.9 |
| Histidine | 1.2 |
| Phenylalanine | 2.9 |
| Tyrosine | 1.8 |
| Threonine | 3.2 |
| Leucine | 5.0 |
| Isoleucine | 2.4 |
| Valine | 3.6 |
| Non-essential amino acids | 69.2 |

$$PM_1 = 361 \text{ mg kg}^{-0.75} \text{ day}^{-1} \tag{13.1}$$

There will be an additional requirement for a source of non-specific nitrogen (90 mg N kg$^{-0.75}$ day$^{-1}$; Moughan *et al.*, 1987a) to offset non-urea nitrogen loss in the urine.

It remains unknown as to whether the efficiency of basal body protein turnover is influenced by the age, gender or genotype of the pig. For the present, it is assumed that the efficiency of basal turnover is a constant, but that the fractional rate of basal protein synthesis declines as the animal grows.

### Integumental losses

Quantitative information on the losses of amino acids from the hooves, hair and cutaneous epithelium of the pig is scarce. The value presented by Moughan (1989), 94 mg protein kg$^{-0.75}$ day$^{-1}$, appears reasonable based on current published information and is close to a published estimate for adult humans from a review by Linder (1991). Equally, information on the amino acid composition of skin and hair is limited. A revised estimate of the amino acid composition of skin and hair in the pig to that given by Moughan (1989), and based on the inclusion of other published data (Mitchell, 1959; Wünsche *et al.*, 1983; MacRae *et al.*, 1993) is given in Table 13.2. For a rapidly growing 50 kg live weight pig, the need to replace lysine loss from the skin and hair would represent only 0.5% of the daily dietary lysine requirement. An estimate of the dietary maintenance requirement (PM$_2$) arising from a loss of amino acids via the skin and hair is:

$$PM_2 = 105 \text{ mg kg}^{-0.75} \text{ day}^{-1} \tag{13.2}$$

*Basal gut endogenous amino acid loss*

Copious quantities of proteinaceous material of endogenous origin enter the alimentary canal of the pig. Much of this material is digested and reabsorbed as amino acids in the small intestine, but some escapes small bowel digestion and enters the large intestine where it is subjected to bacterial degradation. Endogenous protein will also enter the large bowel via the colonic mucosa. An animal given an amount of food dry matter just sufficient to maintain live weight will lose (through urinary and faecal routes) a considerable amount of amino acid nitrogen of gut endogenous origin. It is clear that at higher food dry matter intakes (supporting tissue accretion), gut endogenous amino acid loss is related to food dry-matter intake. This relationship may come about due to a higher degree of cell turnover (a function of metabolic rate and possibly of degree of abrasion) and mucin secretion (lubricant and protective function) at higher food intakes and is likely affected by the nature and amount of fibre in the diet. However, there will be a base level of gut endogenous excretion at low (maintenance and sub-maintenance) food dry-matter intakes which may be a function of body weight rather than dry matter intake (Fuller, 1994).

From the results of a number of recent studies (Souffrant, 1991) it is clear that the traditional approach to determining endogenous amino acid loss from the digestive tract, whereby an animal is given a protein-free diet, will lead to underestimation of the losses pertaining to the use of normal diets. It is important that endogenous losses be determined under protein or peptide alimentation (Boisen and Moughan, 1996b). A recent study (Butts *et al.*, 1993), involving total collection of ileal digesta from the 50 kg live weight pig and employing peptide alimentation (enzyme hydrolysed casein-based diet) followed by ultrafiltration of ileal digesta, provides quantitative information on endogenous ileal amino acid loss as related to food dry-matter intake. The regression relationships from the latter study were extrapolated to predict endogenous loss at a maintenance food dry-matter intake (500 g dry matter per day for a 48 kg live weight pig) and the latter estimates have been expressed on a metabolic body weight basis (Table 13.3).

Endogenous amino acids also enter the large intestine from the colonic mucosa, but it is more difficult to obtain a quantitative appreciation of this component of overall gut endogenous loss as few direct determinations have been made. The main components of small intestinal endogenous protein are digestive enzymes and bile, secreted proteins, mucus and mucosal cells. Moreover, it appears (Fauconneau and Michel, 1970; Sauer, 1976) that the enzymes and secretory proteins are readily digested and that the amino acids are largely reabsorbed, and will thus contribute only a small fraction of total endogenous protein remaining at the end of the ileum. Conversely, mucin protein is largely resistant to digestion in the small intestine (Hashimoto *et al.*, 1963) and along with mucosal cells will form the bulk of endogenous nitrogen excreted from the small intestine (Snook, 1973). Mucin and sloughed epithelial cells are also expected to comprise the main part of large

**Table 13.3.** Estimates[a] of basal endogenous amino acid loss from the terminal ileum of the 50 kg live weight pig.

| Component | Endogenous loss (mg kg$^{-0.75}$ day$^{-1}$) |
| --- | --- |
| Nitrogen | 83.4 |
| Lysine | 28 |
| Methionine | 12 |
| Cysteine | 16 |
| Tryptophan | 12 |
| Histidine | 11 |
| Phenylalanine | 21 |
| Tyrosine | 29 |
| Threonine | 24 |
| Leucine | 31 |
| Isoleucine | 19 |
| Valine | 33 |
| Non-essential amino acids | 285 |

[a] From Butts *et al.* (1993). The estimates for methionine, cysteine and tryptophan are based on the tabulated endogenous loss for lysine and an estimate of the amino acid composition of gut endogenous protein (Boisen and Moughan, 1996b).

intestinal endogenous loss. Mucus-secreting goblet cells are dispersed throughout the mucosal epithelium in both the small and large intestines and cell division and migration occurs in the large as well as the small intestine. It thus seems reasonable to estimate endogenous loss in the large intestine based on determined losses from the upper digestive tract. In earlier work (Moughan, 1989), endogenous loss from the large intestine (caecum plus colon) was estimated by equating large bowel losses to upper-tract losses on an organ tissue weight basis. This approach is crude, however, as mucin and cell loss should be more related to surface area rather than tissue weight, and the different digestive organs have different surface area to weight ratios. An attempt is made here to equate the losses on an organ surface area basis. Information on the linear dimensions of the digestive organs was obtained from Kvasnitskii (1951), Vodovar *et al.* (1964) and Nickel *et al.* (1973) and primary surface areas for the upper and lower tracts were calculated. The estimated primary surface area ratio (SI/LI) for the 50 kg live weight pig (1.3/1) was similar to volumetric capacity ratios reported in the literature. Folding of the tissue occurs in both the small intestine and colon of the pig and this was assumed to increase surface area to a similar extent in the two organs. However, the small intestine contains numerous villi, which increase surface area in the small intestine relative to the large intestine (Warren, 1939; Argenzio, 1984). When the contribution of the small intestinal villi is taken into account, it is estimated that the large intestinal surface area is 0.10 of small intestinal area. Accepting the above arguments, the latter factor can be applied to known small-intestinal losses to obtain estimates of endogenous loss from the large intestine. This estimation relies upon a number of assumptions and there is a considerable degree of uncertainty in some of the values used in the calculations, so it must be stressed that the present estimate of large intestinal endogenous amino acid loss is very much an approximation. Nevertheless, the factor (0.10) is in good agreement with the recent observation of Krawielitzki *et al.* (1990) that large intestinal endogenous secretion is 0.11 total secretion.

Taking the hindgut contribution into account, and for a rapidly growing 50 kg live weight pig, the dietary requirement for lysine to offset the basal endogenous gut lysine loss would be around 3% of the total dietary lysine requirement.

The dietary maintenance requirement ($PM_3$) due to a loss of amino acids via the gut is represented as:

$$PM_3 = 637 \text{ mg kg}^{-0.75} \text{ day}^{-1} \tag{13.3}$$

The basal gut endogenous amino acid loss estimates given here (Table 13.3) are based on observations made at a single live weight (50 kg live weight). Caution should be exercised in extrapolating to other body weights. Information is needed on the effect of body weight on endogenous gut protein losses.

### Synthesis of 'other' compounds

Amino acids may be used by the animal to synthesize 'other' compounds and thus be lost from the metabolic amino acid pool. The compound may be a non-amino-acid nitrogen-containing compound (e.g. synthesis of histamine from histidine or synthesis of serotonin from tryptophan) or the metabolism may represent an irreversible alteration in amino acid structure (e.g. histidine to 3-methyl histidine, lysine to hydroxylysine).

Generally, these conversions are considered to make only a negligible quantitative contribution to amino acid metabolism, and are usually ignored in the modelling of amino acid transactions. In some cases, however, they may be quantitatively significant. For example, methionine is involved in the synthesis of creatine which is lost in the urine of the pig as creatinine in significant quantities. Depending upon the extent of recycling of homocysteine to methionine (which relies upon an adequate supply of choline, betaine or $N^5$-methyltetrahydrofolate) and the animal's cysteine status, the loss of methionine could be significant. There is insufficient information available to allow quantitation of this process.

### Urinary amino acid loss

Amino acids and protein are lost in the urine, but for a healthy animal these losses are negligible (Linder, 1991).

## Overall maintenance amino acid losses

Table 13.4 provides a summary of estimated amounts of amino acids required for the main processes underlying 'protein maintenance' in the 50 kg live weight pig. The final column provides an estimate of the dietary requirement for amino acids for maintenance. A quantitative appreciation of the influence of factors such as breed, sex, age and live weight on these requirement values is generally lacking, but in total and for a rapidly growing animal, the maintenance amino acid requirement is only a small proportion ($< 10\%$) of the total daily amino acid requirement.

## Growth

It is assumed that the basal metabolism, concerned with maintaining vital processes, has a priority for amino acid substrate over the anabolic processes

**Table 13.4.** Estimated amino acid losses associated with 'maintenance' processes in the 50 kg live weight pig and an estimate of the dietary maintenance amino acid requirement.

| | Amino acid loss (mg day$^{-1}$) | | | Dietary |
| --- | --- | --- | --- | --- |
| | Basal turnover[a] | Skin and hair[b] | Gut endogenous[c] | requirement[d] (mg kg$^{-0.75}$ day$^{-1}$) |
| Lysine | 405 | 76 | 579 | 63 |
| Methionine | 119 | 18 | 248 | 23 |
| Cysteine | 79 | 80 | 331 | 29 |
| Tryptophan | 59 | 16 | 248 | 19 |
| Histidine | 173 | 21 | 228 | 25 |
| Phenylalanine | 231 | 51 | 434 | 42 |
| Tyrosine | 153 | 32 | 600 | 46 |
| Threonine | 233 | 57 | 496 | 47 |
| Leucine | 458 | 89 | 641 | 70 |
| Isoleucine | 209 | 43 | 393 | 38 |
| Valine | 278 | 64 | 682 | 61 |
| Nitrogen | 2670[e] | 283 | 1725 | 277 |

[a] Estimated from a basal body protein loss of 0.325 g kg$^{-0.75}$ day$^{-1}$ (6.11 g protein day$^{-1}$) and whole body protein amino acid composition as given by Moughan (1989).
[b] Estimated from an integumental protein loss of 0.094 g protein kg$^{-0.75}$ day$^{-1}$ (1.77 g day$^{-1}$) and the amino acid composition of skin and hair (Table 13.2).
[c] Estimates of basal endogenous ileal losses adjusted for the colonic endogenous contribution.
[d] Assumes an efficiency of utilization of dietary amino acids for maintenance of 0.9.
[e] Inefficiency of basal body protein turnover plus allowance for non-specific urinary nitrogen.

leading to the accretion of new tissue. Moreover, and in addition to the accretion of new tissue, absorbed amino acids may be irreversibly oxidized or may be used for that tissue resynthesis directly associated with the growth processes. These 'other' (non-anabolic) uses of amino acids are referred to here as the 'support costs' for protein accretion (Table 13.1).

## Protein accretion

As the pig receives incremental amounts of balanced amino acids above maintenance and if these are accompanied by sufficient amounts of glucose and other non-amino energy yielding compounds and vitamins and minerals, it accretes incremental amounts of body protein. This will occur until the protein synthetic capacity of the pig has been met. Above this capacity ($P_{dmax}$) balanced amino acids cannot be used for protein synthesis, and as they cannot be stored in the body for prolonged periods of time, will be catabolized. Thus $P_{dmax}$ would appear to be an important constraint on pig growth.

AN UPPER LIMIT TO PROTEIN RETENTION ($P_{dmax}$). The animal's intrinsic impetus for protein growth dominates the regulation of nutrient partitioning (Webster, 1993), and thus $P_{dmax}$ and its quantitative description are of considerable importance. The concept of an intrinsic upper limit to whole-body protein retention, influenced by the animal's genotype, was discussed by Kiela-nowski (1969) and subsequently several studies have been undertaken to characterize $P_{dmax}$ and how it changes with live weight and degree of maturity. Emmans (1981) has argued, quite correctly, that it cannot be proven 'scientifically' that a given growth rate is the potential rate, because it is impossible to demonstrate unequivocally that a given environment (a host of influencing variables, some of which may be unknown) is non-limiting. However, pragmatically, $P_{dmax}$ can be demonstrated and is a valuable con-cept. Numerous dose–response type studies with growing pigs have demonstrated a plateau to body protein deposition which cannot be lifted by the feeding of additional sources of energy and/or balanced protein. Also, several hyperalimentation studies have been conducted with the growing pig (Seddon *et al.*, 1991; Newcomb *et al.*, 1993) which indicate that the supply of a balanced diet at a level exceeding normal *ad libitum* ingestion leads to an increased growth rate but not an increased retention of body protein. These studies imply the operation of a maximal rate of protein deposition. More-over, $P_{dmax}$ is under genetic control (breed, strain, gender) and is influenced by age, live weight and maybe nutritional history of the animal (Whittemore, 1993). $P_{dmax}$ can be manipulated by genetic selection or by administration of exogenous hormones and repartitioning agents (Beerman, 1993) and may be affected by challenges to the immune system (Williams *et al.*, 1993). Pub-lished estimates of $P_{dmax}$ for the growing pig (20 to 100 kg live weight) vary from 80 g day$^{-1}$ to well in excess of 200 g day$^{-1}$.

At a fundamental level protein deposition and, therefore, $P_{dmax}$ is the result of differences between body protein synthesis (PS) and body protein degradation (PD). PS, in turn, is related to the units of RNA in the organ or body and the activity per unit of RNA. Maximal protein retention should equate with maximal PS at minimal PD. PS and PD, the rates of which are much higher than that of protein deposition, appear to be controlled independently of one another, and it is the change in the relative rates of these two processes with age that dictates the pattern of $P_{dmax}$ with age. Both PS and PD increase with food intake but PS increases at a greater rate than PD leading to increases in protein deposition (Reeds and Fuller, 1983). Moreover, it appears that dietary protein and non-protein energy influence protein retention via different mechanisms. Increases in dietary protein stimulate both PS and PD, whereas increases in non-protein energy lead to a small increase in PS and a small decrease in PD (Reeds and Fuller, 1983). It also appears that the effects of dietary protein and non-protein energy on PS and PD are additive rather than interactive. Another example of differential regulation of PS and PD is found in the study of Fuller *et al.* (1987). An increase in body protein retention in growing pigs was brought about by increasing the quantity of protein consumed. The increased protein retention resulted from an increase in PS with a concomitant but smaller increase in PD. A similar sized increase in retention was achieved by altering the quality of the dietary protein rather than increasing the quantity, and PS was not changed but PD decreased substantially.

Both PS and PD are influenced by level of nutrition and the rate of both processes decreases with age. In the adult state PS will normally equal PD. Changes in tissue ribosomal RNA content may be the major factor determining the rate of PS but it is not the only factor as differences in the rate of protein synthesis per unit RNA have been observed (Madsen, 1983a).

Protein retention in the pig has been modelled at the level of PS and PD (Stombaugh and Stombaugh, 1989; Danfaer, 1991), and there is the opportunity here to relate protein retention to changes in cellularity (DNA) during growth and development and to changes in ribosomal RNA per unit DNA and activity per unit RNA. However, there are limited observations on the latter characteristics for the growing pig. Whereas there are some quantitative data for the pig on PS (usually given as fractional synthetic rates, FSR) at the whole body and organ level, there are no data on maximal rates of PS in the pig and there is very little direct information on PD. Because of this, most workers have chosen to emphasize body protein retention *per se* and to describe this relative to an upper limit to deposition.

A central issue in a quantitation of protein metabolism is the description of the relationship between $P_{dmax}$ and age or live weight. In the embryonic and fetal phases of an animal's growth, which is typified by hyperplasia, body protein accretes over time in a disproportionate manner, such that $P_{dmax}$ will increase with age during early growth. Further, in a non-pregnant, non-lactating adult pig there will be no net body protein deposition (i.e. $P_{dmax} = 0$), and it follows that at some stage during the growth phase $P_{dmax}$ must begin

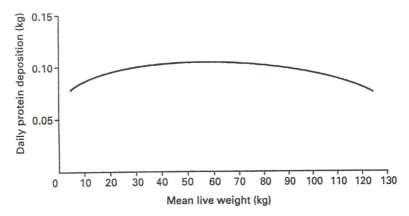

**Fig. 13.1.** Daily protein deposition rate in relation to *mean* live weight. The experiment began at 20 kg and progressed by serial slaughter at increasing live weight points through to about 180 kg. Mean live weight is the mean between 20 kg and the serial slaughter point. (From Whittemore, 1993, published with permission.)

to decline. It has been argued (Whittemore, 1993), however, that for a large part (and possibly the commercially important part) of the postweaning growth phase, $P_{dmax}$ may be rather constant. If this is true it allows for a straightforward description of the parameter, $P_{dmax}$. The latter argument for a constant $P_{dmax}$ over a considerable period of the postnatal growth phase, is consistent with postnatal growth being dominated by hypertrophy rather than hyperplasia (Reeds *et al.*, 1993). The question remains, however, as to when in life is such constancy in $P_{dmax}$ achieved and when does the decline in $P_{dmax}$ begin to occur?

It may be difficult for young pigs (less than 20 kg live weight) to ingest sufficient quantities of nutrients to allow $P_{dmax}$ to be measured, and sometimes in studies purporting to measure $P_{dmax}$, measurement is made of a rate of protein deposition (Pd), which is less than $P_{dmax}$. When particular care has been taken to ensure maximal food intakes, very high growth rates have been reported for young pigs (Whittemore, 1993; Boyd *et al.*, 1995) and thus a constancy in $P_{dmax}$ may occur at earlier live weights than is commonly accepted (Fig. 13.1). The proposition of a constancy in $P_{dmax}$ over a considerable live weight range (20 to 85 kg live weight) receives some recent empirical support (Jacobson, 1992; see also Table 13.5).

If $P_{dmax}$ is constant during part of the postweaning growth period, $P_{dmax}/P_t$ ($P_t$ = whole body protein) must be declining over this period and this implies that PS/$P_t$ must be higher and, or PD/$P_t$ lower at the lower live weights. There is evidence from rats (Goldspink and Kelly, 1984) that the fractional synthetic rate (FSR) of whole body protein is considerably higher at younger ages and body weights and declines with age. Although there will be

**Table 13.5.** Statistical significance of regression equations for the assessment of linearity of the increase in body protein over time for pigs depositing body protein at maximal rates. (Adapted from Jacobson, 1992.)

| | Significance | |
|---|---|---|
| Live weight range | Linear equation | Quadratic equation |
| 25–45 kg | $P < 0.01$ | NS |
| 25–65 kg | $P < 0.01$ | NS |
| 25–85 kg | $P < 0.01$ | NS |
| 25–110 kg | $P < 0.01$ | $P < 0.05$ |

some increase in body RNA during postnatal growth (Novakofski and McCusker, 1993) this may be offset by a declining synthesis rate per unit RNA (Millward *et al.*, 1976). It is interesting that for the growing rat (Millward *et al.*, 1976) it appears that although the FSR ($PS/P_t$) for muscle protein declines with increasing age (29% at 23 days to 5% at 130 days) the total protein synthesized per day is quite constant between 30 g body weight (at 23 days) and 200 g body weight (at 130 days).

The shape of the relationship between $P_{dmax}$ and live weight or age is likely to be influenced by the breed or strain of pig and this will account for some of the considerable variation observed in the literature for the shapes of reported maximal protein deposition curves. It is also possible that in some of these studies, and particularly at the lower live weights, Pd was measured rather than $P_{dmax}$ as there is often no internal evidence that $P_{dmax}$ was obtained. Given the importance of $P_{dmax}$ as a descriptor of pig growth, it is critical that studies, the objective of which is to determine $P_{dmax}$, actually do determine $P_{dmax}$ and that the mathematical relationship between $P_{dmax}$ and age be described appropriately. Ferguson and Gous (1993) have discussed experimental approaches to determining the potential rate of growth of protein in the pig. However, conclusions drawn also differ due to the fitting of quite different mathematical functions. There is debate as to the best procedures to be adopted (Walker and Young, 1993; de Lange, 1995; Black *et al.*, 1995; Emmans, 1995; Whittemore, 1995). Just because a function gives a good fit to the data does not prove validity. The function fitted should also give rise to a growth trajectory which is consistent with the modeller's view of the growth process. The Gompertz function has been shown to be useful for describing the growth trajectory (Whittemore, 1993) and allows simple description of different pig genotypes and sexes. Emmans (1981), who has made extensive use of the Gompertz function in animal growth modelling, makes the point that the function has elegant mathematical properties, appears to fit experimental data well and has parameters which are biologically interpretable.

It is possible that for pigs of high genetic merit, and especially at lower

**Table 13.6.** Amino acid composition (g kg$^{-1}$ crude protein) of whole-body protein in the growing pig. (From Kyriazakis *et al.*, 1993.)

| | Body protein (kg)[a] | |
|---|---|---|
| Amino acid | 1 | 10 |
| Lysine | 68.1 | 72.3 |
| Methionine | 18.0 | 18.0 |
| Cysteine | 11.0 | 9.6 |
| Tryptophan | 8.0 | 8.0 |
| Histidine | 25.7 | 30.7 |
| Phenylalanine | 38.0 | 38.0 |
| Tyrosine | 26.0 | 26.0 |
| Threonine | 38.0 | 38.0 |
| Leucine | 74.0 | 74.0 |
| Isoleucine | 35.0 | 35.0 |
| Valine | 47.0 | 47.0 |

[a] Corresponds approximately to 6 and 60 kg live weight, respectively.

live weights, $P_{dmax}$ may not be reached within the confines of appetite (Bikker, 1994). When this is the case the upper limit to protein retention is no longer regulated by the protein synthetic capacity of the cell, but rather by the upper limit to food intake.

THE AMINO ACID COMPOSITION OF WHOLE-BODY PROTEIN. Several studies have investigated the amino acid composition of protein deposited in the pig's body during growth. The recent work of Kyriazakis *et al.* (1993) has shown that there is a systematic effect of body protein weight on the concentrations of lysine, histidine and cysteine in whole-body protein (Table 13.6). This is likely brought about by changes (Reeds *et al.*, 1993) in the relative proportions (by weight) of different organs and body parts as the pig grows and matures. There are distinct differences in the amino acid compositions of the protein in different body parts (Mitchell, 1959). Evidence that faster growing pigs deposit more protein in the non-lean carcass parts (Cliplef and McKay, 1993; de Greef and Verstegen, 1993; Bikker, 1994) indicates that there may also be an association between the amino acid composition of body protein and rate of protein deposition. This is borne out by other recent studies (Krick *et al.*, 1993; Bikker *et al.*, 1994). Food restriction may also elicit changes in organs and body parts (Seve *et al.*, 1986; Rao and McCracken, 1992; Pekas, 1993) which may in turn alter body protein amino acid composition. For a more accurate quantitative description of body protein metabolism it would be useful to model the growth of different body parts and organs separately. This would allow for the differences in amino acid composition to be taken into account.

Estimation of the pattern of amino acids which can potentially be deposited in the pig's body during a day of growth allows comparison with the estimated utilizable amino acid profile (absorbed dietary amino acids corrected for postabsorptive amino acid losses), and thus prediction of the imbalanced absorbed amino acids. The loss of amino acids due to imbalance can be quantitatively considerable. If non-essential amino acids are limiting relative to the needs for new tissue deposition plus maintenance, then the process of transamination can be described quantitatively (Moughan and Smith, 1984), as can the metabolic conversions of cysteine from methionine, and tyrosine from phenylalanine.

## Support Costs

The growing pig requires a supply of amino acids above that needed for the basal processes and direct body protein deposition. In the main, this need arises because of metabolic processes directly associated with deposition. By way of example, as an animal eats more food and can thus deposit more body protein, there is an associated increase in the loss of endogenous amino acids from the digestive tract. Also, as body protein deposition increases so too does the turnover of body protein and presumably losses from turnover. There will be a greater demand for various metabolites, as the pig grows faster and thus the use of amino acids for synthesis of other compounds will also increase. Moreover, there will be an inevitable loss of amino acids to catabolic pathways, regardless of the animal's energy status (inevitable catabolism). These various costs are described here as the support costs. The gut losses can be described directly, but it is technically difficult, and nor is it necessary from a modelling perspective, to distinguish among the 'other' support costs. These are normally included together for purposes of quantitation.

### Gut amino acid losses

There is now considerable evidence that gut endogenous amino acid losses are positively related to food dry-matter intake. Furthermore, it is also well established that such losses need to be determined experimentally under protein or peptide alimentation rather than by using the traditional protein-free method (de Lange *et al.*, 1990; Moughan and Rutherfurd, 1990). Figure 13.2 depicts endogenous nitrogen output from the terminal ileum of the growing pig related to food dry-matter intake. It appears that as the growing pig ingests more food, the output of endogenous protein increases in a linear manner. This relationship is likely to be dependent upon dietary composition and the range of food dry-matter intakes studied and may be affected by live weight. Gut endogenous amino acid loss is affected by the amount and type of dietary fibre and by the presence of antinutritional factors (ANFs) such as

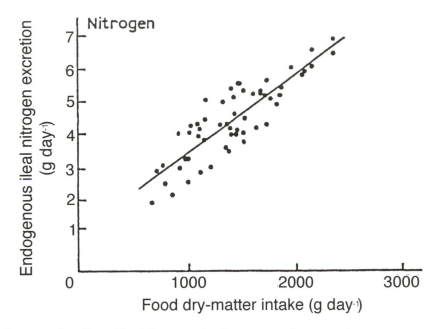

**Fig. 13.2.** The effect of food dry matter intake on gut endogenous nitrogen excretion for the 50 kg live weight pig fed an enzymatically hydrolysed casein-based diet. (From Butts *et al.*, 1993, published with permission.)

lectins and tannins. There is considerable published quantitative data on endogenous protein loss in the pig (Boisen and Moughan, 1996b). When low fibre protein-containing diets are fed to the growing pig, endogenous ileal protein loss is about 20 g kg$^{-1}$ dry matter intake. ANFs usually increase this loss resulting in losses typically ranging from 20 to 40 g kg$^{-1}$ dry matter intake, though losses as high as 100 g kg$^{-1}$ dry matter intake have been reported (Boisen and Moughan, 1996b). The recent studies of Schulze (1994) provide much useful quantitative information on the effects of dietary fibre and ANFs on gut endogenous amino acid loss in the pig.

There will be a secretion of endogenous protein (especially mucus) into the large intestine and this loss also needs to be quantified (p. 308).

### *'Other' support costs*

The main component of the 'other' support costs is the loss of amino acids due to inevitable catabolism. Inevitable catabolism is defined here, in relation to the first-limiting amino acid, as being the degradation of the amino acid through active catabolic pathways, which still occurs when the energy supply is not limiting for protein synthesis. This degradation which will be modulated by blood glucose and fatty acid levels, appears to be an 'inevitable' consequence of the presence in cells of catabolic systems which may be turned down but not shut off completely. The phenomenon of inevitable catabolism has been discussed in some detail (Moughan, 1995b). It appears, based on evidence to date, that at least 10% of the absorbed first-limiting amino acid flow is subject to inevitable oxidation in mammals, with the rate of loss possibly being higher as the flow of absorbed amino acids increases towards the level required to support maximal body protein synthesis. There are two approaches to determining inevitable catabolism. The first direct method involves the use of radioactive tracers to measure amino acid oxidation, while the second indirect approach relies upon estimating the rate of inevitable catabolism as the difference between the amount of absorbed amino acid and that retained plus that used for the basal processes and to replace gut losses. In the latter approach the unexplained loss of absorbed amino acid is assumed to be due to the 'other support costs' and in the main to represent inevitable catabolism. The rate of inevitable catabolism needs to be determined under carefully defined experimental conditions.

A study in the growing pig (20–45 kg live weight) by Batterham *et al.* (1990), in which lysine was clearly the first limiting dietary amino acid, indicates that around 20% of the truly absorbed lysine (for the five lowest lysine diets, 0.1 to 0.5 g lysine $MJ^{-1}$ DE) remains unaccounted for after considering the basal processes, total gut losses and the deposition of body lysine, and this constitutes an estimate of the 'other' support costs. It also appears from this work that over the observed range in body protein deposition, 0.06 to 0.80 $P_{dmax}$, the 'other' support costs are relatively constant (at around 20% of absorbed lysine). The support costs appear to increase linearly with increases in $P_d$. In the latter study, the proportion of absorbed lysine unaccounted for appeared to increase substantially and disproportionately over the last three diets (0.6, 0.7 and 0.8 g lysine $MJ^{-1}$ DE) as $P_{dmax}$ (estimated by the authors to be 121 g per day) was approached. However, the data for these three diets must be interpreted cautiously as there is some uncertainty concerning the actual upper limit to protein retention for these pigs and as to whether or not dietary energy became a limiting factor for protein deposition. In another recent controlled study (P.J. Moughan and S.M. Rutherfurd, unpublished), growing pigs (30 kg live weight) were given a casein/synthetic amino acid based diet over a 19-day period. Lysine was clearly first limiting and the diet contained 15.0 MJ DE $kg^{-1}$. On average the pigs deposited body protein at 97 g $day^{-1}$ (around 0.6 $P_{dmax}$). The true ileal digestibility of dietary reactive lysine was determined as was the retention of lysine in the whole body. After taking into account the lysine deposited and the lysine estimated to be needed for maintenance and gut endogenous loss some 15 ± 7.4% ($\bar{x} \pm$ SD) of the absorbed lysine was found to be unaccounted for. This is in agreement with the finding of Batterham *et al.* (1990). Direct measurement, using isotope infusion, generally indicates that below an animal's amino acid 'requirement', oxidation of the first-limiting dietary amino acid occurs at a relatively low rate (Young *et al.*, 1985; Lobley, 1988;

Benevenga *et al.*, 1993), though in some such studies (Floc'h *et al.*, 1995) quite high rates of oxidation have been reported.

There is also uncertainty concerning the shape of the response of amino acid retention as a function of the absorbed amino acid and thus of 'other' support costs versus amino acid absorption. There is general agreement that for a group of diverse animals the response should approach a plateau curvilinearly (population effect) but there is debate as to whether the response in the individual animal is curvilinear or linear (Gahl *et al.*, 1991; Fuller and Garthwaite, 1993; Susenbeth, 1993; Bikker, 1994; Gahl *et al.*, 1995; Pomar, 1995).

The magnitude of inevitable catabolism may be influenced by factors such as body weight, genotype and the relative dietary concentration of amino acids other than the first-limiting amino acid and may differ for different amino acids and especially for methionine (Baker, 1993; Fuller, 1994). Moreover, the rate of the inevitable catabolism of the first-limiting amino acid may be related to the flux of TCA cycle intermediates arising from carbohydrate or fat dissimilation. It seems to be possible to reduce, but not to eliminate, the rate of inevitable catabolism of an amino acid by giving fat or carbohydrate to an animal.

Inadequate information on the process of inevitable catabolism and the other processes contributing to inefficiency in utilization of the absorbed first-limiting amino acid is viewed as a major weakness in current models of amino acid utilization (Young *et al.*, 1985; Fuller, 1994).

### Preferential amino acid catabolism

Amino acids are primarily used to synthesize proteins and this is an energy (ATP) demanding process, but alternatively, amino acids can be catabolized to provide energy (ATP). Also, an animal needs to maintain its blood glucose level within narrow limits and presumably has a need for a certain amount of fatty acid synthesis. Amino acids may supply precursors for the synthesis of both glucose and fatty acids. This is the basis for the often observed inter-action between dietary energy and protein during animal growth and there is a need in a quantitative description of metabolism to describe the metabolic state whereby absorbed balanced amino acids are expressly catabolized to supply ATP or precursors for glucose or fatty acid synthesis.

It has been clearly demonstrated, in a number of studies, that when a pig is given adequate balanced protein (to achieve its $P_{dmax}$) but a low intake of digestible energy, it breaks down amino acids to directly supply precursors for the synthesis of glucose or fatty acids, and protein deposition is compromised. This catabolism is referred to here as 'preferential' catabolism, which infers that the pig is partitioning amino acids away from protein synthesis towards catabolism, to meet a specific metabolic need. Amino acids can be catabolized preferentially to meet a metabolic demand, as opposed to inevitable catabolism which occurs in situations whereby dietary energy is clearly not limiting and the catabolism is an inevitable consequence of the existence of active catabolic pathways. Two distinct phases for body protein deposition ($P_d$) can be identified. Firstly, $P_d$ may be dependent upon dietary protein intake ($P_i$) and independent of dietary energy ($E_i$). Here preferential catabolism does not occur. Secondly, and at a set $P_i$, $P_d$ is dependent upon $E_i$. In this phase, preferential catabolism occurs and supply of extra energy may lead to a sparing of amino acids and to an increase in $P_d$. There are two considerations of importance in a quantitative description of metabolism: Is the process normally of a significant magnitude and how should it be described?

It appears that the process can have a significant influence on amino acid utilization and with modern breeds of pig which have higher potentials for lean gain but lower voluntary food intakes, preferential catabolism will assume an even greater significance. The potential quantitative significance of the process of preferential catabolism is borne out by data from the recent study of Kyriazakis and Emmans (1992a). Pigs (12 kg live weight) were given, for either 4 or 8 weeks, different allowances of a starch based diet at a set daily intake of a high-protein diet. The amount of the protein diet was increased weekly as were the amounts of the starch diet, except for one treatment group which received no starch. The dietary intake of protein was deemed to be sufficient to meet the requirement for potential lipid-free growth at all stages of the experiment. For the pigs killed at 4 weeks and receiving no additional starch, 3.5 g of lipid day$^{-1}$ were deposited along with 60 g of protein day$^{-1}$. With the highest level of starch supplementation pigs achieved 101 g of protein deposition day$^{-1}$ with 50 g of lipid deposition. For the pigs killed at 8 weeks from 12 kg body weight, and receiving no additional starch 13.5 g of lipid day$^{-1}$ were deposited and 74.5 g of protein. With the highest level of starch supplementation pigs achieved 128 g of daily protein deposition with 80 g day$^{-1}$ of lipid deposition. It is clear that on the lower energy intakes a significant amount of dietary protein (at least that equivalent to supporting 41 and 54 g of body protein deposition day$^{-1}$ for the two time periods, respectively) was degraded preferentially, presumably to supply ATP and to support the observed level of lipid deposition.

There are a number of approaches to a quantitative description of preferential catabolism. It is possible to describe the relationship between body protein deposition and dietary energy intake empirically, for pigs given protein adequate diets (Black *et al.*, 1986). The plateau of such a relationship represents $P_{dmax}$, while the linear increase of protein deposition with increasing energy intake reflects the process of preferential catabolism. The slope of the relationship between protein deposition and metabolizable energy intake is considered to decrease as body weight increases (Fig. 13.3). This effect has been represented (Black *et al.*, 1995) by assuming a declining exponential change in slope ($S$, nitrogen deposition/available energy intake) with body weight ($W$) viz.:

$$S = xe^{-yW} + z \tag{13.4}$$

where $x$, $y$ and $z$ are constants.

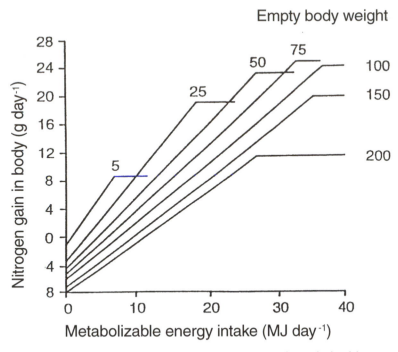

**Fig. 13.3.** Relationship between daily gain in nitrogen and metabolizable energy intake for pigs of the same genotype and increasing body weight. (From Black *et al.*, 1995, published with permission.)

The relationship appears to be affected by genotype and must therefore be derived separately for different populations of pigs. Kyriazakis and Emmans (1992b) have described an empirical relationship between the efficiency of utilization of absorbed ideally balanced protein ($e_p$) and the energy (ME) to protein (DCP) ratio in a feed. At low levels of protein intake, $e_p$ was independent of energy intake, and reflects inevitable amino acid catabolism. At low protein, high energy intakes around 15 to 20% of the absorbed ideal protein was catabolized. It also seems that for these pigs, preferential catabolism began to occur at an ME/DCP ratio of 73 MJ ME kg$^{-1}$. No effect of sex of the pig was found on the relationship between $e_p$ and ME/DCP, which is in contrast to the earlier finding of Campbell and Taverner (1988) and it was assumed to be unaffected by live weight of the pig. A similar approach to that of Kyriazakis and Emmans (1992b) was taken by Miller and Payne (1961) in a model related to dietary protein utilization in humans.

Another approach is to model the process of preferential catabolism based on limits to the outcome of metabolism, tissue deposition. Early pig growth models (Whittemore and Fawcett, 1976) and in accordance with the observation that when energy intake is limiting, pigs generally deposit lipid as well as protein, attempted to predict preferential amino acid catabolism based on a minimum $L_d : P_d$ ratio, which was deemed to vary with sex and genotype. There is experimental evidence for a minimum $L_d : P_d$ ratio (Campbell *et al.*, 1984) for pigs receiving low dietary energy intakes. This concept was subsequently successfully used by Moughan and Smith (1984), but here the modelling objective was to score dietary protein quality for a single day of growth using a standardized pig of defined characteristics. In a later model, the objective of which was to predict growth performance in pigs (Moughan *et al.*, 1987b), the minimum $L_d : P_d$ constraint was replaced by a minimum $L_t : P_t$ constraint (where $L_t$ and $P_t$ are total body lipid and protein, respectively). Such description allows for a net catabolism of lipid, and does not necessarily infer a constant minimum $L_d : P_d$ ratio. It would seem reasonable that a pig would use excess body lipid rather than catabolize absorbed ideal protein. Whittemore (1993) has made the distinction between 'normal' growth and physiological states whereby the animal is severely compromised, and has discussed the concepts of minimal and target fat levels.

There is some indication from the work of Kyriazakis and Emmans (1992a), that the minimum $L_t : P_t$ ratio for the young growing pig may be around 0.3. Black *et al.* (1995) have pointed out that if pigs at zero energy balance gain protein and lose body lipid, the ratio of lipid to protein deposition must be curvilinear with increasing energy intake and therefore cannot be adequately represented by a single minimum lipid to protein deposition ratio. However, the loss of lipid and gain of protein around and below maintenance energy intake may represent a distinct physiological phase, quite independent of the 'normal' growth phase and the former phase may be viewed as an adaptation to metabolic crisis. Moreover, the loss of body lipid at maintenance to support protein gain, may depend upon the animal's body lipid reserves. When adipose stores have become depleted, a constant lipid to protein retention phase may be found. Several recently conducted studies (de Greef, 1992; Kyriazakis and Emmans, 1992a,b; Bikker, 1994) provide some evidence against a single constant minimal $L_d : P_d$ ratio for pigs receiving low energy intakes. De Greef (1992) has argued that rather than a constant minimum $L_d : P_d$ ratio there is a constant ratio of $\Delta L_d : \Delta P_d$ (i.e. marginal ratio) and alternative means of describing preferential catabolism relative to tissue deposition have been discussed (de Greef, 1992; de Greef and Verstegen, 1995; de Lange, 1995).

It is possible to derive the marginal ratio from the empirical relationships between $P_d$ and dietary metabolizable energy intake, and the two approaches to description of preferential catabolism are directly related. A description of preferential amino acid catabolism based around the concept of a minimal body lipid mass has a causal basis and as such is more generalized.

# Conclusion

The view of body protein metabolism as a number of discrete processes (Table 13.1) is an attempt at rationalizing the complexities of the processes leading to lean tissue accretion. Clearer description of these processes physiologically and mechanistically but particularly, causally, and a better quantitation of the processes will lead to an enhanced understanding of growth in the pig and other mammals.

Far too often, there is a well-developed qualitative understanding of the processes underlying growth, but little attention has been paid to quantitation. Hypotheses concerning growth causation need to be developed, tested and discussed. Such a view of growth, rather than the more traditional mechanistic focus, can provide a platform for the development of quantitative relationships which when integrated will provide a quantitative model of whole-body amino acid metabolism. Such models, which can be applied to make quantitative predictions of technical and economic outputs, are enormously useful in practice.

# References

Argenzio, R.A. (1984) Digestion and absorption of carbohydrate, fat and protein. In: Swenson, M.J. (ed.), *Duke's Physiology of Domestic Animals*, 10th edn. Comstock Publishing Associates, Ithaca, New York, pp. 301–310.

Baker, D.H. (1993) Efficiency of amino acid utilization in the pig. In: Batterham, E.S., (ed.), *Manipulating Pig Production IV*. Australasian Pig Science Association, Canberra, Australia, pp. 191–197.

Baldwin, R.L., Calvert, C.C., Hanigan, M.D. and Beckett, J. (1994) Modelling amino acid metabolism in ruminants. In: D'Mello, J.P.F. (ed.), *Amino Acids in Farm Animal Nutrition*. CAB International, Wallingford, UK, pp. 281–306.

Bastianelli, D. and Sauvant, D. (1995) Modelling digestion and absorption in the pig. In: Danfaer, A. and Lescoat, P. (eds), *Proceedings of the IVth International Workshop on Modelling Nutrient Utilisation in Farm Animals*. National Institute of Animal Sciences, Foulum, Denmark, pp. 107–115.

Batterham, E.S. Andersen, L.M. Baigent, D.R. and White, E. (1990) Utilization of ileal digestible amino acids by growing pigs: Effect of dietary lysine concentration on efficiency of lysine retention. *British Journal of Nutrition* 64, 81–94.

Beerman, D.H. (1993) Use of endogenous agents to regulate growth composition. In: Hollis, G.R. (ed.), *Growth of the Pig*. CAB International, Wallingford, UK, pp. 185–213.

Benevenga, N.J., Gahl, M.J. and Blemings, K.P. (1993) Role of protein synthesis in amino acid catabolism. *Journal of Nutrition* 123, 332–336.

Bikker, P. (1994) Protein and lipid excretion in body components of growing pigs: Effects of body weight and nutrient intake. PhD thesis, Wageningen University, The Netherlands, 203 pp.

Bikker, P., Verstegen, M.W.A. and Bosch, M.W. (1994) Amino acid composition of growing pigs is affected by protein and energy intake. *Journal of Nutrition* 124, 1961–1969.

Black, J.L. (1995a) Approaches to modelling. In: Moughan, P.J., Verstegen, M.W.A. and Visser-Reyneveld, M.I. (eds), *Modelling Growth in the Pig*. EAAP Publication No. 78, Wageningen Pers, Wageningen, The Netherlands, pp. 11–22.

Black, J.L. (1995b) Modelling energy metabolism in the pig – critical evaluation of a simple reference model. In: Moughan, P.J., Verstegen, M.W.A. and Visser-Reyneveld, M.I. (eds), *Modelling Growth in the Pig*. EAAP Publication No. 78, Wageningen Pers, Wageningen, The Netherlands, pp. 87–102.

Black, J.L. and de Lange, C.F.M. (1995) Introduction to the principles of nutrient partitioning for growth. In: Moughan, P.J., Verstegen, M.W.A. and Visser-Reyneveld, M.I. (eds), *Modelling Growth in the Pig*. EAAP Publication No. 78. Wageningen Pers, Wageningen, The Netherlands, pp. 33–45.

Black, J.L., Cambell, R.G., Williams, I.H., James, K.J. and Davies, G.T. (1986) Simulation of energy and amino acid utilization in the pig. *Research and Development in Agriculture* 3, 121–145.

Black, J.L., Davies, G.T., Bray, H.J. Giles, L.R. and Chapple, R.P. (1995) Modelling the effects of genotype, environment and health on nutrient utilisation. In: Danfaer, A. and Lescoat, P. (eds), *Proceedings of the IVth International Workshop on Modelling Nutrient Utilisation in Farm Animals*. National Institute of Animal Science, Foulum, Denmark, pp. 85–105.

Boisen, S. and Moughan, P.J. (1996a) Different expressions of dietary protein and amino acid digestibility in pig feeds and their application in protein evaluation – A review. *Acta Agriculturae Scandinavica* 46, 165–172.

Boisen, S. and Moughan, P.J. (1996b) Dietary influences on endogenous ileal protein and amino acid loss in the pig – A review. *Acta Agriculturae Scandinavica* 46, 154–164.

Boyd, R.D., Kensinger, R.S., Harrell, R.J. and Bauman, D.E. (1995) Nutrient uptake and endocrine regulation of milk synthesis by mammary tissue of lactating sows. *Journal of Animal Science* 73 (Suppl. 2), 36–56.

Buttery, P.J. and D'Mello, J.P.F. (1994) Amino acid metabolism in farm animals: An overview. In: D'Mello J.P.F. (ed.), *Amino Acids in Farm Animal Nutrition*. CAB International, Wallingford, UK, pp. 1–10.

Butts, C.A., Moughan, P.J., Smith, W.C., Reynolds, G.W. and Garrick, D.J. (1993) The effect of food dry matter intake on endogenous ileal amino acid excretion determined under peptide alimentation in the 50 kg liveweight pig. *Journal of the Science of Food and Agriculture* 62, 235–243.

Campbell, R.G. and Taverner, M.R. (1988) Genotype and sex effects on the relationship between energy intake and protein deposition in growing pigs. *Journal of Animal Science* 66, 676–686.

Campbell, R.G., Taverner M.R. and Curic D.M. (1984) Effect of feeding level and dietary protein content on the growth, body composition and rate of protein deposition in pigs growing from 45 to 90 kg. *Animal Production* 38, 233–240.

Cliplef, R.L. and McKay, R.M. (1993) Visceral organ weights of swine selected for reduced backfat. *Canadian Journal of Animal Science* 73, 201–206.

Danfaer, A. (1991) Mathematical modelling of metabolic regulation and growth. *Livestock Production Science* 27, 1–18.

de Greef, K.H. (1992) Prediction of production. Nutrition induced tissue partitioning in growing pigs. PhD thesis, Wageningen University, The Netherlands, 117 pp.

de Greef, K.H. and Verstegen, M.W.A. (1993) Partitioning of protein and lipid deposition in the body of growing pigs. *Livestock Production Science* 35, 317-328.

de Greef, K.H. and Verstegen, M.W.A. (1995) Evaluation of a concept on energy partitioning in growing pigs. In: Moughan, P.J., Verstegen, M.W.A. and Visser-Reyneveld, M.I. (eds), *Modelling Growth in the Pig*. Wageningen Pers, Wageningen, The Netherlands, pp. 137-149.

de Lange, C.F.M. (1995) Modelling amino acid utilisation in grower finisher pigs. In: *Proceedings of the Annual Biokyowa Amino Acid Council Meeting 1995*, Toronto. Biokyowa Council, St Louis, Missouri.

de Lange, C.F.M., Souffrant, W.B. and Sauer, W.C. (1990) Real ileal protein and amino acid digestibilities in feedstuffs for growing pigs as determined with the [15]N-isotope dilution technique. *Journal of Animal Science* 68, 409-418.

Dijkstra, J. and France, J. (1995) Modelling and methodology in animal science. In: Danfaer, A. and Lescoat, P. (eds), *Proceedings of IVth International Workshop on Modelling Nutrient Utilisation in Farm Animals*. National Institute of Animal Science, Foulum, Denmark, pp. 9-18.

Emmans, G.C. (1981) A model of the growth and feed intake of *ad libitum* fed animals, particularly poultry. In: Hillyer, G.M., Whittemore, C.T. and Gunn, R.G. (eds), *Computers in Animal Production*, Occasional Publication No. 5. British Society of Animal Production, Edinburgh, pp. 103-110.

Emmans, G.C. (1995) Ways of describing pig growth and food intake using equations. *Pig News and Information* 16(4), 113N-116N.

Fauconneau, G. and Michel, M.C. (1970) The role of the gastrointestinal tract in the regulation of protein metabolism. In: Munro, H.N. and Allison, J.B. (eds), *Mammalian Protein Metabolism*, Vol IV. Academic Press, New York, pp. 481-522.

Felig, P. (1975) Amino acid metabolism in man. *Annual Review of Biochemistry* 44, 933-955.

Ferguson, N.S. and Gous, R.M. (1993) Evaluation of pig genotypes. 1. Theoretical aspects of measuring genetic parameters. *Animal Production* 56, 233-243.

Floc'h, N.Le, Obled, C. and Seve, B. (1995) *In vivo* threonine oxidation rate is dependent on threonine dietary supply in growing pigs fed low to adequate levels. *Journal of Nutrition* 125, 2550-2562.

Folin, O. (1905) A theory of protein metabolism. *American Journal of Physiology* 13, 117-138.

Fox, D.G. and Barry, M.C. (1995) Using whole animal models to evaluate and refine cattle diets under widely varying conditions. In: Danfaer, A. and Lescoat, P. (eds), *Proceedings of IVth International Workshop on Modelling Nutrient Utilisation in Farm Animals*. National Institute of Animal Science, Foulum, Denmark, pp. 143-162.

Fuller, M.F. (1980) Protein and amino acid nutrition of the pig. *Proceedings of the Nutrition Society* 39, 193-203.

Fuller, M.F. (1994) Amino acid requirements for maintenance, body protein accretion and reproduction in pigs. In: D'Mello, J.P.F. (ed.), *Amino Acids in Farm Animal Nutrition*. CAB International, Wallingford, UK, pp. 155-184.

Fuller, M.F. and Garthwaite, P. (1993) The form of response of body protein accretion to dietary amino acid supply. *Journal of Nutrition* 123, 957-963.

Fuller, M.F., Reeds, P.J., Cadenhead, A., Seve, B. and Preston, T. (1987) Effects of quantity and quality of dietary protein on nitrogen metabolism and protein turnover in growing pigs. *British Journal of Nutrition* 57, 287-301.

Gahl, M.J., Finke, M.D., Benevenga, N.J. and Crenshaw T.D. (1991) Use of a four-parameter logistic equation to evaluate the response of growing rats to ten levels of each indispensable amino acid. *Journal of Nutrition* 121, 1720-1729.

Gahl, M.J., Crenshaw, T.D. and Benevenga, N.J. (1995) Diminishing returns in weight, nitrogen and lysine gain of pigs fed six levels of lysine from three supplemented sources. *Journal of Animal Science* 73, 3177-3187.

Goldspink, D.F. and Kelly, F.J. (1984) Protein turnover and growth in the whole body, liver, and kidney of the rat from the fetus to senility. *Biochemical Journal* 217, 507-516.

Hashimoto, Y., Tsuiki, S., Nisizawa, K. and Pigman, W. (1963) Action of proteolytic enzymes on purified bovine submaxillary mucin. *Annals of the New York Academy of Science* 106, 233-246.

Jacobson, L.H. (1992) Determination of the upper genetic limit to body protein growth for an improved pig genotype. Master of Agricultural Science thesis, Massey University, Palmerston North, New Zealand, 132 pp.

Kidder, D.E. and Manners, M.J. (1978) *Digestion in the Pig*. Scientechnica, Bristol, UK, 201 pp.

Kielanowski, S. (1969) Energy and protein metabolism in growing pigs. *Revista Cubana de Ciencia Agricola* 3, 207-216.

Krawielitzki, K., Zebrowska, T., Schadereit, R., Kowalezyk, J., Wünsche, J and Herrmann, U. (1990) Determining nitrogen absorption and nitrogen secretion in different sections of the pig's intestine by digesta exchange between $^{15}$N labelled and unlabelled animals. *Archives of Animal Nutrition* 40, 25-37.

Krick, B.J., Boyd, D.R., Roneker, K.R., Beerman, D.H., Bauman, D.E., Ross, D.A. and Meisinger, D.J. (1993) Porcine somatotropin affects the dietary lysine requirement and net lysine utilization for growing pigs. *Journal of Nutrition* 123, 1913-1922.

Kvasnitskii, A.V. (1951) Voprosy fiziologii pishchevareniya u svinei. *Sel'khozziz, Moscow*.

Kyriazakis, I. and Emmans, G.C. (1992a) The effects of varying protein and energy intakes on the growth and body composition of pigs. 1. The effects of energy intake at constant, high protein intake. *British Journal of Nutrition* 68, 603-613.

Kyriazakis, I. and Emmans, G.C. (1992b) The effects of varying protein and energy intakes on the growth and body composition of pigs. 2. The effects of varying both energy and protein intake. *British Journal of Nutrition* 68, 615-625.

Kyriazakis, I., Emmans, G.C. and McDaniel, R. (1993) Whole body amino acid composition of the growing pig. *Journal of the Science of Food and Agriculture* 62, 29-33.

Linder, M.C. (1991) Nutrition and metabolism of proteins. In: Linder, M.C. (ed.), *Nutritional Biochemistry and Metabolism*. Elsevier, New York, USA, pp. 87-109.

Lobley, G.E. (1988) Protein turnover and energy metabolism in animals: Interactions in leanness and obesity. In: Leclercq, B. and Whitehead, C.C. (eds), *Leanness in Domestic Birds, Genetic, Metabolic and Hormonal Aspects*. Butterworths, London, UK, pp. 331-361.

Madsen, A. (1983a) The molecular basis of animal production: Metabolism in skeletal muscle cells. In: Riis, P.M. (ed.), *Dynamic Biochemistry of Animal Production*. Elsevier World Animal Science Series, A3. Elsevier, Amsterdam, The Netherlands, pp. 9-28.

Madsen, A. (1983b) The molecular basis of animal production: Metabolism in liver cells. In: Riis, P.M. (ed.), *Dynamic Biochemistry of Animal Production*. Elsevier World Animal Science Series, A3. Elsevier, Amsterdam, The Netherlands, pp. 53-74.

MacRae, J.C., Walker, A., Brown, D. and Lobley, G.E. (1993) Accretion of total protein and individual amino acids by organs and tissues of growing lambs and the ability of nitrogen balance techniques to quantitate protein retention. *Animal Production* 57, 237-245.

Miller, D.S. and Payne, P.R. (1961) Problems in the prediction of protein values of diets: The use of food composition tables. *Journal of Nutrition* 74, 413-419.

Millward, D.J. and Rivers, J.P.W. (1988) The nutritional role of indispensable amino acids and the metabolic basis for their requirements. *European Journal of Clinical Nutrition* 42, 367-393.

Millward, D.J., Garlick, P.J., James W.P.T., Sender, P.M. and Waterlow, J.C. (1976) Protein turnover. In: Cole, D.J.A., Boorman, K.N., Buttery, P.J., Lewis, D., Neale, R.J. and Swan, H. (eds), *Protein Metabolism and Nutrition*. European Association of Animal Production Publication No. 16 Butterworths, London, pp. 49-69.

Mitchell, H.H. (1959) Some species and age differences in amino acid requirements. In: Albanese, A.A. (ed.), *Protein and Amino Acid Nutrition*. Academic Press, New York, pp. 11-43.

Moughan, P.J. (1989) Simulation of the daily partitioning of lysine in the 50 kg liveweight pig – A factorial approach to estimating amino acid requirements for growth and maintenance. *Research and Development in Agriculture* 6, 7-14.

Moughan, P.J. (1995a) Modelling protein metabolism in the pig – first principles. In: Moughan, P.J., Verstegen, M.W.A. and Visser-Reyneveld, M.I. (eds), *Modelling Growth in the Pig*. EAAP Publication No. 78. Wageningen Pers, Wageningen, The Netherlands, pp. 59-70.

Moughan, P.J. (1995b) Modelling protein metabolism in the pig – Critical evaluation of a simple reference model. In: Moughan, P.J., Verstegen, M.W.A. and Visser-Reyneveld, M.I. (eds), *Modelling Growth in the Pig*. EAAP Publication No. 78. Wageningen Pers, Wageningen, The Netherlands, pp. 103-112.

Moughan, P.J. and Rutherfurd, S.M. (1990) Endogenous flow of total lysine and other amino acids at the distal ileum of the protein- or peptide-fed rat: The chemical labelling of gelatin protein by transformation of lysine to homoarginine. *Journal of the Science of Food and Agriculture* 52, 179-192.

Moughan, P.J. and Rutherfurd, S.M. (1996) A new method for determining digestible reactive lysine in foods. *Journal of Agricultural and Food Chemistry* 44, 2202-2209.

Moughan, P.J. and Smith, W.C. (1984) Prediction of dietary protein quality based on a model of the digestion and metabolism of nitrogen in the growing pig. *New Zealand Journal of Agricultural Research* 27, 501-507.

Moughan, P.J., Smith, W.C. and Kies, A.K., (1987a) Endogenous urinary metabolite excretion in the growing pig. *New Zealand Journal of Agricultural Research* 30, 183-187.

Moughan, P.J., Smith, W.C. and Pearson, G. (1987b) Description and validation of a model simulating growth in the pig (20 to 90 kg liveweight). *New Zealand Journal of Agricultural Research* 30, 481–489.

Newcomb, M.D., Ott, R.S., van Kempen, T., Lan, Y.H., McKeith, F.K., Novakofski, J.E., Bechtel, P.J. and Easter, R.A. (1993) Effect of hyperalimentation on body composition in swine. *Journal of Animal Science* 71, 144–150.

Nickel, R., Shummer, A., Seiferle, E. and Sack, W.O. (1973) *The Viscera of the Domestic Mammals*. Verlag Paul Parey, Berlin, Germany, 401 pp.

Noblet, J. and Shi, X.S. (1994) Effect of body weight on digestive utilization of energy and nutrients of ingredients and diets in pigs. *Livestock Production Science* 37, 323–338.

Novakofski, J. and McCusker, R.H. (1993) Physiology and principles of muscle growth. In: Hollis, G.R. (ed.), *Growth of the Pig*. CAB International, Wallingford, UK pp. 33–48.

Oldham, J.D. (1988) Amino acid metabolism in ruminants. In: *Proceedings of the 1988 Cornell Nutrition Conference for Feed Manufacturers*. Cornell University, Ithaca, New York, pp. 137–151.

Pekas, J.C. (1993) Maintenance feeding of 100 kg pigs: effect on carcass lean and fat yield and on gastrointestinal organ size. *Animal Production* 57, 455–464.

Pettigrew, J.E. (1995) Metabolic approach to modelling nutrient utilization. In: Moughan, P.J., Verstegen, M.W.A. and Visser-Reyneveld, M.I. (Eds), *Modelling Growth in the Pig*. EAAP Publication No. 78. Wageningen Pers, Wageningen, The Netherlands, pp. 173–183.

Pettigrew, J.E., Gill, M., France, J. and Close, W.H. (1992) A mathematical integration of energy and amino acid metabolism of lactating sows. *Journal of Animal Science* 70, 3742–3761.

Pomar, C. (1995) A systematic approach to interpret the relationship between protein intake and deposition and to evaluate the role of variation in production efficiency in swine. In: M. Ivan (ed.), *Animal Science Research and Development: Moving Toward a New Century*. Centre for Food and Animal Research, Agriculture and Agri-Food, Ottawa, Canada, pp. 361–375.

Rao, D.S. and McCracken, K.J. (1992) Energy: protein interactions in growing boars of high genetic potential for lean growth. 1. Effects on growth, carcass characteristics and organ weights. *Animal Production* 54, 75–82.

Reeds, P.J. (1988) Nitrogen metabolism and protein requirements. In: Blaxter, K. and MacDonald, I. (eds), *Comparative Nutrition*. John Libbey & Co Ltd, London, UK, pp. 55–72.

Reeds, P.J. and Fuller, M.F. (1983) Nutrient intake and protein turnover, *Proceedings of the Nutrition Society* 42, 463–471.

Reeds, P.J., Burrin, D.G., Davis, T.A., Fiorotto, M.A., Mersmann, M.J. and Pond, W.G. (1993) Growth regulation with particular reference to the pig. In: Hollis, G.R. (ed.), *Growth of the Pig*. CAB International, Wallingford, UK, pp. 1–32.

Riis, P.M. (1983a) The pools of tissue constituents and products: Proteins. In: Riis, P.M. (ed.), *Dynamic Biochemistry of Animal Production*. Elsevier World Animal Science Series, A3. Elsevier, Amsterdam, The Netherlands, pp. 75–103.

Riis, P.M. (1983b) The pools of cellular nutrients in amino acids. In: Riis, P.M. (ed.), *Dynamic Biochemistry of Animal Production*. Elsevier World Animal Science Series, A3. Elsevier, Amsterdam, The Netherlands, pp. 151–172.

Sauer, W.C. (1976) Factors affecting amino acid availabilities for cereal grains and their components for growing monogastric animals. PhD thesis, University of Manitoba, USA.

Schoenheimer, R. (1946) *The Dynamic State of Body Constituents*. Harvard University Press, Cambridge, Massachusetts, USA, 81 pp.

Schulz, A.R. (1978) Simulation of energy metabolism in the simple stomached animal. *British Journal of Nutrition* 39, 235-254.

Schulze, H. (1994) Endogenous Ileal nitrogen losses in pigs – dietary factors. PhD thesis, Wageningen University, The Netherlands, 147 pp.

Seddon, I.R., Gos, P.M. and Bayley, H.S. (1991) Hyperalimentation of the pig. *FASEB* 5(5), A1294.

Seve, B., Reeds, P.J., Fuller, M.F., Cadenhead, A. and Hay, S.M. (1986) Protein synthesis and retention in some tissues of the young pig as influenced by dietary protein intake after early-weaning. Possible connection to the energy metabolism. *Reproduction Nutrition Development* 26, 849-861.

Simon, O. (1989) Metabolism of proteins and amino acids. In: Bock, H.D., Eggum, B.O., Low, A.G., Simon, O. and Zebrowska, T. (eds), *Protein Metabolism in Farm Animals*. Oxford University Press, Oxford, and VEB Deutscher Landwirtschaftsverlag, Berlin, pp. 273-366.

Snook, J.T. (1973) Protein digestion – nutritional and metabolic considerations. *World Review of Nutrition and Dietetics* 18, 121-176.

Souffrant, W.B. (1991) Endogenous nitrogen losses during digestion in pigs. In: Verstegen, M.W.A., Huisman, J. and den Hartog, L.A. (eds), *Digestive Physiology in Pigs*, EAAP Publication No. 54. Pudoc, Wageningen, The Netherlands, pp. 147-166.

Stombaugh, D.P. and Stombaugh, I.A. (1989) Modelling protein synthesis and excretion during swine growth. Paper No. 89-4521, *International Winter Meeting of the American Society of Agricultural Engineers (ASAF), December 1989, Louisiana*. ASAE, St Joseph, USA, 17 pp.

Susenbeth, A. (1993) A model to describe the effect of nutrition on growth, body composition and nitrogen excretion in pigs. In: Verstegen, M.W.A., den Hartog, L.A., van Kempen, G.J.M. and Metz, J.H.M. (eds), *Nitrogen Flow in Pig Production and Environmental Consequences*. Pudoc Scientific Publishers, Wageningen, The Netherlands, pp. 90-95.

Vodovar, N., Flanzy, J. and Francois, A.C. (1964) Intestin grêle du porc .1. Dimensions et fonction de l'âge et du poids. Etude de la jonction du canal cholédoque et du canal pancréatique a celuici. *Annales de Biologie Animales Biochimie Biophysique* 4, 27-34.

Walker, B. and Young, B.A. (1993) Prediction of protein accretion, support costs and lipid accretion in the growing female pig and dry sow. *Agricultural Systems* 42, 343-358.

Warren, R. (1939) Serosal and mucosal dimensions at different levels of the dog's small intestine. *Anatomical Record* 75, 427-437.

Waterlow, J.C., Garlick, P.J. and Millward, D.J. (1978) *Protein Turnover in Mammalian Tissues and in the Whole Body*. North Holland Publishing Company, Amsterdam, The Netherlands, 804 pp.

Webster, A.J.F. (1993) Energy partitioning, tissue growth and appetite control. *Proceedings of the Nutrition Society* 52, 69-76

Whittemore, C.T. (1993) *The Science and Practice of Pig Production*. Longman, Harlow, UK, 661 pp.

Whittemore, C.T. (1995) Modelling the requirement of the young growing pig for dietary protein. *Agricultural Systems* 47, 415–425.

Whittemore, C.T. and Fawcett R.H. (1976) Theoretical aspects of a flexible model to simulate protein and lipid growth in pigs. *Animal Production* 22, 87–96.

Williams, N.H., Stahly, T.S. and Zimmerman, D.R. (1993) Impact of immune system activation and dietary amino acid regimen on nitrogen retention of pigs. *Journal of Animal Science* 71 (Suppl. 1), 171.

Wünsche, J., Borgmann, E., Hennig, U., Kreienbring, F. and Bock, H.D. (1983) [The influence of graded protein supply at a high energy level on the fattening performance and the retention and utilisation of feed energy, protein and amino acids by female fattening pigs. 4-Nitrogen and amino acid content in the carcass and the parts of the carcass.] *Archiv fuer Tierernaehrung* 33, 389–413.

Young, V.R., Meredith, C., Hoerr, R., Bier, D.M. and Matthews, D.E. (1985) Amino acid kinetics in relation to protein and amino acid requirements: The primary importance of amino acid oxidation. In: Garrow, J.S. and Halliday, D. (eds), *Substrate and Energy Metabolism in Man*. John Libbey, London, pp. 119–132.

# Carbohydrate and Lipid Metabolism

**14**

## A. Danfaer

*Danish Institute of Agricultural Sciences, Research Centre Foulum, PO Box 50, 8830 Tjele, Denmark*

## Introduction

The aim of this chapter is to present an outline of qualitative and quantitative aspects of carbohydrate and lipid metabolism, mainly in growing pigs.

The significance of carbohydrates in pig nutrition is illustrated by the high feed content (65–75% of dry matter) and the high digestibility (70–90%) of this composite chemical fraction. In normal diets, the lipid content is much less (2–5% of dry matter) and the digestibility is also lower (60–80%) than that of carbohydrates. However, in the production of pig meat, the lipid content of the carcass is an important parameter. The composition of the empty body (0.5–1.0% carbohydrate and 15–25% lipid) is very much different from that of the feed. This means that absorbed carbohydrates are used for lipid synthesis rather than for storage of carbohydrate in the body.

Digested carbohydrates and lipids are absorbed from the small intestine as monosaccharides (mainly glucose) and as long-chain fatty acids (in chylomicron triglycerides), respectively. The fraction of non-starch polysaccharides (NSP) cannot be digested by animal enzymes, but is fermented in the lower part of the small intestine and in the hindgut. The absorbed fermentation products are short-chain fatty acids (SCFA), mainly acetate and propionate. Water-soluble absorbed nutrients like monosaccharides, SCFA and amino acids are transported from the intestinal cells via mesenteric veins and the portal vein to the liver. Chylomicrons enter the peripheral circulation via the lymphatic system.

The main function of carbohydrates and lipids in the body is to provide fuel for energy-requiring processes. In addition to this some fat depots protect internal organs like the kidneys and the intestines, and subcutaneous fat insulates the body against heat loss. Both carbohydrates and lipids have also more specific purposes, e.g. as important components of cellular membranes and secretions, pentoses as constituents of nucleotides and lipids as vehicles for intestinal transport of fat-soluble vitamins. Some polyunsaturated, so-called essential fatty acids, cannot be synthesized in the body and, hence, animals are dependent on dietary supply. These fatty acids have vital functions as part of membrane structures and as precursors for various biologically active compounds.

Carbohydrates are present in the body as free monosaccharides in the extracellular fluid, as components of specific compounds (mucopolysaccharides, glycoproteins and glycolipids), as lactose, and as energy stores in the form of glycogen in liver, muscles and most other tissues. Glucose is the major circulating monosaccharide. Lipids are present as circulating chylomicrons, lipoproteins and non-esterified fatty acids (NEFA), as structural lipids in membranes and nervous tissue, as milk fat, as intramuscular fat and as depot fat in adipose tissue.

In a 60 kg pig, the energy content of free glucose in the extracellular space is about 0.2 MJ while 6.0 MJ is stored in body glycogen. Approximately 75% of the glycogen is found in skeletal muscles and 15–20% in the liver. The glucose entry rate is about 15 MJ day$^{-1}$, i.e. many times higher than the extracellular pool and even higher than the total glycogen stores. Circulating lipids (triglycerides, phospholipids and NEFA) represent 0.7–0.8 MJ, i.e. an insignificant amount of energy in relation to the depot lipids in adipose tissue which can hold 250–500 MJ depending on the fatness of the animal. The plasma NEFA pool (0.005–0.006 MJ) constitutes only a minor fraction of the circulating lipids, but its turnover rate is very high, about 5 MJ per day.

## Transport and Cellular Uptake

Glucose is transported into cells from the interstitial fluid down a concentration gradient by facilitated diffusion. The carrier proteins (glucose transporters) in muscle and adipose tissues are activated by insulin.

Non-esterified long-chain fatty acids mobilized from adipose tissue triglycerides are transported in the blood complexed with the albumin fraction of plasma proteins from which they are released before leaving the blood capillaries (Newsholme and Start, 1973). Cellular uptake of free fatty acids in the interstitial fluid is by means of non-mediated simple diffusion in which the rate of uptake is determined only by the concentration gradient (Madsen, 1983). Glycerol is taken up into cells also by simple diffusion.

Triglycerides are circulating in blood plasma as chylomicrons absorbed from the digestive tract and as very low density lipoproteins (VLDL) synthesized in the liver. Peripheral tissues which utilize triglyceride fatty acids (muscles, adipose tissue and mammary gland) secrete lipoprotein lipases to the endothelial wall of the adjacent blood capillaries. Fatty acids are released from the triglyceride complexes by the action of lipoprotein lipase and then transported into the cells by diffusion. The liberated glycerol is taken up only in tissues with glycerol kinase activity. The activity of lipoprotein lipase from adipose tissue is enhanced by insulin and decreased by catecholamines. After removal of a large part of the triglyceride content in chylomicrons and VLDL by extrahepatic lipoprotein lipases, the remnant particles are taken up and partly degraded in the liver and other tissues (Griglio, 1986; Ganong, 1993). Another extracellular enzyme that may act to hydrolyse triglycerides in lipoproteins is hepatic lipase, which is present in the liver and the kidneys (Bezkorovainy and Rafelson, 1996).

Cholesterol is an essential constituent of cell membranes and is the precursor of various compounds, e.g. bile acids and steroid hormones. It is transported in the blood in chylomicron remnants and in low density lipoproteins (LDL). Chylomicron remnants bind to receptors in the liver. LDL is formed from triglyceride depleted VLDL via IDL (intermediate density lipoproteins) and is the major vehicle for cholesterol transport between body tissues. LDL receptors are present both in the liver and in extrahepatic tissues. After binding of chylomicron remnants and LDL to their receptors at the cell membrane, cholesterol is released intracellularly by endocytosis of the complex, recycling of receptors to the membrane and hydrolysis of cholesteryl esters in lysosomes (Heidemann, 1992).

# An Outline of Carbohydrate and Lipid Metabolism

## Liver

A major role of this organ is to maintain a rather constant supply of circulating nutrients during variations in dietary intake. The major hepatic pathways of carbohydrate and lipid metabolism are shown in Fig. 14.1.

In the absorptive phase after feeding, glucose availability is normally in excess of tissue needs, and the liver acts as a buffer by storing glucose as glycogen (glycogenesis) and to a lesser extent by utilizing glucose and lactate for fatty acid synthesis. NADPH required for fatty acid synthesis is supplied via the pentose phosphate pathway and the pyruvate/malate cycle. To some extent, fatty acids can be taken up from circulating lipoproteins by the action of hepatic lipase (Ganong, 1993; Bezkorovainy and Rafelson, 1996). Fatty acids entering the liver and fatty acids synthesized *de novo* are secreted as triglycerides in VLDL. Glucose is also a substrate for the synthesis of glycerol-P needed for esterification of fatty acids. The liver metabolism in the absorptive stage is influenced by a high insulin : glucagon ratio in the blood plasma. The anabolic hormone insulin stimulates glycogenesis, glycolysis as well as lipid, protein and cholesterol synthesis. On the other hand, glycogen breakdown,

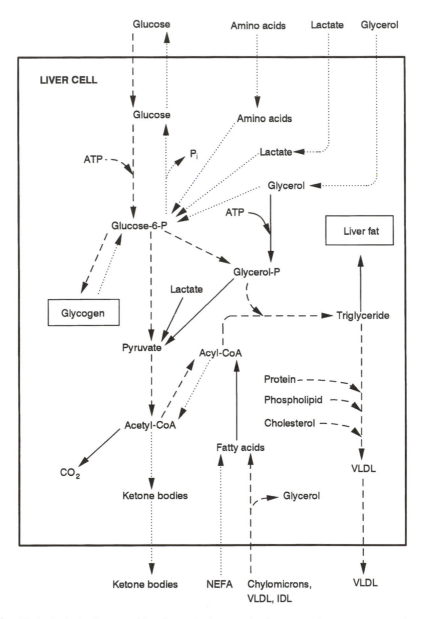

**Fig. 14.1.** Carbohydrate and lipid metabolism in the liver. No distinction is made between cytosolic and mitochondrial pathways.
Prevailing pathways in the fed state: ---------->
Prevailing pathways in the unfed state: ·········>

glucose synthesis and protein breakdown are inhibited by insulin (Bezkorovainy and Rafelson, 1996).

In the postabsorptive phase, the plasma concentrations of glucose and insulin decrease and the reduced blood glucose level causes the concentration of glucagon to increase. This hormone inhibits glycogenesis and glycolysis, but stimulates the degradation of liver glycogen to glucose (glycogenolysis) and also the synthesis of glucose from non-carbohydrate substrates (gluconeogenesis). These substrates are lactate, glucogenic amino acids (primarily alanine), propionate and glycerol. By these metabolic changes, the liver turns from a glucose consumer to a glucose producer in order to maintain glucose availability to peripheral tissues.

In situations with longer periods of food deprivation and in fasting, the plasma glucose level will decrease further and the concentrations of glucagon and epinephrine will be high. The concentration of glucocorticoids could also be elevated. Like glucagon, epinephrine operates by increasing the intracellular level of cAMP and therefore exerts the same effects on glucose metabolism as glucagon. Glucocorticoids stimulate protein breakdown in peripheral tissues and amino acid uptake in the liver, thus providing substrates for glucose synthesis. The glycogen stores are depleted and the supply of blood glucose is totally dependent on gluconeogenesis in the liver and the kidneys. Furthermore, the body seeks to reduce the glucose requirement by using fatty acids and ketone bodies instead of glucose as a fuel in tissues like skeletal muscles, heart, kidneys and also to some extent, in nerve tissue (Bezkorovainy and Rafelson, 1996). Fatty acids mobilized from adipose depots are taken up from the plasma NEFA pool mainly by muscle tissues, but also by the liver in which they are incorporated into VLDL in the cytosol or oxidized in the mitochondria via β-oxidation to acetyl-CoA. This metabolite can be further oxidized to $CO_2$ in the TCA cycle, but the predominant pathway will be ketogenesis and secretion of ketone bodies to the blood.

### Adipose tissue

The function of this tissue is primarily to be a fuel reservoir in the body. The carbohydrate and lipid metabolism is outlined in Fig. 14.2.

In the well-fed state, the uptake of glucose and fatty acids as well as the synthesis of fatty acids and glycerol-P from glucose are stimulated while the rate of lipolysis is decreased under the influence of a high insulin concentration. The conversion of glucose to fatty acids leads to a net production of ATP which could be a limiting factor for the maximum rate of fatty acid biosynthesis from glucose (Flatt, 1970). The surplus of ATP can be reduced if most of the NADPH needed for fatty acid synthesis is generated in the pyruvate/malate cycle. Hence, there seems to be no need for ATP production in the TCA cycle and the carbon flow through this pathway beyond citrate is very low (Martin, 1987). Fatty acids can also be synthesized from acetate which is

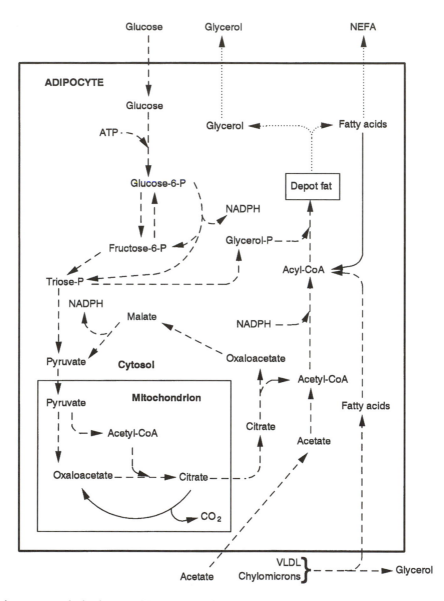

**Fig. 14.2.** Carbohydrate and lipid metabolism in adipose tissue.
Prevailing pathways in the fed state: ---------- >
Prevailing pathways in the unfed state: ·········· >

activated in the cytosol to acetyl-CoA, the starting point of the synthetic
pathway. Adipocytes are able to store some glucose as glycogen. As described
previously, fatty acids from triglycerides in chylomicrons and VLDL are taken

up by the action of lipoprotein lipase, the activity of which is stimulated by insulin. The glycerol moiety of these lipids cannot be activated to glycerol-P in adipose tissue and is returned to the circulation for utilization in the liver. The fatty acids taken up are activated with coenzyme A and esterified with glycerol-P to form triglycerides in the fat stores. Coenzyme A esters of long-chain fatty acids (acyl-CoA) are strong inhibitors of acetyl-CoA carboxylase, a rate-limiting enzyme in the pathway of fatty acid synthesis. This may explain why the capacity of this pathway is significantly decreased by a high lipid content in the diet. Even in the fed state, a fraction of the stored lipid is continuously degraded to fatty acids and glycerol, but normally the rate of lipogenesis is higher than the rate of lipolysis, and as a result lipid is retained.

The rate of lipolysis is regulated by a hormone-sensitive lipase, the activity of which is stimulated via cAMP. The intracellular formation of cAMP is stimulated by epinephrine and glucagon and probably inhibited by insulin (Schade *et al.*, 1979). Glucocorticoids enhance the action of cAMP, and growth hormone (somatotrophin) seems to increase the ability of cate-cholamines to activate cAMP (Ganong, 1993). In the postabsorptive phase, the lower insulin : glucagon ratio leads to a decreased glucose and fatty acid uptake, a decreased lipogenesis and an increased lipolysis. Hence, a net mobilization of fatty acids can occur. Glucose for internal use in the adipo-cytes is also released from the small glycogen stores. During periods of fasting or a substantial energy deficit, e.g. in early lactation of sows, the mobilization of lipid is augmented due to increased plasma levels of epinephrine, somato-trophin and glucocorticoids. In response to this, the oxidative metabolism in most other tissues is adjusted to use fatty acids and ketone bodies in preference to glucose.

## Muscle tissue

The metabolism of carbohydrates and lipids in this tissue (Fig. 14.3) is directed towards a provision of energy for muscle contractions and for protein turnover. The cost of ATP in protein synthesis is rather high and the energy use per unit of net protein accretion in growing animals is therefore highly dependent on the rate of protein turnover.

After a meal when glucose is readily available from absorption, muscles are in an anabolic state. The rates of glucose uptake, glycogenesis, amino acid uptake and protein synthesis are stimulated by insulin while those of glycogenolysis and proteolysis are inhibited. Thus, muscle glycogen stores will be replenished, and in growing animals, muscle protein is retained. Fatty acids are taken up from the circulating NEFA pool as well as from chylomi-crons and VLDL by means of insulin-stimulated lipoprotein lipase. Oxidation of fatty acids provides the predominant source of ATP production if the animal is at rest or performs only moderate exercise. Fatty acids are also

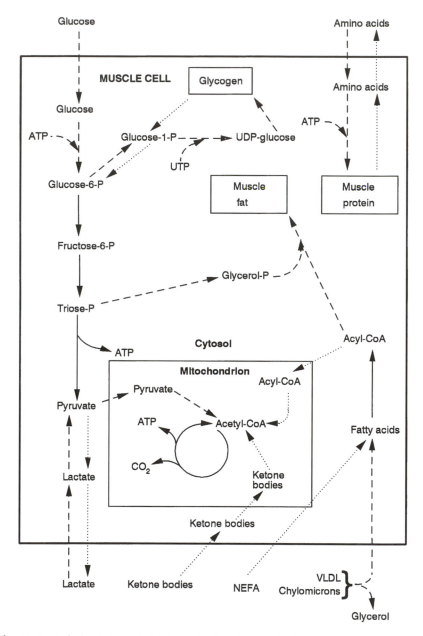

**Fig. 14.3.** Carbohydrate and lipid metabolism in muscle tissue.
Prevailing pathways in the fed state: ----------- >
Prevailing pathways in the unfed state: ·········· >

precursors for the synthesis of intracellular fat. Other substrates for oxidation are glucose and, to some extent, amino acids. In the latter case, amino groups are transferred to pyruvate and the resulting alanine is taken up by the bloodstream and extracted from there by the liver. In this tissue, alanine is transaminated and deaminated back to pyruvate and then metabolized further to glucose in the gluconeogenic pathway. The amino group from alanine is trapped into the synthesis of urea, and the glucose can be returned to the muscles and thus complete the alanine cycle. This cycle is one of the means by which the muscle tissue gets rid of surplus nitrogen from catabolized amino acids.

With bursts of violent exercise, the rate of $O_2$ uptake is too low to provide enough ATP from oxidation of fatty acids and glucose. The increased need of ATP is then met by anaerobic glycolysis whereby glucose is reduced to lactate. This metabolite is exported to the liver and converted back to glucose by gluconeogenesis (the Cori cycle). The pool of ATP in muscle tissue is very small and is used up rapidly in situations with intensive muscular work. At highly violent exercise, ATP is formed from a larger pool of creatine-P, which in turn will be replenished during periods of rest by phosphorylation of creatine with ATP. Skeletal muscles contain different fibre types that are characterized by their contractile properties as well as by their predominant metabolic pathways for the provision of ATP. Slow twitch fibres (type I) have a long latency and are adopted for long, slow, posture-maintaining contractions. These fibres have a high content of myoglobin and mitochondria and, hence, a high oxidative capacity. Fast twitch fibres (type IIB) respond quickly with brief, powerful contractions, have few mitochondria and are mainly glycolytic. According to the myoglobin content, muscles are classified as red muscles (mostly type I fibres) and white muscles (mostly type IIB fibres).

There are no glucagon receptors in skeletal muscles and the increased plasma level of glucagon in the postabsorptive phase has therefore no effect on the carbohydrate metabolism in this tissue. Instead, catecholamines, glucocorticoids and maybe somatotrophin (Dauncey *et al.*, 1994) secreted in response to a declining plasma glucose concentration are the major catabolic hormones regulating muscle metabolism. Epinephrine inhibits glycogenesis and promotes glycogenolysis via cAMP. However, in contrast with the liver, the increased level of cAMP does not inhibit glycolysis (Bezkorovainy and Rafelson, 1996). This seems to be well coordinated with the lack of glucose-6-phosphatase in muscles, meaning that glucose-6-phosphate from glycogen breakdown is not converted to free glucose for export (also unlike in the liver), but is metabolized within the muscle tissue. Glucocorticoids stimulate muscle protein breakdown and transamination of especially branched-chain amino acids. Furthermore, it has been shown that the rate of muscle protein synthesis is much lower just before feeding than at 1–2 h after feeding in minipigs fed twice daily (Danfaer, 1980). The rate of lipolysis is probably low in muscles and the deficit of fuel for aerobic metabolism is met by an increased uptake of fatty acids from the circulation. In starvation (or relative fasting as in early lactation), ketone bodies become an important energy source in muscles, and the rate of proteolysis will exceed the rate of protein synthesis resulting in a net mobilization of amino acids. These are utilized in the liver and the kidneys for gluconeogenesis.

## Mammary gland

Carbohydrate metabolism in this tissue is dominated by the synthesis of lactose and fatty acids from glucose. Lactose synthetase is a two-unit enzyme made up of galactosyl transferase and α-lactalbumin which makes it unique for mammary tissue. Glucose-6-P is metabolized to UDP-galactose which is combined with free glucose to lactose and UDP by the action of lactose synthetase. The synthesis of fatty acids and glycerol-P is as described for adipose tissue. Fatty acids are taken up from circulating chylomicrons and VLDL and to a lesser extent from NEFA depending on the plasma concentration. Fatty acids released from triglycerides by lipoprotein lipase and fatty acids released from the albumin–NEFA complex equilibrate in the interstitial fluid. Therefore, the net uptake of plasma NEFA depends on the relative concentrations of these two pools as well as on the intra- and extracellular concentrations. The milk fat is made of fatty acids derived partly from *de novo* synthesis in the liver and in the mammary gland itself and partly from dietary fatty acids. The fatty acid composition of sow's milk fat differs from that of cow's milk fat by a much lower content of saturated, short- and medium-chain fatty acids, a lower content of stearic acid, and higher contents of palmitoleic, oleic and especially linoleic acid (Jenness, 1974).

There seems to be no internal turnover of the major products (lactose, fat and protein) formed in the mammary gland in contrast to the case in liver, muscle and adipose tissues. The milk products are secreted instead of being stored in the body and subjected to degradation. This makes the formation of milk components energetically more efficient.

## Metabolism of essential fatty acids

Mammals are not able to dehydrogenate long-chain fatty acids at carbon atoms beyond C-9 from the carboxylic end. Linoleic (C18:2, n-6) and α-linolenic acid (C18:3, n-3) are therefore considered as essential fatty acids and must be supplied with the food. These fatty acids are transported in the blood mainly as cholesterol esters and in phospholipids (Leat, 1963).

Linoleic acid of the *n*-6 family is metabolized in the microsomal fraction of the cell into γ-linolenic acid (C18:3, n-6), dihomo-γ-linolenic acid (C20:3, n-6), and arachidonic acid (C20:4, n-6). As structural components of cellular membranes, these fatty acids are important for the stabilization and function of cellular organelles and cell membranes. It has been shown that the ATP formation in mitochondria from liver, cardiac and skeletal muscle tissues is impaired in pigs deficient in linoleic acid (Christensen, 1974, 1986, 1987). Dihomo-γ-linolenic and arachidonic acid are released from membrane phospholipids and oxidized to biologically active substances called eicosanoids: prostaglandins, thromboxanes, leukotrienes and lipoxins. These compounds exert a range of physiological effects, mainly in the tissues in which they are synthesized. Prostaglandins and thromboxanes are potent regulators of vasoconstriction and vasodilation as well as of platelet aggregation (Korbut and Moncada, 1978). Prostaglandins participate in regulation of the female reproductive cycle, and a prostaglandin ($PGF_{2\alpha}$) derived from arachidonic acid has been used to induce parturition in sows (Einarsson, 1981). Leukotrienes stimulate the contraction of smooth muscles and increase the permeability of capillaries (Hedqvist *et al.*, 1980). Eicosanoids are also mediators of inflammatory and immunological reactions (Christensen, 1985; Ganong, 1993). The requirement of linoleic acid (on an energy basis) for maximum daily gain and feed conversion efficiency in growing pigs has been estimated as 0.2% of dietary gross energy although the requirement for some specific biological functions could be higher (Christensen, 1985).

During a series of desaturation and elongation reactions, α-linolenic acid is metabolized to the various members of the n-3 family ending with C22:6, n-3. Eicosapentaenoic acid (C20:5, n-3) is a precursor of the n-3 eicosanoids (Lands, 1992). Fatty acids of the n-3 family are known to reduce the risk of myocardial infarction, ischaemic, inflammatory and immunological disorders in humans (Dyerberg, 1993).

# Quantitative Descriptions of Carbohydrate and Lipid Metabolism

In this section, some data will be presented in order to illustrate the magnitude of nutrient pools and nutrient flow rates in pigs as related to the metabolism of carbohydrates and lipids. The presented data comprise extracellular nutrient concentrations, whole-body entry rates and nutrient metabolism in specific tissues. Except in rather extreme cases, the diet composition does not seem to influence these parameters to any great extent, but most of them are strongly affected by time after feeding (importance of feeding frequency) and by treatment with porcine somatotrophin (pST).

## Extracellular concentrations

### Glucose

Blood plasma concentrations of glucose determined in different experiments for a wide range of live weights are collected in Table 14.1. Values are given for fed animals in their absorptive state (1–2 h postfeeding, mean = 5.9 mM),

**Table 14.1.** Glucose concentration (mM) in blood plasma of pigs. Effects of pST treatment and fasting.

| Live weight (kg) | Fed, controls | Fed, pST | Fasted | (h) | Reference |
|---|---|---|---|---|---|
| 61 | 4.88 | 6.33 | | | Wray-Cahen et al. (1991) |
| 61 | 4.61 | 5.88 | | | Wray-Cahen et al. (1993) |
| 65 | 6.1 | 7.6 | | | Gopinath and Etherton (1989a) |
| 71 | 5.34 | 7.11 | | | Dunshea et al. (1992a) |
| 71 | 6.23 | 7.70 | | | Dunshea et al. (1992b,c) |
| | 5.9 | | | | Ponter et al. (1991) |
| 20–45 | 6.0 | | | | Beech et al. (1991) |
| 30–52 | 6.08 | | | | Malmlöf et al. (1988) |
| 35–50 | 5.96 | | | | Malmlöf et al. (1989) |
| 38–63 | 5.92 | | | | Fuller et al. (1977) |
| 50–60 | 6.16 | | | | Graham and Malmlöf (1993) |
| 6–12 weeks | 6.2 | | 5.3 | (16) | Freeman et al. (1970) |
| 51 | 7.2 | | 5.7 | (16) | Wood et al. (1977) |
| 49–68 | 6.2 | | 5.3 | (24) | Riis and Grummer (1969) |
| | | | 5.0 | (72) | |
| 12 | | | 4.9 | (14) | Cote et al. (1982) |
| 50–75 | | | 5.4 | (12–14) | Gopinath and Etherton (1989a) |
| 84 | | | 4.47 | (12–14) | Gopinath and Etherton (1989b) |
| Mean | 5.9 | 6.9 | 5.2 | | |
| Range | 4.6–7.2 | 5.9–7.7 | 4.5–5.7 | | |

for animals in a postabsorptive state (12-72 h postfeeding, mean = 5.2 mM) and also for fed animals treated with pST (mean = 6.9 mM). There is no relationship between live weight and glucose concentration, but the pST treatment causes the concentration to increase by 25-30%. The decrease in the postabsorptive state is 15-20% and seems to be relatively unaffected by the length of fasting up to 72 h.

### Lactic acid and short-chain fatty acids (SCFA)

The blood content of lactate in pigs seems to vary a great deal due to differences in live weight, diet composition or laboratory procedures. Friend et al. (1964) determined the concentration in both arterial (6.15 mM) and portal venous plasma (6.03 mM) from fed 30 kg pigs, thus indicating no net absorption of lactate from the portal-drained viscera (PDV). Rérat et al. (1993) measured the same parameters in pigs of 61 kg live weight and found much lower levels of L-lactate both in arterial and portal blood, but the concentration in portal plasma (2.6–3.1 mM) was higher than that of arterial plasma (2.2–2.9 mM). This concentration difference, showing a net absorption from PDV, was rather persistent for up to 12 h after a meal, but at this time the concentration had decreased to 1.0–1.2 and 0.8–1.0 mM in portal and arterial plasma, respectively. However, in 20–50 kg pigs fasted for 24 h, the serum concentration increased from 9.1 to 23.7 mM (Mersman and Hu, 1987). The net absorption from PDV of D-lactate is just a small fraction (5–6%) of the net absorption of total lactate (Rérat *et al.*, 1993).

The plasma concentrations of acetate, propionate and butyrate in piglets (20–30 kg) are in the order of 0.60, 0.01 and 0.01 mM, respectively (Freeman *et al.*, 1970; Imoto and Namioka, 1978). Rérat *et al.* (1987) found lower values for acetate (0.27 mM) and higher values for propionate (0.02 mM) in 60 kg pigs and, thus, lower values for total SCFA. The effects of fasting and diet composition on SCFA levels in the blood seem to be small (Freeman *et al.*, 1970), although the plasma acetate concentration was increased on a high-protein, high-fibre diet compared to a low-protein, high-NFE diet, both in the fed state (from 0.37 to 0.66 mM) and after a 15 h fast (from 0.58 to 0.75 mM). Hence, the values were somewhat increased by fasting on both diets (Imoto and Namioka, 1978).

### NEFA, glycerol and triglycerides

Data on NEFA concentrations in blood plasma are shown in Table 14.2. Mean values for fed, fasted and pST-treated pigs are 0.17, 0.82 and 0.13 mM, respectively. Especially for the fed pigs, the range of the values is large (0.04–0.34 mM) with the lowest concentrations determined by Dunshea *et al.* (1992a,c) and the highest by Freeman *et al.* (1970). Some of this variation could be due to differences in live weight, as there seems to be an overall decrease in the NEFA concentration with increasing body weight. This is clearly the case within one experiment in which the live weight ranges from 12 to 73 kg (Mersman and MacNeil, 1985). The concentrations are much higher in the fasted pigs, from 134 to 544% of the values measured on fed pigs. This increase is augmented with longer periods of fasting (Riis and Grummer, 1969; Mersman and MacNeil, 1985) and is caused by an increased net mobilization of lipid from adipose tissues. Within individual experiments, the NEFA concentration is roughly doubled by pST treatment.

Like NEFA, the glycerol content in the blood is increased by fasting as well as by pST. In pigs weighing from 15 to 37 kg, the blood glycerol concentration was 0.043 mM in the fed and 0.086 mM in the fasted state (Mersman, 1986). The same relative increase was found by Mersman and Hu (1987) in 20–50 kg pigs fasted for 24 h. In this experiment, the serum concentrations were 0.187 and 0.353 mM in fed and fasted pigs, respectively. The plasma glycerol concentration was increased by pST treatment from 0.013 to 0.025 mM (Dunshea *et al.*, 1992a) and from 0.009 to 0.010 mM (Dunshea *et al.*, 1992c) in barrows weighing 71 kg.

**Table 14.2.** NEFA concentration (mM) in blood plasma of pigs. Effects of pST treatment and fasting.

| Live weight (kg) | Fed, controls | Fed, pST | Fasted | (h) | Reference |
|---|---|---|---|---|---|
| 61 | 0.059 | 0.137 | | | Wray-Cahen *et al.* (1991) |
| 61 | 0.124 | 0.220 | | | Wray-Cahen *et al.* (1993) |
| 71 | 0.055 | 0.118 | | | Dunshea *et al.* (1992a) |
| 71 | 0.044 | 0.064 | | | Dunshea *et al.* (1992c) |
| 6–12 weeks | 0.337 | | 1.243 | (16) | Freeman *et al.* (1970) |
| 15–37 | 0.186 | | 0.849 | (?) | Mersman (1986) |
| 51 | 0.091 | | 0.337 | (16) | Wood *et al.* (1977) |
| 42–66 | 0.290 | | 0.390 | (18) | Riis and Grummer (1969) |
| | | | 0.600 | (24) | |
| | | | 0.770 | (40) | |
| | | | 1.110 | (72) | |
| 12 | 0.274 | | 1.317 | (30) | Mersman and MacNeil |
| 40 | 0.179 | | 0.973 | (30) | (1985) |
| 73 | 0.154 | | 0.823 | (30) | |
| 24–88 | 0.200 | | 0.480 | (12) | Mersman and MacNeil |
| | | | 0.790 | (24) | (1985) |
| | | | 0.900 | (48) | |
| | | | 0.900 | (72) | |
| Mean | 0.17 | 0.13 | 0.82 | | |
| Range | 0.04–0.34 | 0.06–0.22 | 0.34–1.32 | | |

The concentration of triglycerides in blood plasma is 0.33–0.54 mM (Faidly *et al.*, 1990; Beech *et al.*, 1991) and a little higher in serum, 0.5–0.8 mM (Thacker and Bowland, 1981; Lepine *et al.*, 1993) in pigs fed a range of different diets. The plasma concentration is higher when the pigs are fasted. In 24 kg pigs, the concentration was increased from 0.37 mM in the fed state to 0.50, 0.52, 0.62 and 0.79 mM after 12, 24, 48 and 72 h of fast, respectively (Mersman and MacNeil, 1985). The corresponding figures for larger pigs (88 kg LW) were 0.37, 0.39, 0.43, 0.43 and 0.50 mM (Mersman and MacNeil, 1985). In another group of pigs from the same experiment, the increase in triglyceride plasma concentration after 30 h of fasting was also higher in small than in large pigs. Values in the fed and fasted state for pigs weighing 12, 40 and 73 kg were 0.93 and 1.59, 0.58 and 0.88, and 0.69 and 0.62 mM, respectively (Mersman and MacNeil, 1985).

## Insulin and glucagon

These pancreatic hormones are the major regulators of the blood glucose level and are strongly involved in the control of carbohydrate and lipid metabolism. As shown in Table 14.3, the plasma concentration of insulin in

**Table 14.3.** Insulin concentration (ng ml$^{-1}$) in blood plasma of pigs. Effects of pST treatment and fasting.

| Live weight (kg) | Fed, controls | Fed, pST | Fasted | (h) | Reference |
|---|---|---|---|---|---|
| 61 | 1.60 | 3.90 | | | Wray-Cahen *et al.* (1991) |
| 61 | 0.87 | 2.79 | | | Wray-Cahen *et al.* (1993) |
| 65 | 2.20 | 5.20 | | | Gopinath and Etherton (1989a) |
| 71 | 0.61 | 3.33 | | | Dunshea *et al.* (1992a) |
| 71 | 0.73 | 2.89 | | | Dunshea *et al.* (1992b,c) |
| | 1.30 | | | | Ponter *et al.* (1991) |
| 30–52 | 0.47 | | | | Malmlöf *et al.* (1988) |
| 35–50 | 0.66 | | | | Malmlöf *et al.* (1989) |
| 38–63 | 0.77 | | | | Fuller *et al.* (1977) |
| 50–60 | 1.03 | | | | Graham and Malmlöf (1993) |
| 51 | 2.36 | | 0.88 | (16) | Wood *et al.* (1977) |
| 51 | 1.45 | | 0.70 | (16) | |
| Mean | 1.2 | 3.6 | 0.8 | | |
| Range | 0.5–2.4 | 2.8–5.2 | 0.7–0.9 | | |

untreated, fed pigs (mean = 1.2 ng ml$^{-1}$) is higher than in fasted, but much lower than in pST-treated pigs (3.6 ng ml$^{-1}$). In the individual experiments, the concentration is decreased by 50–60% 16 h postfeeding (Wood *et al.*, 1977) and increased by two- to fivefold by pST treatment (Gopinath and Etherton, 1989a; Wray-Cahen *et al.*, 1991, 1993; Dunshea *et al.*, 1992a,b,c). The presented values in fed pigs do not reflect the steep increase within 1 h after feeding, peaking at 5–7 ng ml$^{-1}$ (Rérat *et al.*, 1985, 1993). Intravenous infusion of a glucose dose (2.5 mmol kg$^{-1}$) increases plasma insulin within 15 min in female piglets (about 25 kg), whereas similar infusions of SCFA have very small effects on plasma insulin (Sano *et al.*, 1995). In pST-treated gilts, the serum insulin concentration is positively correlated with the live weight (from 20 to 60 kg), but there is no relation between live weight and insulin concentration in untreated pigs (Caperna *et al.*, 1993).

The concentration of glucagon is in the order of ten times lower than the insulin concentration and, contrary to insulin, the plasma glucagon level is not increased during the absorptive phase (Ostaszewski and Nissen, 1988; Scholz-Ahrens *et al.*, 1990) except on diets with a very high protein content (Rérat *et al.*, 1985). In female piglets, plasma glucagon is decreased by intravenous dosage of glucose, and increased by similar doses of acetate and n-butyrate within 15 min (Sano *et al.*, 1995). The glucagon concentration is increased by fasting (from 75 to 90 pg ml$^{-1}$, 12 h postfeeding) (Rérat *et al.*, 1993) and decreased by pST treatment (from 68 to 35 pg ml$^{-1}$) (Wray-Cahen

**Table 14.4.** Glucose entry rate (mmol (kg MBW)$^{-1}$ min$^{-1}$) in pigs. Effects of pST treatment and fasting.

| Live weight (kg) | Fed, controls | Fed, pST | Fasted | (h) | Reference |
|---|---|---|---|---|---|
| 61 | 0.191 | 0.055 | | | Wray-Cahen *et al.* (1993) |
| 71 | 0.172 | 0.111 | | | Dunshea *et al.* (1992b) |
| 22 | 0.116 | | | | Danfær (1980) |
| 72 | 0.134 | | | | Dunshea *et al.* (1992b) |
| 6–12 weeks | 0.114 | | 0.044 | (16) | Freeman *et al.* (1970) |
| 18 | 0.198 | | 0.041 | (13) | Danfær (1980) |
| 48–68 | 0.193 | | 0.089 | (17) | Riis and Grummer (1969) |
| | | | 0.052 | (24) | |
| | | | 0.028 | (72) | |
| 12 | | | 0.031 | (14) | Cote *et al.* (1982) |
| 16 | | | 0.043 | (24) | Trayhurn *et al.* (1981) |
| | | | 0.041 | (24) | |
| | | | 0.043 | (24) | |
| 14–21 | | | 0.039 | (42) | Wolf and Brenner (1988) |
| 84 | | 0.073 | | (12–14) | Gopinath and Etherton (1989b) |
| Mean | 0.16 | 0.083 | 0.048 | | |
| Range | 0.11–0.20 | 0.055–0.11 | 0.028–0.089 | | |

*et al.*, 1993). In gilts of 20, 40 and 60 kg live weight, the daily integrated serum level of glucagon is not affected by pST treatment, but increases (like insulin) with the live weight in the pST-treated animals (Caperna *et al.*, 1993).

## Whole body entry rates

### Glucose

Turnover rates of extracellular glucose expressed as mmol per kg metabolic live weight per min are presented in Table 14.4. The mean values are 0.16, 0.048 and 0.083 in fed, fasted and pST-treated animals, respectively. Increasing the length of the fasting period (up to 72 h) decreases the glucose entry rate (Riis and Grummer, 1969).

### NEFA and glycerol

The entry rate of NEFA is increased by fasting as well as by pST treatment. Freeman *et al.* (1970) estimated turnover rates in fed and fasted (16 h) piglets as 44 and 50 μmol (kg MBW)$^{-1}$ min$^{-1}$, respectively. In larger pigs (42–64 kg), the entry rate was 26 in the fed state and 26, 49, 71 and 101 μmol (kg MBW)$^{-1}$ min$^{-1}$ after 18, 24, 40 and 72 h fast, respectively (Riis and Grummer, 1969). Values determined in a more recent experiment (Dunshea *et al.*, 1992c) were much lower, viz. 3.9 in fed controls and 4.7 µmol (kg MBW)$^{-1}$ min$^{-1}$ in fed, pST-treated pigs of 71 kg live weight. However, as the turnover rate constant ($k$) was similar (0.5–0.7 min$^{-1}$) both in the work of Riis and Grummer (1969) and in that of Dunshea *et al.* (1992c), the big difference in entry rates obtained in the two studies can largely be explained by differences in plasma NEFA concentrations (see Table 14.2).

Entry rates of individual fatty acids in fed and 16 h fasted piglets have been determined by Freeman *et al.* (1970). Estimates in the fed and fasted state were 197 and 153, 22 and 21, 4.8 and 11, and 3.0 and 9.3 µmol (kg MBW)$^{-1}$ min$^{-1}$ for acetate, palmitate, stearate and oleate, respectively. These results suggest that a 16 h fast decreases the flux rate of acetate and increases that of stearate and oleate. Wood *et al.* (1977) found a similar value (2.8) as did Freeman *et al.* (1970) for the oleate flux rate in 51 kg pigs, whereas the values estimated by Dunshea *et al.* (1992c) in 71 kg pigs were somewhat lower, viz. 1.08 in fed controls and 1.55 µmol oleate (kg MBW)$^{-1}$ min$^{-1}$ in fed, pST-treated pigs. In the same study, glycerol entry rates were 1.96 and 2.18 µmol (kg MBW)$^{-1}$ min$^{-1}$ in untreated and treated pigs, respectively (Dunshea *et al.*, 1992c).

### Extracellular pool sizes, turnover rate constants and entry rates

Water-soluble substances like glucose, lactate, SCFA and glycerol are assumed to be uniformly distributed in the total extracellular space, i.e. blood plasma and interstitial fluid. Non-esterified long-chain fatty acids are bound to albumin and therefore assumed to be distributed only in the blood plasma. The concentration of free LCFA in the interstitial fluid is extremely low (< 0.001 mM) and this pool represents a negligible amount in relation to the plasma NEFA pool (Newsholme and Start, 1973). The volume of the extracellular fluid is 20% and the plasma volume is 5% of the body weight (Ganong, 1993).

In Table 14.5, extracellular pool sizes, rate constants and whole-body entry rates of glucose, fatty acids and glycerol are given for fed, fasted and pST-treated pigs of 60 kg live weight. Calculation of the tabulated values is based on data from Riis and Grummer (1969), Freeman *et al.* (1970), Dunshea *et al.* (1992c) and Wray-Cahen *et al.* (1993). Treatment with pST decreases both the glucose entry rate and the rate constant, but increases LCFA entry rates and decreases the corresponding rate constants. The entry rate of glycerol is increased, but the rate constant is not affected. Fasting decreases rate constants and entry rates of glucose and acetate, but contrary to what is the case with pST treatment, the glucose pool size is also decreased. The NEFA pool size and entry rate are increased by fasting, whereas the rate constant is decreased. This is also true for some individual fatty acids. Thus, rate constants for palmitate, stearate and oleate have been estimated as 0.87, 0.59 and 0.78 min$^{-1}$, respectively, in fed pigs and as 0.36, 0.34 and

**Table 14.5.** Extracellular pool sizes, entry rates and turnover rate constants in 60 kg pigs. Effects of pST treatment and fasting.

| | Pool size | | Turnover rate constant | | Entry rate | |
|---|---|---|---|---|---|---|
| | mmol | (%) | min$^{-1}$ | (%) | mmol min$^{-1}$ | (%) |
| **Glucose**[a] | | | | | | |
| Fed, controls | 70.0 | (100) | 0.052 | (100) | 3.62 | (100) |
| Fed, pST | 88.2 | (126) | 0.019 | (37) | 1.70 | (47) |
| Fasted, 12–24 h | 59.5 | (85) | 0.020 | (38) | 1.19 | (33) |
| **Acetate**[a] | | | | | | |
| Fed, controls | 7.56 | (100) | 0.57 | (100) | 4.30 | (100) |
| Fasted, 12–24 h | 6.66 | (88) | 0.50 | (89) | 3.30 | (77) |
| **Total NEFA**[b] | | | | | | |
| Fed, controls | 0.501 | (100) | 0.64 | (100) | 0.321 | (100) |
| Fed, pST | 0.726 | (145) | 0.53 | (83) | 0.388 | (121) |
| Fasted, 12–24 h | 0.858 | (171) | 0.54 | (84) | 0.462 | (144) |
| **Oleate**[b] | | | | | | |
| Fed, controls | 0.037 | (100) | 0.63 | (100) | 0.0233 | (100) |
| Fed, pST | 0.063 | (170) | 0.53 | (84) | 0.0334 | (143) |
| **Glycerol**[a] | | | | | | |
| Fed, controls | 0.103 | (100) | 0.41 | (100) | 0.0423 | (100) |
| Fed, pST | 0.115 | (112) | 0.41 | (100) | 0.0470 | (111) |

[a] Assumed distributed in extracellular fluid (20% of LW).
[b] Assumed distributed in blood plasma (5% of LW).

0.34 min$^{-1}$, respectively, in fasted pigs (Freeman *et al.*, 1970; Wood *et al.*, 1977).

Turnover rate constants are estimated by isotopic labelling of the nutrient in question on the assumption that the disappearance from the blood (i.e. the rate of cellular uptake) follows first-order mass-action kinetics. This assumption seems to be reasonable in the case of fatty acids and glycerol, as these compounds move across cell membranes by passive diffusion in the direction of a concentration gradient. By a slight modification of Fick's law of diffusion, the rate of cellular uptake can be described as $d\,(C_{ex} - C_{int})\,V$, where $d$ is the diffusion coefficient, $C_{ex}$ and $C_{int}$ are the extracellular and intracellular concentrations, respectively, of the nutrient in question, and $V$ is the extracellular volume. Using the turnover rate constant, $k$, the rate of cellular uptake is described as $k\,C_{ex}\,V$, and hence the relationship between $k$ and $d$ depends on the ratio of $C_{int}$ to $C_{ex}$. It is not known whether the fatty acid diffusion coefficient is constant in specific tissues or if it is under the influence of metabolic or hormonal control. However, as judged by the relatively small changes in the fatty acid turnover rate constant when pigs are treated with pST or fasted (Table 14.5) and given that the ratio of $C_{int}$ to $C_{ex}$ is rather constant, the diffusion coefficient does not seem to be strongly regulated.

The cellular uptake of glucose in most tissues (apart from the small intestine and the renal tubules) is by means of facilitated diffusion. This means that glucose transport across the cell membrane is directed by the concentration gradient, but the passage is dependent on specific carrier proteins located in the cell membrane. The rate of glucose uptake then follows saturation kinetics and can be described as $V_{max} C_{ex} (K_m + C_{ex})^{-1}$, where $C_{ex}$ is the extracellular glucose concentration, $V_{max}$ is the rate at infinitely high $C_{ex}$, and $K_m$ is the affinity constant. $V_{max}$ expresses the capacity of the cellular uptake and is dependent on the number of active carrier proteins. There is a family of glucose transport proteins (GLUT 1–5) located in various tissues (Ganong, 1993). In muscle, adipose and some other tissues, the number of active transporters (GLUT 4) is increased by insulin and, hence, the estimated turnover rate constant cannot be expected to be fixed in a range of situations where the tissue exposure to insulin is changed. The observed decrease in the turnover rate constant in pST-treated and fasted pigs (Table 14.5) may be explained by a reduced tissue sensitivity to insulin and by a decreased insulin concentration, respectively, and thereby a lower number of transporters. Using the information in Table 14.5 about extracellular concentrations and entry rates of glucose, and assuming that the affinity constant ($K_m$) is 5 mM (Ganong, 1993), $V_{max}$ for the cellular uptake can be calculated as 6.7, 2.9 and 2.4 mmol min$^{-1}$ for normally fed, pST-treated and fasted pigs, respectively.

## Tissue metabolism

### Liver

Rates of net absorption (= net portal appearance) can be determined as the portal–arterial blood concentration difference of a nutrient times the rate of portal blood flow. In fed pigs, the blood flow rates in the portal vein and in the hepatic artery are measured as 0.118 and 0.028 l (kg MBW)$^{-1}$ min$^{-1}$, respectively (Simões Nunes *et al.*, 1989). The portal blood flow in a 60 kg pig is then 2.54 l min$^{-1}$ comprising 80% of the total blood flow through the liver. The rate of net glucose absorption in pigs of about 60 kg live weight fed an 800 g test meal after a 18–24 h fast is 3.3–4.1 mmol min$^{-1}$ (Giusi-Périer *et al.*, 1989; Simões Nunes *et al.*, 1989; Simões Nunes and Malmlöf, 1992; Rérat *et al.*, 1993). The hepatic balance of glucose, i.e. the use of glucose in the liver for glycogen storage, lipid synthesis, oxidation, etc., is estimated as 41–57% of the net absorption (Simões Nunes *et al.*, 1989; Simões Nunes and Malmlöf, 1992). Hence, the amount of glucose available for extrahepatic tissues is approximately 50% of the net absorption, viz. 1.7–2.1 mmol min$^{-1}$ in a 60 kg pig. This parameter would correspond to the glucose entry rate which is estimated as 3.6 mmol min$^{-1}$ in the fed state (Table 14.5). However, the estimated net hepatic output of approximately 2 mmol min$^{-1}$ is probably too low for full-fed pigs because the net absorption from a small meal after 18–24 h of fasting is likely to be an underestimate and the hepatic balance is likely to be overestimated (an increased need for replenishment of glycogen stores) compared to the situation in normally fed animals.

The rates of lactate net absorption and lactate hepatic balance have been measured by Simões Nunes and Malmlöf (1992) as 0.35 and 0.28 mmol min$^{-1}$, respectively, leaving 0.07 mmol min$^{-1}$ for use in extrahepatic tissues.

## Muscle tissue

The rate of glucose net uptake in the hindlimb (66% muscle and 8% fat) of 55 kg pigs is 0.155 mmol min$^{-1}$ (Wray-Cahen et al., 1992). If one hindlimb as a rough estimate contains 12–15% of the muscles in the body, then the total glucose uptake in muscle tissues of a fed 60 kg pig would be 1.1–1.4 mmol min$^{-1}$, i.e. 30–40% of the entry rate (see Table 14.5). In the same study (Wray-Cahen et al., 1992), the net hindlimb NEFA uptake was measured as 0.006 mmol min$^{-1}$. With the same assumptions as used above, the NEFA uptake in whole-body muscle tissues is 0.043–0.053 mmol min$^{-1}$ corresponding to about 15% of the NEFA entry rate (0.32 mmol min$^{-1}$; Table 14.5).

Glucose recycling due to the Cori cycle (glucose → lactate → glucose) is about 15% of the glucose entry rate (Trayhurn et al., 1981; Wolf and Brenner, 1988). From this figure, the rate of anaerobic glucose metabolism to lactate can be calculated as 0.54 mmol min$^{-1}$ equal to 40–50% of the estimated glucose uptake in muscles.

## Adipose tissue

Dunshea et al. (1992b, c) determined glucose entry rate, glucose utilization in adipose tissue, fatty acid and glycerol synthesis, fatty acid uptake from the blood as well as mobilization of fatty acids and glycerol in control pigs and in pST-treated pigs. Their results for the control pigs are recalculated in order to apply to animals of 60 kg live weight and presented in Fig. 14.4. All rates are given as mmol C min$^{-1}$. According to these data, about half of the glucose entry rate (11.33) is taken up by adipose tissue and 62% of this is incorporated into lipid (6.99) together with fatty acids taken up from the circulation (3.82) and re-esterified fatty acids from lipolysis of fat depots (4.05 − 1.68 = 2.37). Hence, the total lipid synthesis is 13.18. Lipolysis is 4.28 and thus 8.90 mmol lipid C min$^{-1}$ is retained. Assuming that the pig contains 10.8 kg lipid (18% of body weight), that the average molecular mass of lipid triglyceride is 860, and that the average chain length of triglyceride fatty acids is 17.4 mol C, the total lipid pool size can be calculated as 690 mol C. The fractional rates of lipid synthesis, lipolysis and lipid retention are then 2.75, 0.90 and 1.85% of the pool per day. In pigs treated with pST, the rate of lipid mobilization is twice as high compared with control pigs, but the rate of lipid synthesis from glucose is much more affected and is only 6% of the rate in the untreated pigs (Dunshea et al., 1992b). Hence, the rates of glucose uptake and utilization in

**Blood**

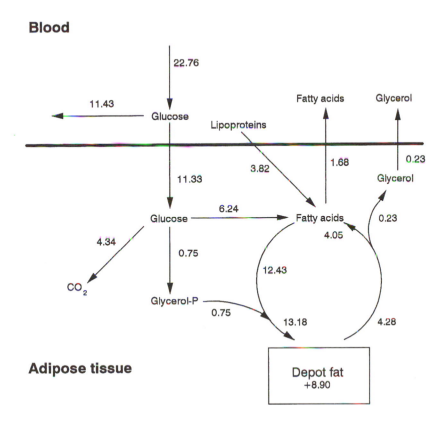

**Fig. 14.4.** Rates of adipose tissue glucose and lipid metabolism in a fed 60 kg pig. All rates are presented as mmol C min$^{-1}$.

adipose tissue are decreased in spite of increased plasma levels of both glucose and insulin in pST-treated pigs (see Tables 14.1 and 14.3). This apparent paradox can be explained, not by a decreased insulin binding, but by a decreased insulin sensitivity in adipose tissue exposed to pST (Magri *et al.*, 1990; Etherton *et al.*, 1993). Even when the glucose total entry rate is decreased by pST (see Table 14.4), the glucose availability for non-adipose tissues is unchanged because of the strongly reduced utilization in adipose tissue (Dunshea *et al.*, 1992b). These *in vivo* observations are confirmed by results from *in vitro* studies. Decreased *in vitro* rates of lipogenesis and glucose oxidation were associated with much lower activities of key lipogenic and NADPH-generating enzymes as well as lower concentrations of glucose transporter (GLUT4) mRNA and fatty acid synthase mRNA in adipocytes from pST-treated pigs (Magri *et al.*, 1990; Etherton *et al.*, 1993; Harris *et al.*, 1993).

**Table 14.6.** Rate of total $CO_2$ production and oxidation rates of glucose, acetate and NEFA in fed and fasted pigs of 60 kg live weight. (From Riis and Grummer, 1969; Freeman *et al*, 1970.)

| Parameter | Fed | | Fasted 12–24 h | |
|---|---|---|---|---|
| | mmol min$^{-1}$ | % | mmol min$^{-1}$ | % |
| Total $CO_2$ production | 20 | | 14 | |
| $CO_2$ from glucose oxidation | 6.00 | 30 | 2.24 | 16 |
| Glucose entry rate | 2.97 | | 1.19 | |
| Glucose oxidation to $CO_2$ | 1.00 | 34 | 0.37 | 31 |
| $CO_2$ from acetate oxidation | 2.40 | 12 | 2.94 | 21 |
| Acetate entry rate | 4.30 | | 3.30 | |
| Acetate oxidation to $CO_2$ | 1.20 | 28 | 1.47 | 45 |
| $CO_2$ from NEFA oxidation | 2.20 | 11 | 5.04 | 36 |
| NEFA entry rate | 0.69 | | 0.93 | |
| NEFA oxidation to $CO_2$[a] | 0.13 | 18 | 0.29 | 31 |

[a] Average chain length of plasma NEFA: 17.5 C atoms (Dunshea *et al.*, 1992c).

## Oxidations

The proportions of the total $CO_2$ production derived from oxidations of glucose, acetate and NEFA as well as the entry rates of these substrates have been determined by Riis and Grummer (1969) and by Freeman *et al.* (1970) in fed and fasted pigs. In the experiment of Riis and Grummer (1969), the total $CO_2$ production was also measured. From these data, it can be calculated how much of the entry rates is oxidized to $CO_2$ and how much is utilized for syntheses and other transactions. This is shown in Table 14.6 in which the calculated rates are adjusted to pigs of 60 kg live weight. The total $CO_2$ production is 20 mmol min$^{-1}$ in the fed and 14 mmol min$^{-1}$ in the fasted state. In fed pigs, 30, 12 and 11% of $CO_2$ is derived from glucose, acetate and NEFA, respectively. In fasted pigs, the priority of substrates for oxidation is changed and the corresponding figures are 16, 21 and 36%, respectively. The remainder of the $CO_2$ production (47% in the fed and 27% in the fasted state) is mainly from oxidation of amino acids, other SCFA, ketone bodies and triglycerides. The percentage of the glucose entry rate oxidized to $CO_2$ is slightly decreased by fasting (from 34 to 31%) whereas the proportions of acetate and NEFA oxidized to $CO_2$ are increased from 28 to 45% and from 18 to 31%, respectively, by fasting. The proportions of total $CO_2$ production derived from glucose, acetate and NEFA are determined by the use of $^{14}C$ labelling methods (Riis and Grummer, 1969; Freeman *et al.*, 1970). The resulting figures could be too low because any exchange of labelled with unlabelled carbon in the substrates will lead to an underestimation of the transfer quotients.

# A Static Model of Carbohydrate and Lipid Metabolism

Based on some of the quantitative data presented in the foregoing sections, a simple model of carbohydrate and lipid metabolism in a growing pig of 60 kg live weight is constructed (Fig. 14.5). The model is static and illustrates the fed state only. Therefore, the rates are presented as mmol C min$^{-1}$ and cannot be extrapolated to daily flow rates in pigs fed twice a day.

The entry rates of glucose and NEFA are taken from Table 14.5 and represent hepatic glucose output (21.72) and mobilization of NEFA from adipose tissues (5.60), respectively. It is assumed that the average fatty acid chain length is 17.5 carbon atoms (Dunshea *et al.*, 1992c). The rate of glucose output from the liver is assumed to be 60% of the net absorption rate from PDV (Simões Nunes and Malmlöf, 1992). Hence, the net absorption and the hepatic balance of glucose are 36.20 and 14.48, respectively. The following rates of metabolism in adipose tissue are adopted from Fig. 14.4: glucose uptake (11.33), glucose oxidation (4.34), lipogenesis from glucose (6.99), lipid retention (8.90) and fatty acid re-esterification (4.05). The rate of lipolysis is the sum of re-esterified and mobilized fatty acids (4.05 + 5.60 = 9.65) as well as the accompanying glycerol (9.65/17.5 = 0.55). By balancing the triglyceride pool, the uptake of blood fatty acids is found as 8.90 + 5.60 + 0.55 − 6.99 = 8.06. The corresponding transport of triglyceride in chylomicrons and VLDL is 8.06 × (1 + 17.5)/17.5 = 8.52 and the recycling of glycerol to the liver is 0.55 + 8.52 − 8.06 = 1.01.

The absorption of lipid is calculated from an assumed daily intake of 2.5 kg feed containing 2.3% fat (Dunshea *et al.*, 1992a) with an ileal digestibility of 65% (Jørgensen, 1991). Thus, 2.5 × 23 × 0.65 = 37 g per day is absorbed equivalent to 1.67 mmol triglyceride C min$^{-1}$. The export of triglyceride in VLDL from the liver is then 8.52 − 1.67 = 6.85 (6.48 in fatty acids and 0.37 in glycerol). The surplus of glycerol in the liver, 1.01 − 0.37 = 0.64, is assumed to be oxidized via the glycolytic pathway and the TCA cycle. The rate of glucose uptake in muscles and other tissues is 21.72 − 11.33 = 10.39. According to Table 14.6, the rate of whole-body glucose oxidation is 34% of the entry rate (21.72 × 0.34 = 7.38), and 7.38 − 4.34 = 3.04 is then oxidized in non-adipose tissues.

The rate of glycogen synthesis in the liver is 5–10 mg h$^{-1}$ g$^{-1}$ liver after fasting and re-feeding (Bergman, 1983). Assuming an average value of 7.5 mg and a liver weight of 1.5 kg (2.5% of LW), the synthesis rate can be calculated as 7.5 × 1500 × 6/(162 × 60) = 6.94. Hence, glucose available in the liver for fatty acid synthesis and oxidation is 14.48 − 6.94 = 7.54. The requirement of NADPH in fatty acid synthesis is 14 mol mol$^{-1}$ palmitate ($C_{16}$). The pyruvate/malate cycle furnishes 8 mol and the remaining 6 mol NADPH is assumed to be supplied from the complete oxidation of 0.5 mol glucose in the pentose phosphate pathway. This means that for each mole fatty acid carbon synthesized, 3/16 mol glucose carbon must be oxidized. The rate of glucose oxidation is then 7.54 × 3/(16 + 3) = 1.19, and the rate of fatty acid synthesis

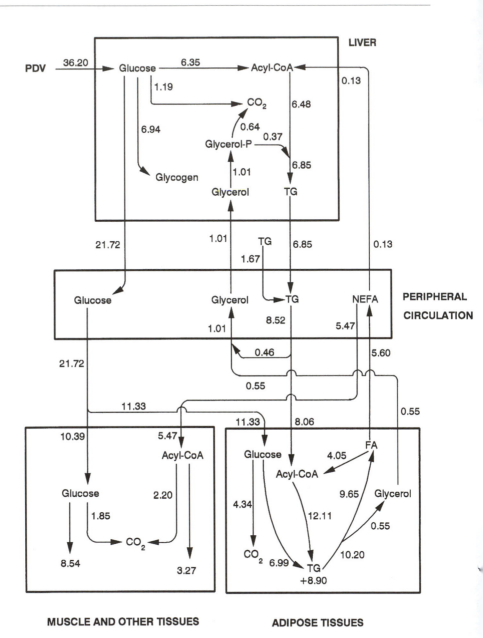

**Fig. 14.5.** Rates of whole body glucose and lipid metabolism in a fed 60 kg pig. All rates are presented as mmol C min$^{-1}$.

in the liver is $7.54 - 1.19 = 6.35$. Then, the hepatic uptake of fatty acids from the blood NEFA pool is $6.48 - 6.35 = 0.13$.

Glucose oxidation in muscles and other tissues is $3.04 - 1.19 = 1.85$ leaving 8.54 for glycogen synthesis and other purposes. The rate of fatty acid uptake is $5.60 - 0.13 = 5.47$ and the rate of whole-body fatty acid oxidation is taken as 2.20 from Table 14.6. This is assumed to take place in muscles and other tissues and, hence, the use of fatty acids for other purposes (e.g. lipogenesis) is $5.47 - 2.20 = 3.27$ mmol C min$^{-1}$.

As the quantitative data used in this model are derived from several experiments using different animals, feeding protocols, analytical methods, etc., the calculated whole-body flux rates are somewhat tentative and should be considered with caution. For example, the rates of lipid turnover in the model are considerably higher than found by Dunshea *et al.* (1992b,c) because their estimated fatty acid mobilization was very low (1.7 mmol C min$^{-1}$; Fig. 14.4) compared to the fatty acid entry rate used in the model (5.6 mmol C min$^{-1}$; Table 14.5).

# Perspectives

In this chapter, a brief review of carbohydrate and lipid metabolism in growing pigs is presented in both qualitative and quantitative terms. Also, a few examples of interaction between carbohydrate, lipid and protein metabolism are pointed out, i.e. gluconeogenesis, amino acid oxidation and provision of energy for protein synthesis. A static model of whole-body glucose and lipid flow rates is constructed. The model applies to fed pigs of 60 kg live weight, but should be considered as tentative. The purpose of the model is to give an impression of the magnitude of glucose and lipid flow rates in a 60 kg pig and also to serve as a starting point for the development of a dynamic, mechanistic simulation model. A model of this kind could be constructed in many ways depending on its purpose, e.g. as a research model for testing of alternative hypotheses concerning the intermediary metabolism in growing pigs or as a tool for an improved feed evaluation system by which the actual value of a given feed (expressed on the basis of the obtained animal performance) in a given situation can be predicted (Danfaer, 1990a,b).

In order to develop a dynamic and mechanistic simulation model of nutrient digestion, absorption and metabolism in growing pigs, more quantitative information than presented in this chapter is needed. The key data required to build an animal model according to the principle of rate-state formalism are masses of nutrient pools and rates of transaction (France and Thornley, 1984). Hence, there is a need for more experimental research to extend our knowledge of intracellular nutrient pool sizes, nutrition partition between alternative pathways in various tissues as well as causes of variation in these parameters. The use of multicatheterized animals to measure nutrient balances across a specified organ or tissue bed combined with isotopic labelling of nutrients seems to be a suitable method to provide such information (Danfaer *et al.*, 1995).

# References

Beech, S.A., Elliott, R. and Batterham, E.S. (1991) Sucrose as an energy source for growing pigs: a comparison of the effects of sucrose, starch and glucose on energy and protein retention. *Animal Production* 53, 383–393.

Bergman, E.N. (1983) The pools of tissue constituents and products: carbohydrates. In: Riis, P.M. (ed.), *Dynamic Biochemistry of Animal Production*. Elsevier, Amsterdam, pp. 137–149.

Bezkorovainy, A. and Rafelson, M.E. Jr. (1996) *Concise Biochemistry*, International edn. Marcel Dekker, Inc., New York, 624pp.

Caperna, T.J., Steele, N.C., Evock-Clover, C.M., Brocht, D., McMurtry, J.P. and Rosebrough, R.W. (1993) Acute effects of administration of porcine growth hormone on circulating levels of hormones and metabolites in 20-, 40-, and 60-kilogram gilts. *Journal of Animal Science* 71, 897–905.

Christensen, K. (1974) Different levels of linoleic acid in the diet of pigs. 2. The effect on the fatty acid composition of mitochondrial phospholipids and mitochondrial function. *Zeitschrift für Tierphysiologie, Tierernährung und Futtermittelkunde* 33, 10–20.

Christensen, K. (1985) *Determination of Linoleic Acid Requirements in Slaughter Pigs*. Report 577 from the National Institute of Animal Science, Copenhagen, Denmark, 158pp.

Christensen, K. (1986) Influence of different dietary concentrations of linoleic acid on the essential fatty acid (EFA) status and functional characteristics of porcine hepatic and cardiac mitochondria. *Comparative Biochemistry and Physiology* 85B, 419–425.

Christensen, K. (1987) Influence of different dietary concentrations of linoleic acid on the essential fatty acid (EFA) status and functional characteristics of porcine skeletal muscle mitochondria. *Journal of Animal Physiology and Animal Nutrition* 57, 278–290.

Cote, P.J., Wangsness, P.J., Varela-Alvarez, H., Griel, L.C., Jr and Kavanaugh, J.F. (1982) Glucose turnover in fast-growing, lean and in slow-growing, obese swine. *Journal of Animal Science* 54, 89–94.

Danfaer, A. (1980) Protein- og aminosyreomsætning hos voksende grise. Effekt af anaboliske steroider samt fodringshyppighed [Protein turnover and amino acid metabolism in growing pigs. Effects of anabolic steroids and feeding frequency.] PhD thesis, The Royal Veterinary and Agricultural University, Copenhagen, Denmark.

Danfaer, A. (1990a) *A Dynamic Model of Nutrient Digestion and Metabolism in Lactating Dairy Cows*. Report 671 from the National Institute of Animal Science, Foulum, Denmark, 511pp.

Danfaer, A. (1990b) Lactation models in cattle. In: Robson, A.B. and Poppi, D.P. (eds), *Proceedings of Third International Workshop on Modelling Digestion and Metabolism in Farm Animals*. Lincoln University, Canterbury, New Zealand, pp. 305–318.

Danfaer, A., Tetens, V. and Agergaard, N. (1995) Review and an experimental study on the physiological and quantitative aspects of gluconeogenesis in lactating ruminants. *Comparative Biochemistry and Physiology* 111B, 201–210.

Dauncey, M.J., Burton, K.A., White, P., Harrison, A.P., Gilmour, R.S., Duchamp, C. and Cattaneo, D. (1994) Nutritional regulation of growth hormone receptor gene expression. *The FASEB Journal* 8, 81–88.

Dunshea, F.R., Bauman, D.E., Boyd, R.D. and Bell, A.W. (1992a) Temporal response of circulating metabolites and hormones during somatotropin treatment of growing pigs. *Journal of Animal Science* 70, 123–131.

Dunshea, F.R., Harris, D.M., Bauman, D.E., Boyd, R.D. and Bell, A.W. (1992b) Effect of porcine somatotropin on in vivo glucose kinetics and lipogenesis in growing pigs. *Journal of Animal Science* 70, 141–151.

Dunshea, F.R., Harris, D.M., Bauman, D.E., Boyd, R.D. and Bell, A.W. (1992c) Effect of somatotropin on nonesterified fatty acid and glycerol metabolism in growing pigs. *Journal of Animal Science* 70, 132–140.

Dyerberg, J. (1993) The 'discoveries' of the $\Omega$-3 fatty nutrition and their relation to human health. In: *A Challenge to Future Food Production. Proceedings of a Minisymposium organized in connection with the 44th Annual Meeting of EAAP*. Research Centre Foulum, pp. 7–18.

Einarsson, S. (1981) Comparative trial with natural prostaglandin and an analogue (Cloprostenol) in inducing parturition in sows. *Acta Veterinaria Scandinavica* (Suppl. 77), 321–326.

Etherton, T.D., Louveau, I., Sørensen, M.T. and Chaudhuri, S. (1993) Mechanisms by which somatotropin decreases adipose tissue growth. *American Journal of Clinical Nutrition* 58 (Suppl.), 287S–295S.

Faidly, T.D., Luhman, C.M., Galloway, S.T., Foley, M.K. and Beitz, D.C. (1990) Effect of dietary fat source on lipoprotein composition and plasma lipid concentrations in pigs. *The Journal of Nutrition* 120, 1126–1133.

Flatt, J.P. (1970) Conversion of carbohydrate to fat in adipose tissue: an energy-yielding and, therefore, self-limiting process. *Journal of Lipid Research* 11, 131–143.

France, J. and Thornley, J.H.M. (1984) *Mathematical Models in Agriculture*. Butterworths, London, 335pp.

Freeman, C.P., Noakes, D.E. and Annison, E.F. (1970) The metabolism of glucose, acetate, palmitate, stearate and oleate in pigs. *The British Journal of Nutrition* 24, 705–716.

Friend, D.W., Nicholson, J.W.G. and Cunningham, H.M. (1964) Volatile fatty acid and lactic acid content of pig blood. *Canadian Journal of Animal Science* 44, 303–308.

Fuller, M.F., Weekes, T.E.C., Cadenhead, A. and Bruce, J.B. (1977) The protein-sparing effect of carbohydrate. 2. The role of insulin. *The British Journal of Nutrition* 38, 489–496.

Ganong, W.F. (1993) *Review of Medical Physiology*, 16th edn. Prentice-Hall, London, 774pp.

Giusi-Périer, A., Fiszlewicz, M. and Rérat, A. (1989) Influence of diet composition on intestinal volatile fatty acid and nutrient absorption in unanesthetized pigs. *Journal of Animal Science* 67, 386–402.

Gopinath, R. and Etherton, T.D. (1989a) Effects of porcine growth hormone on glucose metabolism of pigs: I. Acute and chronic effects on plasma glucose and insulin status. *Journal of Animal Science* 67, 682-688.

Gopinath, R. and Etherton, T.D. (1989b) Effects of porcine growth hormone on glucose metabolism of pigs: II. Glucose tolerance, peripheral tissue insulin sensitivity and glucose kinetics. *Journal of Animal Science* 67, 689-697.

Graham, H. and Malmlöf, K. (1993) A note on the effect of potato fibre on blood glucose and insulin response in pigs. *Swedish Journal of Agricultural Research* 23, 149-152.

Griglio, S. (1986) Lipoprotein metabolism in isolated and cultured hepatocytes. Part 2: Lipoprotein catabolism by liver cells. In: Guillouzo, A. and Guguen-Guillouzo, C. (eds), *Research in Isolated and Cultured Hepatocytes*. John Libbey Eurotext Ltd/INSERM, London, pp. 135-153.

Harris, D.M., Dunshea, F.R., Bauman, D.E., Boyd, R.D., Wang, S.-Y., Johnson, P.A. and Clarke, S.D. (1993) Effect of in vivo somatotropin treatment of growing pigs on adipose tissue lipogenesis. *Journal of Animal Science* 71, 3293-3300.

Hedqvist, P., Dahlén, S.-E., Gustafsson, L., Hammarström, S. and Samuelsson, B. (1980) Biological profile of leukotrienes $C_4$ and $D_4$. *Acta Physiologica Scandinavica* 110, 331-333.

Heidemann, S. (1992) The molecular and cellular basis of physiological regulation. In: Cunningham, J.G. (ed.), *Textbook of Veterinary Physiology*. W.B. Saunders Company, Philadelphia, pp. 3-31.

Imoto, S. and Namioka, S. (1978) VFA metabolism in the pig. *Journal of Animal Science* 47, 479-487.

Jenness, R. (1974) The composition of milk. In: Larson, B.L. and Smith, V.R. (eds), *Lactation: A Comprehensive Treatise. Vol. III. Nutrition and Biochemistry of Milk/Maintenance*. Academic Press, New York, pp. 3-107.

Jørgensen, H. (1991) Fedt og fedtsyrers fordøjelighed hos svin [Digestibility of fat and fatty acids in pigs.] PhD thesis, The Royal Veterinary and Agricultural University, Copenhagen, Denmark.

Korbut, R. and Moncada, S. (1978) Prostacyclin ($PGI_2$) and thromboxane $A_2$ interaction in vivo. Regulation by aspirin and relationship with antithrombotic theraphy. *Thrombosis Research* 13, 489-500.

Lands, W.E.M. (1992) Biochemistry and physiology of n-3 fatty acids. *The FASEB Journal* 6, 2530-2536.

Leat, W.M.F. (1963) Fatty acid composition of the serum lipids of pigs given different amounts of linoleic acid. *The Biochemical Journal* 89, 44-51.

Lepine, A.J., Garleb, K.A., Reinhart, G.A. and Kresty, L.A. (1993) Plasma and tissue fatty acid profiles of growing pigs fed structured or non-structured triacylglycerides containing medium-chain and marine oil fatty acids. *Journal of Nutritional Biochemistry* 4, 362-372.

Madsen, A. (1983) The molecular basis of animal production: Metabolism in skeletal muscle cells. In: Riis, P.M. (ed.), *Dynamic Biochemistry of Animal Production*. Elsevier, Amsterdam, pp. 9-28.

Magri, K.A., Adamo, M., Leroith, D. and Etherton, T.D. (1990) The inhibition of insulin action and glucose metabolism by porcine growth hormone in porcine adipocytes is not the result of any decrease in insulin binding or insulin receptor kinase activity. *Biochemical Journal* 266, 107-113.

Malmlöf, K., Simões Nunes, C. and Örberg, J. (1988) Effects of a high dietary fibre level on postprandial porto-arterial differences in the plasma concentrations of immunoreactive insulin, glucose and free amino acids in the growing pig. *Swedish*

*Journal of Agricultural Research* 18, 67–75.

Malmlöf, K., Simões Nunes, C. and Askbrandt, S. (1989) Effects of guar gum on plasma urea, insulin and glucose in the growing pig. *The British Journal of Nutrition* 61, 67–73.

Martin, B.R. (1987) *Metabolic Regulation: A Molecular Approach*. Blackwell Scientific Publications, Oxford, 299pp.

Mersman, H.J. (1986) Comparison of plasma free-fatty-acid and blood-glycerol concentrations with measurement of lipolysis in porcine adipose tissue in vitro. *Journal of Animal Science* 63, 757–769.

Mersman, H.J. and Hu, C.Y. (1987) Factors affecting measurements of glucose metabolism and lipolytic rates in porcine adipose tissue slices in vitro. *Journal of Animal Science* 64, 148–164.

Mersman, H.J. and MacNeil, M.D. (1985) Relationship of plasma lipid concentrations to fat deposition in pigs. *Journal of Animal Science* 61, 122–128.

Newsholme, E.A. and Start, C. (1973) *Regulation in Metabolism*. John Wiley, London, 349pp.

Ostaszewski, P. and Nissen, S. (1988) Infusion of nicotinic acid stimulates leucine oxidation and inhibits protein synthesis in pigs before and during a meal. *Hormone and Metabolic Research* 20, 154–157.

Ponter, A.A., Salter, D.N., Morgan, L.M. and Flatt, P.R. (1991) The effect of energy source and feeding level on the hormones of the entero-insular axis and plasma glucose in the growing pig. *The British Journal of Nutrition* 66, 187–197.

Rérat, A., Chayvialle, J.A., Kandé, J., Vaissade, P., Vaugelade, P. and Bourrier, T. (1985) Metabolic and hormonal effects of test meals with various protein contents in pigs. *Canadian Journal of Physiology and Pharmacology* 63, 1547–1559.

Rérat, A., Fiszlewicz, M., Giusi, A. and Vaugelade, P. (1987) Influence of meal frequency on postprandial variations in the production and absorption of volatile fatty acids in the digestive tract of conscious pigs. *Journal of Animal Science* 64, 448–456.

Rérat, A., Giusi-Périer, A. and Vaissade, P. (1993) Absorption balances and kinetics of nutrients and bacterial metabolites in conscious pigs after intake of maltose- or maltitol-rich diets. *Journal of Animal Science* 71, 2473–2488.

Riis, P.M. and Grummer, R.H. (1969) The relationship between glucose and fatty acid metabolism in pigs under various feeding conditions. *Acta Agriculturæ Scandinavica* 19, 11–17.

Sano, H., Nakamura, E., Takahashi, H. and Terashima, Y. (1995) Plasma insulin and glucagon responses to acute challenges of acetate, propionate, *n*-butyrate and glucose in growing gilts (*Sus scrofa*). *Comparative Biochemistry and Physiology* 110A, 375–378.

Schade, D.S., Woodside, W. and Eaton, R.P. (1979) The role of glucagon in the regulation of plasma lipids. *Metabolism* 28, 874–886.

Scholz-Ahrens, K.E., Hagemeister, H., Unshelm, J., Agergaard, N. and Barth, C.A. (1990) Response of hormones modulating plasma cholesterol to dietary casein or soy protein in minipigs. *The Journal of Nutrition* 120, 1387–1392.

Simões Nunes, C. and Malmlöf, K. (1992) Glucose absorption, hormonal release and hepatic metabolism after guar gum ingestion. *Reproduction, Nutrition, Development* 32, 11–20.

Simões Nunes, C., Rérat, A., Galibois, I., Vaugelade, P. and Vaissade, P. (1989) Hepatic and gut balances of glucose, amino-nitrogen, ammonia and urea in the pig after ingestion of casein or rapeseed proteins. *Nutrition Reports International* 40, 901–907.

Thacker, P.A. and Bowland, J.P. (1981) Effects of dietary propionic acid on serum lipids and lipoproteins of pigs fed diets supplemented with soybean meal or canola meal. *Canadian Journal of Animal Science* 61, 439–448.

Trayhurn, P., Dauncey, M.J. and Ingram, D.L. (1981) Determination of glucose turnover and recycling with $^{14}C$ and $^{3}H$ labelled tracers in pigs exposed to thermally neutral and cold environments. *Comparative Biochemistry and Physiology* 69B, 69–73.

Wolf, H.P.O. and Brenner, K.V. (1988) The effect of etomoxir on glucose turnover and recycling with [1-$^{14}C$], [3-$^{3}H$]-glucose tracer in pigs. *Hormone and Metabolic Research* 20, 204–207.

Wood, J.D., Gregory, N.G., Hall, G.M. and Lister, D. (1977) Fat mobilization in Pietrain and Large White pigs. *The British Journal of Nutrition* 37, 167–186.

Wray-Cahen, D., Ross, D.A., Bauman, D.E. and Boyd, R.D. (1991) Metabolic effects of porcine somatotropin: nitrogen and energy balance and characterization of the temporal pattern of blood metabolites and hormones. *Journal of Animal Science* 69, 1503–1514.

Wray-Cahen, D., Bell, A.W., Boyd, R.D. and Bauman, D.E. (1992) Effect of porcine somatotropin and insulin on nutrient uptake by the hindlimb of growing pigs. *The Proceedings of the Nutrition Society* 51, 35A.

Wray-Cahen, D., Boyd, R.D., Bauman, D.E. and Ross, D.A. (1993) Effect of porcine somatotropin on the response of growing pigs to acute challenges of glucose, insulin and epinephrine and during a hyperinsulinemic–euglycemic clamp. *Domestic Animal Endocrinology* 10, 103–115.

# Energy Flows

## 15

## G.C. Emmans

*Animal Biology Division, Scottish Agricultural College, West Mains Road, Edinburgh EH9 3JG, UK*

## Introduction

During the latter part of the eighteenth, and the early part of the nineteenth, centuries the idea of 'energy' as a unifying idea gradually emerged. It was accepted that energy could take different forms and that these could be converted into each other, at least to some extent. The idea of an equivalence of different forms was established, with the quantification of the heat equivalent of work by Joule in Manchester, UK, being particularly elegant. Animal energetics was of importance from an early stage with Lavoisier – later unfortunately guillotined in the French Revolution – one of the founding fathers. The subsequent history of animal energetics, and the related problem of feed evaluation, is extensive and an important part of the history of animal science.

It was realized that it was useful to consider the diet of an animal as supplying a resource called energy as well as material nutrients. The energy yielded by the diet was not independent of the material of the diet, but arose from it. Armsby, working in Pennsylvania in the USA in the early years of the twentieth century, stressed the value of seeing the diet as supplying both material nutrients and energy and made clear the relationships between them. Increasingly, the problem of assigning a value to a foodstuff came to be seen as that of assigning it an energy value which expressed its usefulness to the animal.

Kellner and Kohler (1900), working in Germany on ruminants, proposed that the values of feeds could be described as their value relative to starch – a starch equivalent, or *SE*, value. Armsby (1903) and Armsby and Fries (1915) approached the problem of energy directly and developed the idea of a net energy, $NE$, value of a foodstuff. The $NE$ value (MJ kg$^{-1}$) was the metabolizable energy ($ME$, MJ kg$^{-1}$) value after subtracting the heat increment ($HI$, MJ kg$^{-1}$). Armsby emphasized the importance of accounting for the extra heat produced as the result of feeding another 1 kg of an ingredient. He assumed, and had some experimental results to support this assumption, that the $HI$ value of an ingredient was independent of the level of feeding. Later work at Pennsylvania by Forbes and his colleagues (Forbes *et al.*, 1928, 1930) showed that there could be no single $NE$ value for an ingredient as the heat produced by feeding was not a linear function of the amount fed.

Kellner's approach was extended to pigs by Fingerling in Germany in the 1920s and 1930s (e.g. Fingerling *et al.*, 1938a,b). Just *et al.* (1983), working in Denmark, and Noblet *et al.* (1994), working in France, have used the approach of Fingerling to try to produce a net energy, $NE$, system for pigs. The problems raised by Forbes *et al.* (1928, 1930) appear to have been overlooked. In this chapter the problems raised by attempts to move to an $NE$ system are presented against a background of the traditional way of seeing the problem.

## Background and methods

All of the energy that the pig gets which is available for work comes from its diet. The pig needs to do work in order to carry a range of functions which will be described below. The pig can get radiant heat from its environment, which can help to keep it warm, but this source of energy gain is excluded here as it cannot be used to do work. It is of importance in environments that are not thermally neutral (see Chapter 4). The potential energy of the food exists in the form of chemical bonds. When these are broken, either by combustion or by metabolism, the potential energy that is released is either stored in other chemical bonds, or is used to do work. When work has been done heat is produced according to Joule's 'heat equivalent of work'.

The overall flow of energy in the pig, and in other animals, can be measured as heat either by measuring it directly, or by estimating it indirectly. The direct measurement of heat in a calorimeter involves very accurate measurements of temperature using a large number of thermocouples. The direct method is now used infrequently. The two indirect methods widely used for animals in calorimeters are: (i) measuring gas flows to estimate oxygen consumption and carbon dioxide production; (ii) measuring the terms of the carbon and nitrogen balances and hence estimating energy retention. Blaxter (1989) reviews the methods. Experiments on animals in calorimeters can give information on energy flows over quite short time periods and can be seen as measuring instantaneous rates. As the change in the state of the animal is either small, or can be balanced across animals in the design of the experiment, there are no problems about differences in maintenance. The heat increments, and energy retentions, of different diets can be estimated.

To estimate energy flows over longer periods of time the animal must either be kept in a calorimeter for a longer period, which is expensive, or the alternative technique of serial slaughter must be used. In interpreting the data from slaughter experiments there is the problem of integrating maintenance expenditure over time. The serial slaughter technique estimates energy retentions as lipid and protein over a period of time. Animals are killed at the beginning and end of the period and the retentions calculated. The efficiency of the design is increased as the ratio of the energy contents of the final animals to that of the initial animals is large. However, as the period is made longer, so the problem of estimating cumulative maintenance becomes greater. An example of how this can be done, using data from a serial slaughter experiment in broiler chickens, is in Emmans (1994). The use of data from serial slaughter experiments is illustrated in Kyriazakis and Emmans (1992a, b). That similar results can be obtained from calorimeter and slaughter experiments is indicated by the analysis of Kyriazakis and Emmans (1992a, b) where data from a serial slaughter experiment were compared with the predictions of energy requirements of ARC (1981) which were based on calorimetry experiments.

The classical quantities of energy need to be defined and described. Digested energy is related to the digestible components of the organic matter to connect with the material in Chapter 3. The energy exchanges in the growing animal are built up by considering, in succession: the pig in a fasting state in a thermally neutral environment; the pig fed to be at maintenance; the immature pig eating a diet such that it is retaining protein and lipid. The importance of distinguishing energy flows in cold and non-cold environments connects with the material in Chapter 4. The breeding sow is not considered here as it is dealt with in Chapter 7.

## Energy as an Integrating Concept

As with matter, there is a Principle of the Conservation of Energy. This means that the energy which enters the animal either stays there or comes out. The energy lost from the body is in the form of the potential energy of the physical excretal – faeces, urine and combustible gases – or as heat. The energy that stays in the body is as the potential energy of the change in the weights of the combustible components. Conventionally these are taken to be only protein and lipid, with the small amount of carbohydrate ignored.

There are thus only a few overall processes that need to be considered in order to get a complete description of the pig in energy terms. The energy supplied by the diet is its heat of combustion or gross energy, $GE$. Deducting the energy losses in the faeces, $FE$, urine, $UE$, and combustible gases, conventionally considered as methane, *CE*, yields the metabolizable energy, *ME*:

$$GE - (FE + UE + CE) = ME \qquad (15.1)$$

The *ME* can be seen to be a calculated quantity, rather than one that is measured directly. All errors in the measured quantities accumulate into the estimation of the *ME* value of the diet. The *ME* yielded to the pig by the diet eaten is either lost as heat, *H*, or retained in the body as chemical energy, *ER*:

$$ME = H + ER \qquad (15.2)$$

The heat produced by the pig needs to be lost to the environment. A thermally neutral environment is one to which the heat produced can be comfortably lost. Consideration of the energy relationships thus connect the pig with its climatic environment as is considered in more detail in Chapter 4. As the *ER* depends on the rates of retention of protein and lipid, and these are, at least in part, genetic values the consideration of energy leads to a consideration of the potential of the pig for retaining protein and lipid as considered in Chapters 8 and 10.

The gross energy of the diet eaten by the pig (*GE*, MJ day$^{-1}$) is the heat of combustion of the food that it eats. It is only the organic matter of the food that yields heat when burnt in a bomb calorimeter. The convention of expressing energy contents of foods and diets on an 'as is', or dry matter, basis will inevitably lead to imprecision. Expressing energy values per unit weight of organic matter, rather than on an 'as is' or dry matter basis, is to be encouraged.

The energy yielded as heat when a food is totally combusted in a bomb calorimeter can be seen as arising from the combustion of the three major components of the organic matter – protein, fat and carbohydrate. While these components are not true chemical entities their heats of combustion across the kinds of raw materials fed to pigs can be taken as quasi-constants. The approximate heats of combustion are: fat, 39.6 kJ g$^{-1}$, protein, 23.8 kJ g$^{-1}$ and carbohydrate 17.3 kJ g$^{-1}$. Different amino acids have different heats of combustion as do fatty acids of different chain length. The value for carbohydrate depends on the proportions of starch, sugars and non-starch polysaccharides. For fine calculations such effects need to be allowed for.

A pig eating a diet will usually produce faeces that will yield energy when burnt. Again the value of the heat of combustion of the faeces, expressed per unit of organic matter, will arise from the heats of combustion of the components of the organic matter in the faeces. But there is little value for our present purposes in considering the composition of the faeces in these terms as it is only the energy that is of importance. For predicting the outputs of pollutants such as P and N the composition of the faeces in these terms will be relevant. It is conventional to assume that the faeces produced have been derived from the diet which is being eaten. There is debate about the distinction between the faeces and the faecal energy, which arises directly from the food eaten, and energy which arises from the body as a consequence of the food having been eaten. This problem is discussed elsewhere (Chapter 13). For the purposes in this chapter all quantities are 'apparent'.

As the faecal energy (*FE*, kJ day$^{-1}$) is clearly not available to the pig it is subtracted from the *GE* to estimate a quantity called the digestible energy (*DE* kJ day$^{-1}$) as $DE = GE - FE$. The *DE* is a widely used measure of the energy supplied by its diet to the pig (ARC, 1981). For this definition to be useful it is not necessary that the organic matter in the faeces is actually derived directly from the food, i.e. it does not need to be feed residues. In fact it is the normal case that the faecal organic matter consists, at least in part, of tissue from the gut and microbes that grow in the gut, as well as true feed residues. The definition of *DE* is a formal one and makes no assumptions about the mechanics by which the combustible material gets to the faeces. The important point is that if the energy is lost in the faeces it is not available to the animal.

In ruminants a considerable proportion of the gross energy eaten may appear as gases which are combustible. In the normal case all of this gas is methane; hydrogen may appear as a rare event. The gases are the by-product of fermentation in the rumen. In the pig similar processes can take place in the hindgut but the scale is very much less. It is rare for more than 2% of the GE to appear as combustible gases even on feeds of high fibre content. Kirchgessner *et al.* (1991) measured the effect of the level of fermentable material in the diet on the production of methane. Demeyer and Degraeve (1991) discuss the differences between fermentations in pigs and ruminants. Zhu *et al.* (1993) discuss the stoichiometry of fermentation in pigs.

When protein is burnt in a bomb calorimeter it is completely combusted to the oxides of C, N and S. In the body the oxidation of protein is incomplete and some compounds containing N and energy are lost in the urine. The energy in the urine, *UE*, and in the combustible gases, *CE*, are unavailable to the pig. When these quantities are subtracted from the *DE* the result is the metabolizable energy (*ME*, kJ day$^{-1}$). The definition is:

$$ME = DE - UE - CE \tag{15.3}$$

As an example the energy flows (excluding the combustible gases which were not measured and which were probably negligible on this diet) for five pigs of Fingerling *et al.* (1938a) are given in Table 15.1. They were given 1200 g barley, 100 g fishmeal, 50 g blood meal and 5 g salt per day.

# Fasting

The conventional starting point for considering energy flows is the case where the pig is given no food. This may seem strange when pigs are never going to be fasted in agricultural practice, but there is sense in it. When it is

**Table 15.1.** The flows of energy through the five pigs of
Fingerling *et al.* (1938a).

| Measure | Mean | SD |
|---|---|---|
| Gross energy (kJ day$^{-1}$) | 22,978 | 6 |
| Faeces energy (kJ day$^{-1}$) | 3,334 | 175 |
| Urine energy (kJ day$^{-1}$) | 1,264 | 26 |
| Metabolizable energy (*ME*, kJ day$^{-1}$) | 18,380 | 165 |
| 6.25 × N retention (*PR*, g day$^{-1}$) | 42.2 | 10.2 |
| Carbon retention (*CR*, g day$^{-1}$) | 75.6 | 13.0 |
| Lipid retention (g day$^{-1}$)$^a$ | 69.9 | 21.0 |
| Energy retention (*ER*, kJ day$^{-1}$)$^b$ | 3,771 | 689 |
| Heat loss (*H*, kJ day$^{-1}$)$^c$ | 14,609 | 812 |
| Live weight (kg) | 156.4 | 2.5 |
| Heat (kg$^{-0.75}$ day$^{-1}$) | 330.3 | 17.5 |
| Energy retention (kJ kg$^{-0.75}$ day$^{-1}$) | 85.3 | 15.8 |

$^a$ Lipid retention calculated as $(CR - 0.526 \times PR)/0.765$
which assumes that the C content of protein is 0.526 and of
lipid is 0.765.
$^b$ $ER = 23.8\ PR + 39.6\ LR$.
$^c$ $H = ME - ER$.

fed the animal's energy flows depend on what it is fed. But when fasted they
do not depend on what it is not fed! The fasting state is thus independent of
food.

In principle the fasting condition is straightforward – the pig is given no
food. But in practice there are two problems: (i) at what temperature should
the measurements be made? (ii) when should they be made after the initiation
of the fast? In pigs the second question is not of as much importance as it is in
ruminants as the rate of heat production changes little after two days of
fasting, and only to the extent expected from the change in the size of the pig.
But the first question – At what temperature should the fasting pig be kept? –
is of great importance as it affects the interpretation of the data. Its importance has often been overlooked (Tess *et al.*, 1984).

If the fasting pig is kept at a low temperature its rate of heat production
(*HP*, kJ day$^{-1}$) will include a term for cold thermogenesis (*CT*, kJ day$^{-1}$). As
the temperature is increased so *CT* will decrease to become zero at the lower
critical temperature of the fasted pig. When $CT = 0$ then $HP = FHP$, kJ day$^{-1}$,
which is the thermoneutral fasting heat production. As pigs are poorly
insulated, and usually have no bedding in a calorimeter where their heat
production is measured or estimated, the temperature at which the fasted pig
will be thermally neutral is likely to be above 30°C (see Chapter 4). Heat
productions in the fasting pig which are measured at temperatures lower
than this will contain a proportion as *CT* and will not have the meaning

attributed to them which is that they estimate energy maintenance. There appear to be no estimates of *FHP* in pigs over a range in weight which can be shown to have been measured unequivocally at thermoneutrality. The way in which *FHP* varies with size, and possibly between genotypes of pig, is therefore not yet clear and appropriate well-designed experiments are needed.

The thermoneutral fasting heat production is a measure of the basal metabolism of the pig. It will vary with the level of physical activity. As the pig will be catabolizing protein a part of the *FHP* will arise from the work of excreting the N-containing compounds in the urine which arises from this catabolism (Emmans, 1994). The *FHP* when reduced by this amount is called the maintenance heat (*MH*, kJ day$^{-1}$). It is to be expected that *MH* will vary between pig genotypes at maturity, and between degrees of maturity for a given genotype. It is conventional to scale *FHP*, and maintenance generally, to the size of the pig measured as its live weight. There are, however, obvious problems with such a scaling.

To deal with differences between genotypes and degrees of maturity Emmans and Fisher (1986) proposed the general scaling rule for maintenance heat that:

$$MH = (M_E/P_m^{0.27}). P \qquad \text{MJ day}^{-1} \tag{15.4}$$

where $P_m$ is the mature, and $P$ the current, protein weight of the pig. The value of $M_E$ is assumed to be a quasi-constant the value of which was estimated as 1.63 MJ unit$^{-1}$ (Emmans, 1987).

The more usual way to scale maintenance in the pig is to make it some function of live weight, *W*. The most common form is to make it proportional to $W^b$ where the value of the exponent '*b*' is assumed not to vary between types of pig – different genotypes and sexes. With the value of '*b*' fixed at 0.75 the value of the constant of proportionality is usually estimated to be about 420 kJ kg$^{-0.75}$ day$^{-1}$ (ARC, 1981). It is worth noting that the total heat production of the pigs in Table 15.1, which were fed a diet such that they were retaining appreciable amounts of protein and lipid, was much lower than this estimate of maintenance when scaled in this way.

## Feeding at Maintenance

In a mature, non-productive pig maintenance is the condition when the rates of protein and lipid retention are both zero. Such a condition is, at least in principle, able to be achieved. As a consequence the rate of energy retention is also zero and the rate of heat production at maintenance, $P_m$, is the *ME* intake, $ME_m$. Where the environment is thermally neutral it is to be expected that $ME_m$ will exceed *FHP*. The ratio $FHP/ME_m$ is called $k_m$ which is the energetic efficiency of using the ME from the feed for maintenance. It is to be expected that the value of $k_m$ will be different for different feeds although, in pigs, such differences are generally ignored.

It has been proposed (Emmans, 1994) that $ME_m$ will exceed $MH$ for three reasons and, in the absence of any change in physical activity, for only three reasons. The reasons for the extra heat, or the causes of it, arise from the fact that the consumption of feed increases the work that the pig does and that the extra work leads to an increment in heat production. The extra work is assumed to be the consequence of: (i) producing urinary N; (ii) carrying out fermentation; and (iii) the direct effect of eating feed. The amount of work done to eat feed is assumed to depend on its digestibility. The details are in Emmans (1994). Because of the problems of pigs often being cold at low levels of feeding the proposals have not yet been tested specifically for pigs but they have been well tested for poultry and ruminants. The $FHP/ME_m = k_m$ could be used as a general efficiency number for a feed or foodstuff. The $NE$ of a diet for maintenance is the product of $ME$ and $k_m$.

There are practical problems in measuring the value of $k_m$ in pigs. These include the difficulty of keeping small pigs thermally neutral at low levels of feeding, the high $k_m$ values of the usual foodstuffs, which means that the heat increment is small below maintenance, and making the proper allowance for different levels of activity which can vary systematically with the level of feeding.

## Positive Protein and Lipid Retentions

In order to retain either protein or lipid work must be done and heat will be produced as a consequence. Kielanowski (1965) pointed out that the energetic efficiencies of using feed $ME$ for retaining energy as protein and lipid above maintenance need not necessarily be the same as each other. It is generally accepted (Pullar and Webster, 1977; ARC 1981; Emmans, 1994) that the energetic efficiency of using $ME$ for protein retention is less than that for using it for lipid retention. Where this is the case it is not sensible to add the retentions together and to attempt to quantify the efficiency of using feed $ME$ for overall energy retention. If energy retention is treated as a single quantity then inevitable errors will result. This will be the case both for feed evaluation and for estimating energy requirements.

## Energy Systems and Feed Evaluation

The main function of an energy system is to allow energy values to be assigned to foodstuffs which can be used to estimate the amount of a given mixed feed needed to meet a particular animal performance (Blaxter and Boyne, 1978). It has become conventional to describe animal performance in terms of thermoneutral maintenance (either the conventional *FHP*, or *MH* as described above), the rates of retention of protein and lipid (*LR* and *PR* or, as has often been the case, *ER*) and the energy needed for cold thermogenesis, *CT*, if any. The two conditions of interest are a cold environment and one that is not cold.

## Cold environments

When the environment is cold then *CT* is positive and the relevant energy scale is *ME* (ARC, 1981). It is assumed that the efficiency with which *ME* meets *CT* is unity. The relevant equation is then:

$$ME = EHD + ER \tag{15.5}$$

where *EHD* is the environmental heat demand, which includes *CT*, and *ER* is the total energy retention, assumed to be as lipid and protein. If the *ME* supply is the classical value, i.e. that measured with no correction for N-retention, then *ER* is that with the protein retention converted to energy with its full heat of combustion, which is agreed to be close to 23.8 kJ g$^{-1}$. If, however, the *ME* is corrected to a value for zero N-retention, by subtracting close to 35.2 kJ g$^{-1}$ N in the urine, then the *ER* value for protein retention should also be so corrected. If this is not done then the apparent equations do not balance. The equations are:

$$\text{Classical: } ME_c = EHD + ER \tag{15.6}$$
$$\text{N-corrected: } ME_n = EHD + ER_n \tag{15.7}$$

where $ME_n = ME_c - (35.2/6.25) \times PR$ and $ER_n = ER - (35.2/6.25) \times PR$ where *PR* is the rate of protein retention.

## Thermally neutral and hot environments

The second condition is when the environment is such that $CT = 0$, which will occur because it is thermoneutral or hot. Only if the *ME* from different diets is used with the same efficiencies for the functions of maintenance, protein and lipid retention is *ME* then the suitable energy scale. But this is widely recognized not to be the case because the amount of *ME* needed to support a unit of a particular function varies between foodstuffs and diets. This has often been ignored in pigs and *ME* has been treated as a sufficient energy scale; even *DE* has sometimes been seen as a sufficient energy scale. For thermally neutral and hot environments the problem is to predict the heat production of the animal when eating different amounts of different diets (Emmans, 1994).

Since the efficiency with which the *ME* from the diet replaces *FHP* will

**Table 15.2.** The effects of adding sugar to a basal diet. Data of Fingerling *et al.* (1938b).

| Increments due to sugar | Mean increment (kJ day$^{-1}$) |
|---|---|
| Gross energy (kJ day$^{-1}$) | 6453 |
| Metabolizable energy (kJ day$^{-1}$) | 6315 |
| Weight correction (kJ day$^{-1}$) | −341 |
| Corrected *ME* (kJ day$^{-1}$) | 6656 |
| Energy retention (kJ day$^{-1}$) | 4648 |
| Heat (kJ day$^{-1}$) | 2008 |

normally be greater than the efficiency of retaining positive energy (as either protein or lipid) there is no sense in adding positive energy retention to the *FHP* and calling the result 'net energy', *NE*. If this is done the *NE/ME* value of the diet will vary systematically with the level of feeding. As was pointed out earlier the term *NE* comes from Armsby (1903) and was used to reflect his view at that time that heat production was a simple linear function of feed intake for a given feed. Once this was shown not to be the case (Forbes *et al.*, 1928, 1930) the term became a confusing one with no biological merit. It is long past the time when *NE* should be used.

The situation is not helped, in the sense of producing misleading conclusions, when the energy retained as protein is added to that retained as lipid to calculate the *NE* gained. It would be proper to add the two energy retentions to each other only if the energetic efficiencies of the retentions were the same. Suppose a pig on a given diet, its basal diet, is retaining both protein and lipid at positive rates. It is now given 1 kg of another feed in addition to its basal diet. The *NE* of the additional feed will be calculated as the increase in energy retention apparently resulting from it. But the NE value will depend on what the pig apparently does with the extra feed, i.e. on the proportion of the extra energy retained as protein. This proportion is not only a characteristic of the feed being added but will depend on the basal diet being given – both composition and quantity – and on the genotype and state of the pig. Where *NE* values are determined in this way it will almost inevitably be concluded that the *NE* values of a feed ingredient depend on these variables. It is likely that a complex, and inaccurate, system of feed evaluation will be the result.

## The heat increment of feeding

The usual way of estimating the *NE* value of an ingredient is to add it to a basal diet. The increases in the relevant quantities are then measured. The example in Table 15.2 is taken from Emmans (1994). Pigs were given an increment of sugar to a basal diet and virtually all of the extra *GE* given appeared as extra *ME*.

Of the extra *ME* resulting from the addition of sugar, after a small correction for a difference in the weight of the pigs, 4648 kJ day$^{-1}$ were retained and 2008 kJ day$^{-1}$ were lost as heat. It is possible to state that, in this particular case, the extra *NE* provided by the sugar was 4648 kJ day$^{-1}$, or 12.62 kJ g$^{-1}$ of sugar fed. The energetic efficency of the addition is thus 4648/(4648 + 2008) = 0.698. (Muller *et al.*, 1991, found that sows used a mixture of 0.75 acetic and 0.25 propionic acids with an energetic efficiency of 0.699 $\pm$ 0.032 which is the same as that found for sugar). But there is no certainty that either value would be found if the experiment were to be repeated with a different basal diet, or a different kind of pig, or the same kind of pig in a different state. Even with a protein-free feed, such as starch, additions to a basal diet can lead to an increase in protein retention as found by Kyriazakis and Emmans (1992a, b) who gave growing pigs increments of starch to basal diets of different protein content. They found that where the basal diet was of low protein content the rate of protein retention was unaffected by the starch addition. But, on a basal diet of high protein content, the addition of starch led to an appreciable, and highly significant, increase in the rate of protein retention. The experiment was not intended to look at the efficiency of starch as an energy source but it does show that, if protein and lipid are retained with different energetic efficiencies, then the *NE* value of a material will vary with the basal diet being given.

There is another way of seeing the problem. Instead of asking how much extra energy will be retained we can ask how much extra heat will be produced by an addition of a food. It has been proposed (Emmans, 1994) that there are only a few functions that need work to be done, and hence heat to be produced. For pigs on diets fed above maintenance where fermentation is negligible these are food intake, N excretion, protein retention and lipid retention. The work of food intake is directly related to the indigestibility of the food so that the heat produced is directly related to amount of organic matter appearing in the faeces which is called faecal organic matter, FOM. The estimate of the value, in both steers and broiler chickens, was 3.80 kJ g$^{-1}$. To excrete the N-containing products of protein catabolism 4.67 kJ g$^{-1}$ of heat is produced. This estimate is close to that of Whittemore and Fawcett (1976). The heat produced by the retention of 1 g of protein was 36.5 kJ. For lipid retention the estimates from data on pigs were 16.4 kJ g$^{-1}$ when the retention was not from feed lipid and 4.4 kJ g$^{-1}$ when it was. In order to predict the extra heat that will be produced by an addition of a feed it is thus necessary to predict the extra retentions of both protein and lipid as well as the extra heat arising from other causes.

In order to disentangle the two related, but separate, problems of assigning an energy value to a foodstuff and predicting the performance of a pig from a knowledge of its diet the causes of the heat increment of feeding need to be distinguished.

## Expected heat increments

At maintenance, when this is defined strictly as zero rates of retention of both protein and lipid, there are no heat increments due to retentions. In the scheme proposed by Emmans (1994) there are just three causes of the heat increment at maintenance – excretion of N, production of FOM and fermentation. The consequence is that the value of $k_m$ will decrease as the protein content of the feed increases, as the digestibility decreases and, probably, as the fibre content increases. By allowing for these effects the effectiveness of the $ME$ of the feed, when fed at maintenance, can be calculated.

The effective energy value of the feed, $EE$ MJ kg$^{-1}$, is given by:

$$EE = ME_n - 4.67 \times DCP - 3.80 \times FOM - 0.616 \times CE \qquad (15.8)$$

where $ME_n$ is the $ME$ yield of the diet at maintenance, MJ kg$^{-1}$; $DCP$ is the digestible crude protein content, kg kg$^{-1}$; FOM is the output of faecal organic matter, kg kg$^{-1}$; and $CE$ is the yield of combustible gases, MJ kg$^{-1}$. If these quantities are known for a foodstuff, or can be estimated from other information, then the $EE$ value can be calculated.

It was possible to estimate the $EE$ values of diets from their $ME_n$ and protein contents only for both ruminants and poultry. The equations, taken from Emmans (1994) were:

$$\text{Ruminants: } EE = 1.15ME_n - 3.84 - 4.67DCP$$
$$\text{(MJ kg}^{-1}\text{ organic matter)} \qquad (15.9)$$
$$\text{Poultry: } EE = 1.17ME_n - 2.44 - 4.20CP$$
$$\text{(MJ kg}^{-1}\text{ dry matter)} \qquad (15.10)$$

where $CP$ is the crude protein content and $DCP$ the digestible crude protein content, both in kg kg$^{-1}$. It is likely that similar equations could be derived from suitable experiments on pigs. The requirement of the growing animal for $EE$, $EERQ$, is the same for ruminants, poultry and pigs:

$$EERQ = MH + 50PR + 56LR \qquad \text{(MJ day}^{-1}\text{)} \qquad (15.11)$$

# Discussion

The detailed processes involved in the energetics of pigs are of considerable complexity. But, because of the laws of thermodynamics, particularly the Principle of the Conservation of Energy and Hess's law, considerable simplification can be achieved at the level of the whole animal. The problem to be solved is to account for the heat increment of feeding, i.e. why do pigs produce more heat when they are given more of a particular feed in a thermally neutral environment?

The problem of accounting for the heat increment of feeding is closely connected with the two further problems of assigning an energy value to a foodstuff and estimating the animal's energy requirement to sustain some level of performance. The first step in solving the problems is to account for the losses of energy in the material excreta from the diet eaten – faeces, urine and combustible gases. After allowing for these losses the metabolizable energy, *ME*, of the diet is known. As the *ME* is either retained in the animal or is lost as heat, the problem in finding the *ME* needed to sustain particular rates of retention of protein and lipid is that of predicting the heat increment.

It has been tempting to try to solve the two separate problems of feed evaluation and animal requirements in one step as if they were the same problem. When overcome by this temptation, as has happened frequently over the past century, researchers have tried to measure the increase in energy retention resulting from the addition of a feed and have called that value the *NE* of the ingredient. When, as is inevitable, the value has been found to vary with the detailed conditions of the experiment – the state of the animal and the basal diet are probably most influential – it has then been suggested that the values need to be adjusted for these variables.

It is much easier to recognize that the energetic efficiencies of maintenance, lipid retention and protein retention are different and not to get involved in trying to collapse these functions into one measure called *NE*. The *ME* of the diet appears to be a sufficient measure of its ability to do work. But it needs to be recognized that some of the work that needs to be done is a direct consequence of eating the food. When appropriate allowances are made for the work associated with eating indigestible material, excreting N in the urine and fermentation, the remaining energy appears to be equally effective in meeting the energy needs of the animal. For this reason the *ME* after making these allowances has been called the effective energy, *EE* (Emmans, 1994). The amounts of *EE* needed to replace 1 MJ of lipid loss (1 MJ) and to sustain positive retentions of 1 kg of protein (50 MJ) and 1 kg of lipid (56 MJ) are then constant across diets.

Seeing the pig as an energy processor and converter can only be a partial view, but it is a useful one nevertheless. It helps to connect it with its environment – through its heat production that must be lost – and with the composition of its diet and its feed intake when fed *ad libitum*. Finally, the energy needs, and dietary balance, of different genotypes at different degrees of maturity connect with descriptions of different kinds of pigs, their potential to grow and fatten and their rules for partitioning scarce resources.

# References

Agricultural Research Council (ARC) (1981) *The Nutrient Requirements of Pigs*. Commonwealth Agricultural Bureau, Farnham Royal, Slough, UK.

Armsby, H.P. (1903) *The Principles of Animal Nutrition*. Macmillan, New York.

Armsby, H.P. and Fries, J.A. (1915) Net energy values of feeding-stuffs for cattle. *Journal of Agricultural Research* 3, 435–491.

Blaxter, K.L. (1989) *Energy Metabolism in Animals and Man*. Cambridge University Press, Cambridge, UK.

Blaxter, K.L. and Boyne, A.W. (1978) The estimation of the nutritive value of feeds as energy sources for ruminants and the derivation of feeding systems. *Journal of Agricultural Science* (Cambridge) 90, 47-68.

Demeyer, D.I. and Degraeve, K. (1991) Differences in stoichiometry between rumen and hindgut fermentation. *Zeitschrift fur Tierphysiologie Tierernahrung und Futtermittelkunde* 22, 50-61.

Emmans, G.C. (1987) Growth, body composition and feed intake. *World's Poultry Science Association Journal* 43, 208-227.

Emmans, G.C. (1994) Effective energy: a concept of energy utilisation applied across species. *British Journal of Nutrition*, 71, 801-821.

Emmans, G.C. and Fisher, C. (1986) Problems of nutritional theory. Fisher, C. and Boorman, K.N. (eds), *Nutritional Requirements and Nutritional Theory*. Butterworths, London.

Fingerling, G., Eisenkolbe, P., Heintzsch, B., Just, M. and Knauth, G. (1938a) Untersuchungen uber den Stoff- und Energie-umsatz wachsender Schweine. *Landwirtschaft Versuchstationen* 129, 193-234.

Fingerling, G., Eisenkolbe, P., Heintzsch, B., Just, M. and Knauth, G. (1938b) Untersuchungen uber den Stoff- und Energie-umsatz wachsender Schweine. *Zeitschrift fur Tierernahrung und Futtermittelkunde* 1, 193-234.

Forbes, E.B., Braman, W.W. and Kriss, M. (1928) The energy metabolism of cattle in relation to the plane of nutrition. *Journal of Agricultural Research* 37, 253-300.

Forbes, E.B., Braman, W.W. and Kriss, M. (1930) Further studies of the energy, metabolism of cattle in relation to the plane of nutrition. *Journal of Agricultural Research*, 40, 37-78.

Just, A., Jorgensen, H. and Fernandez, J.A. (1983) Maintenance requirement and the net energy value of different diets for growth in pigs. *Livestock Production Science* 10, 487-506.

Kellner, O. and Kohler, A. (1900) Untersuchungen uber den Stoff- und energieumsatz volljahrigen Ochsen bei Erhaltungsfutter. *Landwirtschaft Versuchstationen* 47, 1-474.

Kielanowski, J. (1965) Estimates of the energy cost of protein deposition in growing animals. In: *Proceedings of the Symposium on Energy Metabolism*. EAAP Publication No. 11. Troon, UK.

Kirchgessner, M., Kreuzer, M., Muller, H.I. and Windisch, W. (1991) Release of methane and of carbon-dioxide by the pig. *Zeitschrift fur Agrarbiologie Agrikulturchemie Okologie* 44, 103-113.

Kyriazakis, I. and Emmans, G.C. (1992a) The effects of varying protein and energy intakes on the growth and body composition of pigs. 1. The effects of energy intake at constant high protein intake. *British Journal of Nutrition* 68, 603-613.

Kyriazakis, I. and Emmans, G.C. (1992b) The effects of varying protein and energy intakes on the growth and body composition of pigs. 2. The effects of varying both energy intake and protein intake. *British Journal of Nutrition* 68, 615-625.

Muller, H.I., Kirchgessner, M. and Roth, F.X. (1991) Energetic efficiency of a mixture of acetic and propionic-acid in sows. *Zeitschrift fur Tierphysiologie Tierernahrung und Futtermittelkunde* 65, 140-145.

Noblet, J., Fortune, H., Shi, X.S. and Dubois, S. (1994) Prediction of net energy value of feeds for growing pigs. *Journal of Animal Science* 72, 344-354.

Pullar, J.D. and Webster, A.J.F. (1977) The energy cost of fat and protein deposition in the rat. *British Journal of Nutrition* 37, 355–363.

Tess, M.W., Dickerson, G.E., Nienaber, J.A. and Ferrell, C.L. (1984) The effects of body composition on fasting-heat production in pigs. *Journal of Animal Science* 58, 99–110.

Whittemore, C.T. and Fawcett, R.H. (1976) Theoretical aspects of a flexible model to simulate protein and lipid growth in pigs. *Animal Production* 22, 87–96.

Zhu, J.Q., Fowler, V.R. and Fuller, M.F. (1993) Assessment of fermentation in growing pigs given unmolassed sugar-beet pulp – a stoichiometric approach. *British Journal of Nutrition* 69, 511–525.

# Conclusions

# Future Directions for Models in Pig Biology

**16**

## I. Kyriazakis

*Animal Biology Division, Scottish Agricultural College, West Mains Road, Edinburgh EH9 3JG, UK*

## Introduction

In the introductory chapter of this book, the need for a quantitative approach to pig biology, and the advantages that will arise from it in relation to simulation modelling were briefly addressed. Almost all the chapters that followed have emphasized these advantages in one way or another, and therefore I do not feel the need to address them further. There is no doubt that the development of quantitative theories, and the construction of simulation models in pig biology will increasingly yield benefits both in the advancement of pig science and in its application to practice.

I have also viewed theories and, by extension, models which deal with biological processes as both evolving and dynamic, so that they can eventually be replaced by better ones; this is the nature of progress in science. For this reason I would like to discuss here what I see as the future direction for models in pig biology, and consequently to make some suggestions on where future efforts should be put. Prior to doing this, there is the need to emphasize the problem areas which might hinder the development of theories and models. I see these mainly arising from the third stage of development of a theory (Chapter 1): the identification and parameterization of the variables and constants of the theory's functional forms. There is one further major issue related to the development of future models in pig science, which is that of the problems involved in getting models into practice. This did not seem to me to fall within the aims of this book, and therefore will not be discussed further (see Black *et al.*, 1993; Emmans, 1995).

© CAB INTERNATIONAL 1999. *A Quantitative Biology of the Pig*
(ed. I. Kyriazakis)

## Problems Arising from System Descriptions

It was suggested that in order to be able to predict the behaviour of any system a predictive theory and a sufficient description of the system are required. For theories developed in pig biology, the elements of the system which call for descriptions are usually the pigs, the food offered to them and the environment in which they are kept (Chapter 1). These descriptions usually represent the values of the variables in the theory's functional forms, and can be seen as the inputs required by the theory or model to produce the quantitative predictions or outputs. Examples of such inputs which may be needed by different models are the initial live weight of the pig, the amino acid content of the food it is offered and the ambient temperature of the environment in which it is kept. All these have ranges and limits which must be known in order to ask questions on how the system will behave if, for example, the size or the kind of the pig changes.

The above seems a rather straightforward process. However, I would argue that the process of parameterization and sufficient description in the terms required by the theory represent a major bottleneck in the development of predictive theories, and importantly in their application. A trivial example would be the development of a model which requires the description of the food offered to pigs in terms which are either not available to food chemists, or where there is a considerable doubt over the methodology used to obtain the measurements. This problem usually reflects the fact that people who develop models and those who do experiments tend to be different sets of people. Furthermore, it is not unusual that data are generated, stored or presented in such a form that they cannot be used for the purposes of a model. To emphasize this point, I will use the example of the description of the potential growth of the pig (Chapter 8); it represents a real problem, where some progress has been made to resolve it.

It was suggested in Chapter 8 that the sufficient set of parameters which describes the *potential* chemical growth of the pig are: mature protein weight, $P_m$, the mature fatness, $LPR_m$, and the Gompertz rate parameter, $B$. Tests of this description have been made (Ferguson and Gous, 1993a; see also Chapter 8). There are estimates of the values of these variables for some kinds of pig (Whittemore *et al.*, 1988; Ferguson and Gous, 1993b) from experiments which either were set out to estimate them, or are estimated from data presented in sufficient detail (Doornenbal, 1971; Kyriazakis *et al.*, 1993). However, estimates are not currently available for many of the genotypes of pigs available in Europe and the United States. Ferguson and Gous (1993a) have described a protocol which could be used to obtain values of the above growth parameters for a given population of pigs. The next step would be to convince those people whose job it is to provide descriptions of pigs, i.e. the pig breeders, that information needs to be obtained and hence provided in such terms. This task is different from that of convincing breeders about the usefulness of a predictive theory, and will represent a major challenge. This is especially the case as pig breeders are mostly interested, for selection purposes, in the *actual* growth of pigs in a certain environment, rather than the *potential* for growth of the pigs they breed.

In conclusion, sufficient descriptions of the system still represent a major challenge to the development of theories in pig biology. For the three elements of the system raised above, it is perhaps the chemical and physical descriptions of the food offered to pigs that are the least problematic. The problem here is to make descriptions of the foods consistent with what is presented in the digestive tract of the pig (Chapter 3). The necessary relationship between theory and system description is illustrated by the proposal of Kyriazakis and Emmans (1995) that it is the water-holding capacity (WHC) of foods which is responsible for constraining intake. To use this theory to predict the intake of bulky foods the description of a food must now include its WHC value. Regarding the description of the environment, some new dimensions, mainly of the infectious and the social conditions under which pigs are kept, will be needed to improve the accuracy of prediction (Chapters 4 and 5). The sufficient description of the pig represents a major challenge, especially in the light of different theories that can be developed, which could have different requirements (Chapters 2 and 8). An important point is that the values for describing the pig in a model are derived from populations and hence they are not singular as values for food lysine content, for example, may be. They will have variation and covariation and these have to be estimated if theories are to be developed for populations of pigs.

## Theories and models about individuals and about populations

The majority of the theories developed in this book have mainly dealt with problems developed at the level of an individual animal, usually *the average pig*, although this was not stated explicitly. I presume that this was done for the sake of simplicity. In most cases, however, theories that predict the responses of a group or a population of pigs are of interest; nutritionists, for example, are interested in the prediction of the response of a population of animals as a herd to a nutritional treatment, rather than the response of individuals. It could, therefore, be argued that similarly theories should be either developed at a level of populations, or made flexible for adaptation to that extent. The variation between the individuals in a population may be of direct interest where grading schemes are non-linear for weight and fatness.

The problem of including stochastic elements in models of pig biology, and animal science in general, has not yet received a lot of attention; neither have its implications for the development of theories. Knap (1995) has summarized the advantages arising from the introduction of stochasticity into models, and these are related to the fact that there would be a difference in the response of the average individual in the population and the mean population response, which is the average of all individuals. There are instances where these differences are likely to be systematic and large enough to be of importance (Emmans and Fisher, 1986). In order to make stochastic predictions in simulation models, one needs to repeat the simulation for a number of individuals representative of the population and then determine the mean and within-population variation from the results (Ferguson *et al.*, 1997). This would require good estimates of the parameters in the model that vary between animals, the nature of their distribution and the possible correlations that exist between individual parameters. These requirements put at a great disadvantage theories which either do not have precise measurement of such parameters, or which require a large number of parameters. In the last instance, the problem ceases to be tractable, and the task of developing models that predict the responses of populations becomes unsolvable (Chapter 8).

## What Kind of Models in Pig Biology?

Here I will follow the suggestion of Whittemore (1993) to classify theories and models in biology into two classes: *empirical* and *deductive*. Empirical is the term used for models which are based on equations that describe the relationship between two or more variables, where these are the result of statistical curve-fitting exercises; statistical could be a better term. Strictly speaking, empirical relationships can only be euphemistically called 'theories'. On the other hand, deductive models are those where underlying biological processes are represented by biologically interpretable functions and parameters.

In Chapter 10, where the prediction of voluntary food intake was discussed, some examples of empirical equations that related food intake to pig live weight were given. It was argued there that such approaches have very little chance of successfully predicting the food intake of pigs in any general way, since they have no theoretical basis and they can be misleading when extrapolated beyond the bounds of the data from which they are derived. A similar view was emphasized in Chapter 8, where the modelling approaches to *actual* and *potential* growth were discussed. It was stressed that it is the biological concept of potential growth that is of importance in making predictions into the future and which needs to be described, rather than simply collecting data on the growth of real pigs to try to see what relationships exist. Such relationships can be very complex and, therefore, attractive to their inventor. Nevertheless, this complexity neither should nor could imply generality.

Given the preference for deductive models, the question that arises then is at what level underlying biological principles of such models should be represented. For example, models that aim to predict the growth of a pig could have representations made at whole animal, organ, tissue or cellular level. There is the view that deductive modelling has a greater chance of success if the underlying biological processes are reduced to that of chemical or physical descriptions, i.e. mechanistic modelling, although the justification for this is not always made clear (Black, 1995; Dijkstra and France, 1995). If, for example, one is to simulate protein growth in the pig, then modelling of the processes of protein synthesis and protein degradation would be required at least. There are two counter-arguments to this view. Firstly, adherence only to the laws of chemistry and physics to simulate biological processes usually ignores physiological controls and the modifying effects these might have on such processes. I will expand on this point below. Secondly, even if a satisfactory representation of the biological processes is achieved in such terms, information on the variables and constants of the functional forms does not currently exist and it might be doubtful whether it would ever be obtained (Baldwin and Sainz, 1995; Pettigrew, 1995). The implications of this have been elaborated upon previously.

Reductionist approaches to developing 'mechanistic' models in pig science are rather scarce. The two notable exceptions are those of Pomar *et al.* (1991) and Pettigrew, *et al.* (1992) which model the growth of pigs and lactation in sows, respectively, with a variable degree of success. This is done by tracing the flow of specific energy containing metabolites in the body of the pig, in terms of substrate kinetics. It must be emphasized that neither of these models can be classified as purely mechanistic, since some representations are made at higher levels. Neither attempts to predict food intake. A third attempt to model mechanistically the metabolism of carbohydrate and lipid is made in Chapter 14 of this book. These three attempts contrast with the plethora of mechanistic models in ruminant biology (see for example the book by Forbes and France, 1993); some explanations for this will be offered below. In the model of Pettigrew *et al.* (1992) the relationships between different substrate (nutrient) pools in the body of the sow are represented by simple enzyme kinetics equations. However, in a subsequent paper by the same author it is admitted that: 'In some cases, the body's control mechanisms [i.e. hormone concentrations] produce a different relationship between substrate concentrations and rate than is reflected in the [enzyme kinetics] equations. ... Attempts to estimate hormone concentrations, and tissue sensitivity to those hormones, will be difficult because of the complexity of the endocrine system' (Pettigrew, 1995).

In the light of the above view, I am tempted to suggest that future developments in models in pig biology, while will be useful in practice, will be unlikely to lie in 'mechanistic' approaches. Deductive models developed at higher levels are more likely to succeed in future developments. Such higher levels include 'nutrient partitioning' models, which for example trace the overall flows of energy and amino acids in the pig's body. Representations at this level seem to meet the requirement that 'if accurate predictions are required at one level, it is essential that the model is constructed at some level below the one that is required' (Black, 1995). However, ultimately the decision of what kind of models to develop will depend on their perceived uses. Models developed to address specific research issues will be able to afford a high degree of reductionism, since they will not necessarily require a high degree of accuracy in their predictions and will not need to be used by other people. The vast majority of models which are developed in ruminant biology fall precisely in this category, and hence have limited applications. On the other hand, commercial pig production has a greater history of reliance on models of pig biology, e.g. for decision-making purposes. For this, a high degree of accuracy in the predictions is paramount. Such models will have representations in them that might never be able to be reduced into the levels of mechanism, although they will be making perfect biological sense.

## Concluding Remarks

It is the characteristic of good science that scientific and practical issues are, or should be, different aspects of the same thing. Inventors of theories and models in pig biology should therefore be aspiring to developments which have a wider applicability and are capable of solving problems in practical pig science. 'Mechanistic' models are unlikely to achieve this aim and their future use will probably be limited to the research environment in which they have been developed. Deductive models developed at slightly higher levels, e.g. at the level of partitioning rules (Kyriazakis, 1996), appear to have the quality of ready applicability. Their ability to be transferred into practice will, however, frequently depend on factors which are independent of their underlying science (Black *et al.*, 1993).

A major bottleneck in the development and application of all models in pig science is the process of their parameterization. The sufficient description of the pig represents a major challenge, especially in the light of different theories that can be developed, which call for different descriptions. This will arise from the close co-operation of people who develop models and pig breeders (Emmans and Kyriazakis, 1997). Emmans (1995) noted that 'a pig food that is sold comes, and has come for many years, with a description that has relevance for its use. The same needs to be the case in the future for pigs that are sold by breeders.' A better description of the pig would also enable easier transition to models which predict the responses of a population of pigs to their treatment. The issue of stochasticity does not seem to be part of the thoughts of the people who are currently developing theories and models in pig biology. These are currently developed at the level of the 'average' pig.

It can be concluded from the above that the process of model developing should be a concerted effort of the inventor(s) and the users. It is only through this exchange that theories and hence models are tested, criticized and eventually replaced.

# References

Baldwin, R.L. and Sainz, R.D. (1995) Bottlenecks for improvement of future models in animal nutrition science. In: Danfaer, A. and Lescoat, P. (eds), *Proceedings of IVth International Workshop on Modelling Nutrient Utilisation in Farm Animals*. National Institute of Animal Science, Foulum, Denmark, pp. 163–171.

Black, J.L. (1995) The evolution of animal growth models. In: Moughan, P.J., Verstegen, M.W.A. and Visser-Reyneveld, M.I. (eds), *Modelling Growth in the Pig*. Wageningen Pers, Wageningen, The Netherlands, pp. 3–9.

Black, J.L., Davies, G.T. and Fleming, J.F. (1993) Role of computer simulation in the application of knowledge to animal industries. *Australian Journal of Agricultural Research* 44, 541–555.

Dijkstra, J. and France, J. (1995) Modelling and methodology in animal science. In: Danfaer, A. and Lescoat, P. (eds), *Proceedings of IVth International Workshop on Modelling Nutrient Utilisation in Farm Animals*. National Institute of Animal Science, Foulum, Denmark, pp. 9–18.

Doornenbal, H. (1971) Growth, development and chemical composition of the pig. I. Lean tissue and protein. *Growth* 35, 281–295.

Emmans, G.C. (1995) Problems in applying models in practice. In: Moughan, P.J., Verstegen, M.W.A. and Visser-Reyneveld, M.I. (eds), *Modelling Growth in the Pig*. Wageningen Pers, Wageningen, The Netherlands, pp. 223–227.

Emmans, G.C. and Fisher, C. (1986) Problems in nutritional theory. In: Fisher, C. and Boorman, N. (eds), *Nutrient Requirements of Poultry and Nutritional Research*. Butterworths, London, UK, pp. 9–36.

Emmans, G.C. and Kyriazakis, I. (1997) Models of pig growth: problems and proposed solutions. *Livestock Production Science* 51, 119–129.

Ferguson, N.S. and Gous, R.H. (1993a) Evaluation of pig genotypes. I. Theoretical aspects of measuring genetic parameters. *Animal Production* 56, 233–243.

Ferguson, N.S. and Gous, R.H. (1993b) Evaluation of pig genotypes. II. Testing experimental procedure. *Animal Production* 56, 245–249.

Ferguson, N.S., Gous, R.H. and Emmans, G.C. (1997) Modelling the growth and nutrition of pigs. *Animal Science* 64, 513–522.

Forbes, J.H. and France, J. (1993) *Quantitative Aspects of Ruminant Digestion and Metabolism*. CAB International, Wallingford, Oxon, UK, pp. 1–10.

Knap, P.W. (1995) Aspects of stochasticity: variation between animals. In: Moughan, P.J., Verstegen, M.W.A. and Visser-Reyneveld, M.I. (eds), *Modelling Growth in the Pig*. Wageningen Pers, Wageningen, The Netherlands, pp. 165–172.

Kyriazakis, I. (1996) A solution to the problem of predicting the response of an animal to its diet. *Proceedings of the Nutrition Society* 55, 155–166.

Kyriazakis, I. and Emmans, G.C. (1995) The voluntary food intake of pigs given feeds based on wheat bran, dried citrus pulp and grass meal, in relation to measurements in feed bulk. *British Journal of Nutrition* 73, 191–207.

Kyriazakis, I., Leus, K., Emmans, G.C., Haley, C.S. and Oldham, J.D. (1993) The effect of breed (Large White X Landrace vs purebred Meishan) on the diets selected in pigs given a choice between two foods that differ in their crude protein contents. *Animal Production* 56, 121–128.

Pettigrew, J.E. (1995) Metabolic approach to modelling nutrient utilisation. In: Moughan, P.J., Verstegen, M.W.A. and Visser-Reyneveld, M.I. (eds), *Modelling*

*Growth in the Pig*. Wageningen Pers, Wageningen, The Netherlands, pp. 173–183.

Pettigrew, J.E., Gill, H., France, J. and Close, W.H. (1992) A mathematical integration of energy and amino acid metabolism of lactating sows. *Journal of Animal Science* 70, 3742–3761.

Pomar, C., Harris, D.L. and Minvielle, F. (1991) Computer simulation of swine production systems. *Journal of Animal Science* 69, 1468–1488.

Whittemore, C.T. (1993) *The Science and Practice of Pig Production*. Longman, Harlow, Essex, UK, pp. 572–575.

Whittemore, C.T., Tullis, J.B. and Emmans, G.C. (1988) Protein growth in pigs. *Animal Production* 45, 437–445.

# Index